Lecture Notes in Computer Science　　10115

Commenced Publication in 1973
Founding and Former Series Editors:
Gerhard Goos, Juris Hartmanis, and Jan van Leeuwen

More information about this series at http://www.springer.com/series/7412

Shang-Hong Lai · Vincent Lepetit
Ko Nishino · Yoichi Sato (Eds.)

Computer Vision – ACCV 2016

13th Asian Conference on Computer Vision
Taipei, Taiwan, November 20–24, 2016
Revised Selected Papers, Part V

 Springer

Editors
Shang-Hong Lai
National Tsing Hua University
Hsinchu
Taiwan

Ko Nishino
Drexel University
Philadelphia, PA
USA

Vincent Lepetit
Graz University of Technology
Graz
Austria

Yoichi Sato
The University of Tokyo
Tokyo
Japan

ISSN 0302-9743   ISSN 1611-3349 (electronic)
Lecture Notes in Computer Science
ISBN 978-3-319-54192-1   ISBN 978-3-319-54193-8   (eBook)
DOI 10.1007/978-3-319-54193-8

Library of Congress Control Number: 2017932642

LNCS Sublibrary: SL6 – Image Processing, Computer Vision, Pattern Recognition, and Graphics

Printed on acid-free paper

This Springer imprint is published by Springer Nature
The registered company is Springer International Publishing AG
The registered company address is: Gewerbestrasse 11, 6330 Cham, Switzerland

Preface

Welcome to the 2016 edition of the Asian Conference on Computer Vision in Taipei. ACCV 2016 received a total number of 590 submissions, of which 479 papers went through a review process after excluding papers rejected without review because of violation of the ACCV submission guidelines or being withdrawn before review. The papers were submitted from diverse regions with 69% from Asia, 19% from Europe, and 12% from North America.

The program chairs assembled a geographically diverse team of 39 area chairs who handled nine to 15 papers each. Area chairs were selected to provide a broad range of expertise, to balance junior and senior members, and to represent a variety of geographical locations. Area chairs recommended reviewers for papers, and each paper received at least three reviews from the 631 reviewers who participated in the process. Paper decisions were finalized at an area chair meeting held in Taipei during August 13–14, 2016. At this meeting, the area chairs worked in threes to reach collective decisions about acceptance, and in panels of nine or 12 to decide on the oral/poster distinction. The total number of papers accepted was 143 (an overall acceptance rate of 24%). Of these, 33 were selected for oral presentations and 110 were selected for poster presentations.

We wish to thank all members of the local arrangements team for helping us run the area chair meeting smoothly. We also wish to extend our immense gratitude to the area chairs and reviewers for their generous participation in the process. The conference would not have been possible without this huge voluntary investment of time and effort. We acknowledge particularly the contribution of 29 reviewers designated as "Outstanding Reviewers" who were nominated by the area chairs and program chairs for having provided a large number of helpful, high-quality reviews. Last but not the least, we would like to show our deepest gratitude to all of the emergency reviewers who kindly responded to our last-minute request and provided thorough reviews for papers with missing reviews. Finally, we wish all the attendees a highly simulating, informative, and enjoyable conference.

January 2017

Shang-Hong Lai
Vincent Lepetit
Ko Nishino
Yoichi Sato

Organization

ACCV 2016 Organizers

Steering Committee

Michael Brown	National University of Singapore, Singapore
Katsu Ikeuchi	University of Tokyo, Japan
In-So Kweon	KAIST, Korea
Tieniu Tan	Chinese Academy of Sciences, China
Yasushi Yagi	Osaka University, Japan

Honorary Chairs

Thomas Huang	University of Illinois at Urbana-Champaign, USA
Wen-Hsiang Tsai	National Chiao Tung University, Taiwan, ROC

General Chairs

Yi-Ping Hung	National Taiwan University, Taiwan, ROC
Ming-Hsuan Yang	University of California at Merced, USA
Hongbin Zha	Peking University, China

Program Chairs

Shang-Hong Lai	National Tsing Hua University, Taiwan, ROC
Vincent Lepetit	TU Graz, Austria
Ko Nishino	Drexel University, USA
Yoichi Sato	University of Tokyo, Japan

Publicity Chairs

Ming-Ming Cheng	Nankai University, China
Jen-Hui Chuang	National Chiao Tung University, Taiwan, ROC
Seon Joo Kim	Yonsei University, Korea

Local Arrangements Chairs

Yung-Yu Chuang	National Taiwan University, Taiwan, ROC
Yen-Yu Lin	Academia Sinica, Taiwan, ROC
Sheng-Wen Shih	National Chi Nan University, Taiwan, ROC
Yu-Chiang Frank Wang	Academia Sinica, Taiwan, ROC

Workshops Chairs

Chu-Song Chen	Academia Sinica, Taiwan, ROC
Jiwen Lu	Tsinghua University, China
Kai-Kuang Ma	Nanyang Technological University, Singapore

Tutorial Chairs

Bernard Ghanem   King Abdullah University of Science and Technology,
                            Saudi Arabia
Fay Huang   National Ilan University, Taiwan, ROC
Yukiko Kenmochi   Université Paris-Est, France

Exhibition and Demo Chairs

Gee-Sern Hsu   National Taiwan University of Science and
                            Technology, Taiwan, ROC
Xue Mei   Toyota Research Institute, USA

Publication Chairs

Chih-Yi Chiu   National Chiayi University, Taiwan, ROC
Jenn-Jier (James) Lien   National Cheng Kung University, Taiwan, ROC
Huei-Yung Lin   National Chung Cheng University, Taiwan, ROC

Industry Chairs

Winston Hsu   National Taiwan University, Taiwan, ROC
Fatih Porikli   Australian National University, Australia
Li Xu   SenseTime Group Limited, Hong Kong, SAR China

Finance Chairs

Yong-Sheng Chen   National Chiao Tung University, Taiwan, ROC
Ming-Sui Lee   National Taiwan University, Taiwan, ROC

Registration Chairs

Kuan-Wen Chen   National Chiao Tung University, Taiwan, ROC
Wen-Huang Cheng   Academia Sinica, Taiwan, ROC
Min Sun   National Tsing Hua University, Taiwan, ROC

Web Chairs

Hwann-Tzong Chen   National Tsing Hua University, Taiwan, ROC
Ju-Chun Ko   National Taipei University of Technology, Taiwan,
                            ROC
Neng-Hao Yu   National Chengchi University, Taiwan, ROC

Area Chairs

Narendra Ahuja   UIUC
Michael Brown   National University of Singapore
Yung-Yu Chuang   National Taiwan University, Taiwan, ROC
Pau-Choo Chung   National Cheng Kung University, Taiwan, ROC
Larry Davis   University of Maryland, USA

Sanja Fidler	University of Toronto, Canada
Mario Fritz	Max Planck Institute for Informatics, Germany
Yasutaka Furukawa	Washington University in St. Louis, USA
Bohyung Han	Pohang University of Science and Technology, South Korea
Hiroshi Ishikawa	Waseda University, Japan
C.V. Jawahar	IIIT Hyderabad, India
Frédéric Jurie	University of Caen, France
Iasonas Kokkinos	CentraleSupélec/Inria, France
David Kriegman	UCSD
Ivan Laptev	Inria, France
Kyoung Mu Lee	Seoul National University
Jongwoo Lim	Hanyang University, South Korea
Liang Lin	Sun Yat-Sen University, China
Tyng-Luh Liu	Academia Sinica, Taiwan, ROC
Huchuan Lu	Dalian University of Technology, China
Yasuyuki Matsushita	Osaka University, Japan
Francesc Moreno-Noguer	Institut de Robòtica i Informàtica Industrial
Greg Mori	Simon Fraser University, Canada
Srinivasa Narasimhan	CMU
Shmuel Peleg	Hebrew University of Jerusalem, Israel
Fatih Porikli	Australian National University/CSIRO, Australia
Ian Reid	University of Adelaide, Australia
Mathieu Salzmann	EPFL, Switzerland
Imari Sato	National Institute of Informatics, Japan
Shin'ichi Satoh	National Institute of Informatics, Japan
Shiguang Shan	Chinese Academy of Sciences, China
Min Sun	National Tsing Hua University, Taiwan, ROC
Raquel Urtasun	University of Toronto, Canada
Anton van den Hengel	University of Adelaide, Australia
Xiaogang Wang	Chinese University of Hong Kong, SAR China
Hanzi Wang	Xiamen University
Yu-Chiang Frank Wang	Academia Sinica, Taiwan, ROC
Jie Yang	NSF
Lei Zhang	Hong Kong Poly University, SAR China

Contents – Part V

Video Understanding

3D Vision

Divide and Conquer: Efficient Density-Based Tracking of 3D Sensors in Manhattan Worlds

Yi Zhou[1,2(✉)], Laurent Kneip[1,2], Cristian Rodriguez[1,2], and Hongdong Li[1,2]

[1] Research School of Engineering, The Australian National University,
Canberra, Australia
{yi.zhou,laurent.kneip,cristian.rodriguez,hongdong.li}@anu.edu.au
[2] Australian Centre for Robotic Vision, Canberra, Australia

Abstract. 3D depth sensors such as LIDARs and RGB-D cameras have become a popular choice for indoor localization and mapping. However, due to the lack of direct frame-to-frame correspondences, the tracking traditionally relies on the iterative closest point technique which does not scale well with the number of points. In this paper, we build on top of more recent and efficient density distribution alignment methods, and notably push the idea towards a highly efficient and reliable solution for full 6DoF motion estimation with only depth information. We propose a divide-and-conquer technique during which the estimation of the rotation and the three degrees of freedom of the translation are all decoupled from one another. The rotation is estimated absolutely and drift-free by exploiting the orthogonal structure in man-made environments. The underlying algorithm is an efficient extension of the mean-shift paradigm to manifold-constrained multiple-mode tracking. Dedicated projections subsequently enable the estimation of the translation through three simple 1D density alignment steps that can be executed in parallel. An extensive evaluation on both simulated and publicly available real datasets comparing several existing methods demonstrates outstanding performance at low computational cost.

1 Introduction

3D depth sensors are a powerful alternative to cameras when it comes to automated localization and mapping. They perform especially well in man-made indoor environments, which are often composed of homogeneously colored planar pieces, and thus provide sufficient well-defined 3D structures for depth sensors, but insufficient texture for a reliable application of classical image-based localization techniques. Further advantages of active sensing are given by absolute (metric) scale operation (and therefore absence of scale drift) and resilience against illumination or appearance changes in the environment, ultimately even permitting operation in complete darkness. Depth sensors are an

Electronic supplementary material The online version of this chapter (doi:10. 1007/978-3-319-54193-8_1) contains supplementary material, which is available to authorized users.

© Springer International Publishing AG 2017
S.-H. Lai et al. (Eds.): ACCV 2016, Part V, LNCS 10115, pp. 3–19, 2017.
DOI: 10.1007/978-3-319-54193-8_1

engineering answer to the inverse problem of structure-from-motion, and ubiquitous success is demonstrated by numerous successful applications in robotics [1,2], autonomous driving (e.g. *Google Chauffeur*), and—more recently—consumer electronics (e.g. *Google Tango, Meta Glass*).

Depth sensors produce point cloud measurements. The fundamental problem behind incremental motion estimation with depth sensors therefore is the registration of two 3D point sets A and B. The most popular technique by far is given by the Iterative Closest Point (ICP) method [3]. The basic idea is straightforward: We find approximate correspondences between pairs of points between A and B by simply associating the spatially nearest neighbor of set B to each point of set A. We then minimize the sum of squared distances over a euclidean transformation in closed form. We finally iterate over these two steps until convergence. The complexity of the algorithm is an immediate consequence of the need to find the closest point for each point in each iteration. Even the fastest implementations [4,5] therefore fail to deliver real-time performance on CPU as soon as we consider modern sensors returning dense depth images at VGA resolution. Distance-transform based ICP variants such as the ones used in KinectFusion [6] and Kintinuous [7] achieve real-time performance, however only by leveraging the power of a GPU.

A more efficient alternative registration principle transforms the data into lower dimensional, spatial density distribution functions [8]. The general advantage of density alignment based methods is that they do no longer depend on the establishment of one-to-one or even weighted, fuzzy one-to-many point correspondences [9]. Our work lifts this concept to a general, real-time motion estimation framework for 3D sensors. The key of our approach consists of exploiting the structure of man-made environments, which often contain sets of orthogonal planar pieces. We furthermore rely on efficient dense surface normal vector computation in order to estimate the rotation independently of the translation. As we will show, the exploitation of this prior furthermore allows us to split up the translational alignment of the density distribution functions into three independent steps, namely one along each direction in the corresponding cartesian coordinate frame.

In summary, we present a highly efficient motion estimation framework for popular 3D sensors such as the Microsoft Kinect, based on alignment of density distribution functions. Our contributions are listed as follows:

- Efficient, decoupled estimation of camera rotation using mean-shift for multimode tracking in surface normal vector distributions.
- Estimation of absolute rotation by exploiting the properties of Manhattan Worlds, thus resulting in manifold-constrained multi-mode tracking.
- Efficient decoupled estimation of individual translational degrees of freedom through 1D kernel density estimates.
- Integration into a real-time framework able to process dense depth images with VGA resolution at more than 50 Hz on a laptop with only CPU resources. The result is an attractive 6 DoF tracker for autonomous mobile systems, which often have limited computational resources or energy supply.

We conclude the introduction by reviewing related work. Section 2 then introduces our main idea for motion estimation in Manhattan Worlds based on 3D sensors. The decoupled estimation of rotation and translation are presented in Sects. 3 and 4, respectively. Section 5 finally presents our extensive experimental evaluation on both simulated and real data. We test and evaluate our algorithm against existing alternatives on publicly available datasets, showcasing outstanding performance at the lowest computational cost.

Related Work: 3D Point set registration is a traditional problem that has been investigated extensively in the computer vision community. We are limiting the discussion to methods that process mainly rigid, geometric information. The most commonly used method is given by the ICP algorithm [3], which performs registration through iterative minimization of the SSD distance between spatial neighbors in two point sets. The costly repetitive derivation of point-to-point correspondences can be circumvented by representing and aligning point clouds using density distribution functions. The idea goes back to [10,11], who represent point clouds as explicit Gaussian Mixture Models (GMM) or implicit Kernel Density Estimates (KDE), and then find the relative transformation (not necessarily Euclidean) by aligning those density distributions. [8] summarizes the idea of using GMMs for finding the aligning transformation, and notably derives a closed-form expression for computing the L2 distance between two GMMs. Yet another alternative which avoids the establishment of point-to-point correspondences is given by [12], which utilizes a distance transformation in order to efficiently and robustly compute the cost of an aligning transformation. The distance transformation itself, however, is again computationally intensive.

Classical ICP or even density alignment based methods are prone to local minima as soon as the displacement is too large. In order to tackle situations of large view-point changes, [13] investigated globally optimal solutions to the point set registration problem. This method is however inefficient and thus not suited for real-time applications, where the frame-to-frame displacement anyway remains small enough for a successful application of local methods.

From a more modern perspective, the ICP algorithm and its close derivatives [4–7] still represent the algorithm of choice for real-time LIDAR tracking. The upcoming of RGB-D cameras has however led to a new generation of 2D-3D registration algorithms that exercise a hybrid use of both depth and RGB information. [14] for instance uses the depth information along with the optimized relative transformation to warp the image from one frame to the next, thus permitting direct and dense photometric error minimization. [15–18] apply a similar idea to RGB camera tracking. More recently, [19] even applied ICP and distance transforms to semi-dense 2D-3D registration.

The special structure of man-made environments can be exploited to simplify or even robustify the formulation of motion estimation with exteroceptive sensors. [20,21] introduce planar surfaces into the mapper which are often contained in our man-made environments. [22] combines point and plane features towards fast and accurate 3D registration. In our work, we additionally exploit the fact that indoor environments such as corridors frequently contain orthogonal structure in the surface arrangement. [23] coined the term *Manhattan World* (MW) to denote such an environment, and they estimated the camera orientation through Bayesian vanishing point estimation in a single RGB image. [24] presents a video compass using a similar idea. Tracking the *Manhattan Frame* (MF) can be regarded as absolute orientation estimation, and thus leads to significant reduction or even complete elimination of the rotational drift. Silberman et al. [25] improve VP-based MW orientation estimation by introducing depth and surface normal information obtained from 3D sensors. More recently, [26] proposes the inference of an explicit probabilistic model to describe the world as a mixture of Manhattan frames. They employ an adaptive Markov-Chain Monte-Carlo sampling algorithm with Metropolis-Hasting split/merge moves to identify von-Mises-Fisher distributions of the surface normal vectors. In [27], they adapt the idea to a more computationally friendly approach for real-time tracking of a single, dominant MF. Our work is closely related, except that our mean-shift tracking scheme [28] is simpler and more computationally efficient than the MAP inference scheme presented in [27], which depends on approximations using the Karcher mean in order to achieve real-time performance. We furthermore extend the idea to full 6DoF motion estimation.

2   Overview of the Proposed Algorithm

Our method is summarized in Fig. 1, and consists of three main steps. Note again that we use only depth information:

– We first start by extracting surface normal vectors $\mathbf{n}_i$ from the measured point clouds, which later allows us to compute the orientation of the sensor independently of the translation. Our method is a hyper-threaded CPU implementation of the approach presented in [29], which can efficiently return

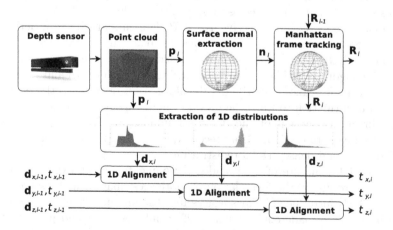

Fig. 1. Overview of the proposed, decoupled motion estimation framework for 3D sensors in Manhattan World.

normal vectors for every pixel in a dense depth image. In order to get smooth and regularized surface normal vectors, the depth map is pre-processed by a smoothing guided filter [30].

– We then rely on the assumption that there is a dominant MF in the environment. This allows us to simply track a number of modes in the density distribution of the surface normal vectors, which can be done in a non-parametric way by employing the mean shift algorithm on the unit sphere. It prevents us from having to identify the parameters of a complete explicit model of the density distribution function. We present a manifold-constrained mean-shift algorithm that takes the orthogonality prior into account. Note that the optimization of the rotation is not a classical registration step, but a simple tracking procedure that uses information of a single frame only to produce a drift-free estimate of the absolute orientation.

– By knowing the absolute orientation in each frame, we can easily unrotate the point clouds of a frame pair and assume that the transformation that separates them is a pure translation. A further benefit is that the principal directions of a Gaussian Mixture Model of the point cloud can be constrained to align with the basis axes. In other words, the covariance matrices become diagonal by which the purely translational alignment cost can effectively be split up into three independent terms, namely one for each dimension. We are therefore allowed to simply solve for each translational degree of freedom independently. We notably do so by extracting kernel density distributions of the point clouds projected onto the basis axes, and by performing three simple 1D alignments. Again note that—due to the unrotation—the obtained relative displacement is immediately expressed in the world frame.

We will in the following explain the details of the rotation and translation alignment.

3 Absolute Orientation Estimation Based on Manifold-Constrained Mean-Shift Tracking

We estimate the absolute orientation by tracking a dominant MF in the surface normal vector distribution of each frame. We will start by introducing the mean-shift tracking scheme that operates under the assumption that a sufficiently close initialization point is known. We then conclude by explaining the initialization in the very first frame, which builds on top of our mean-shift extension.

3.1 Basic Idea

For structures that obey the MW assumption, the surface normal vectors $\mathbf{n}_i$ have an organized distribution on the unit sphere $\mathbb{S}^2$, which can be exploited for recognizing the MW orientation. It is reasonable to assume that the unit vectors $\mathbf{n}_i$ are samples of a probability density function, as they are more likely to be distributed around the basis axes of the MW (in both directions). The process of finding the dominant axes is therefore equivalent to mode seeking in this density distribution (i.e. finding local maxima in the density distribution function). The modes are additionally constrained to be orthogonal with respect to each other. We therefore express the MF by a proper 3D rotation matrix $\mathbf{R} \in SO(3)$ of which each column $\mathbf{r}_j$ captures the direction of one of the dominant axes of the MF. Special care however needs to be taken in order to deal with the non-uniqueness of the representation, as each $\mathbf{r}_j$ could in principle be replaced by its negative (although we ensure that $\mathbf{R}$ always remains a right-handed matrix).

A popular, fast, and notably non-parametric method to seek modes is given by the mean shift algorithm [31]. Given an approximate location for a mode, the algorithm applies local Kernel Density Estimation (KDE) to iteratively take steps in the direction of increasing density. We apply this idea to our unit normal vectors on the manifold $\mathbb{S}^2$ using a Gaussian kernel over conic section windows of the unit sphere. The result is optimal under the assumption that the angles between the normal vectors and their corresponding mode centre have a Gaussian distribution. We independently compute one mean shift vector for each basis vector $\mathbf{r}_j$, which potentially results in a non-orthogonal updated MF $\hat{\mathbf{R}}$. We therefore finish each overall iteration by reprojecting $\hat{\mathbf{R}}$ onto the nearest $\mathbf{R} \in SO(3)$. The following explains the update of each mode within a single mean-shift iteration, as well as the projection back onto $SO(3)$.

3.2   Mean Shift on the Unit Sphere

The core of our method is a single mean shift iteration for a dominant axis given a set of normal vectors on $\mathbb{S}^2$. It works as follows:

- We start by finding all normal vectors that are within a neighbourhood of the considered centre $\mathbf{r}_j$. The extent of this neighbourhood is notably defined by the kernel-width of our KDE. In our case, the window is a conic section of the unit sphere and the apex angle of the cone θ_{window} defines the size of the local window. Relevant normal vectors for mode j need to lie inside the respective cone, and thus satisfy the condition

$$\|\mathbf{n}_i \times \mathbf{r}_j\| < \sin(\frac{\theta_{\text{window}}}{2}). \tag{1}$$

Let us define the index i_j which iterates through all $\mathbf{n}_i$ that fulfill the above condition. Note that—if choosing $\theta_{\text{window}} < \frac{\pi}{2}$—every $\mathbf{n}_i$ contributes to at most one mode.

- We then project all contributing $\mathbf{n}_{i_j}$ into the tangential plane at $\mathbf{r}_j$ in order to compute a mean shift. Let

$$\mathbf{Q} = \begin{bmatrix} \mathbf{r}_{mod(j+1,3)} & \mathbf{r}_{mod(j+2,3)} & \mathbf{r}_{mod(j+3,3)} \end{bmatrix}. \tag{2}$$

Then

$$\mathbf{n}'_{i_j} = \mathbf{Q}^T \mathbf{n}_{i_j} \tag{3}$$

represents the normal vector rotated into the MF, with a cyclic permutation of the coordinates such that the last coordinate is along the direction of axis j. In order for the distances in the tangential plane to represent proper geodesics on $\mathbb{S}^2$ (or equivalently angular deviations), we apply the Riemann logarithmic map. The rescaled coordinates in the tangential plane are given by

$$
\mathbf{m}'_{i_j} = \frac{\sin^{-1}(\lambda)\,\text{sign}(n'_{i_j,z})}{\lambda} \begin{bmatrix} n'_{i_j,x} \\ n'_{i_j,y} \end{bmatrix}, \tag{4}
$$

where $\lambda = \sqrt{n'^2_{i_j,x} + n'^2_{i_j,y}}$.

Note that this projection has the advantage of correctly projecting normal vectors from either direction into the same tangential plane.

– We compute the mean shift in the tangential plane

$$
\mathbf{s}'_j = \frac{\sum_{i_j} e^{-c\|\mathbf{m}'_{i_j}\|^2}\,\mathbf{m}'_{i_j}}{\sum_{i_j} e^{-c\|\mathbf{m}'_{i_j}\|^2}}. \tag{5}
$$

where c is a design parameter that defines the width of the kernel.

– To conclude, we transform the mean shift back onto the unit sphere using the Riemann exponential map

$$
\mathbf{s}_j = \overline{\left[\frac{\tan(\|\mathbf{s}'_j\|)}{\|\mathbf{s}'_j\|}\mathbf{s}'^T_j \; 1 \right]}^T, \tag{6}
$$

where $\overline{[\cdot]}$ returns the input 3-vector divided by its norm.

– The updated direction $\hat{\mathbf{r}}_j$ is finally obtained by reapplying the current rotation with permuted axes

$$
\hat{\mathbf{r}}_j = \mathbf{Q}\mathbf{s}_j. \tag{7}
$$

3.3 Maintaining Orthogonality

After computing a mean shift for each mode $\mathbf{r}_j$, we effectively obtain an expression for the updated "rotation matrix"

$$
\hat{\mathbf{R}} = \begin{bmatrix} \hat{\mathbf{r}}_0 & \hat{\mathbf{r}}_1 & \hat{\mathbf{r}}_2 \end{bmatrix}. \tag{8}
$$

This update may however violate the orthogonality constraint on our rotation matrix. We easily circumvent this problem by re-projecting $\hat{\mathbf{R}}$ onto the closest matrix on $SO(3)$ under the Frobenius norm. Each column of $\hat{\mathbf{R}}$ is re-weighted by a factor λ_i which describes how certain the observation of a direction is. In order to determine the weighting factors, we introduce a non-parametric variance approximation by utilizing a double parzen-widow-based KDE. The method is detailed in the supplemental material. The updated rotation matrix is finally given by

$$
\mathbf{R} = \mathbf{U}\mathbf{V}^T, \text{ where} \tag{9}
$$

$$
[\mathbf{U}, \mathbf{D}, \mathbf{V}] = \text{SVD}\left(\begin{bmatrix} \lambda_0\hat{\mathbf{r}}_0 & \lambda_1\hat{\mathbf{r}}_1 & \lambda_2\hat{\mathbf{r}}_2 \end{bmatrix} \right). \tag{10}
$$

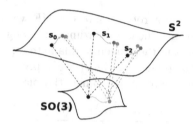

Fig. 2. Illustration of our cascaded manifold-constrained mean-shift implementation. We first compute updates s_j for each mode on $\mathbb{S}^2$, which brings us from the black to the blue modes. The blue modes however do no longer represent a point on the underlying manifold SO(3). We find the nearest rotation through a projection onto the manifold (green arrow), thus returning the red modes which are closest and at the same time fulfill the orthogonality constraint. (Color figure online)

As illustrated in Fig. 2, our method thus represents a double, cascaded manifold-constrained mean-shift extension, where the update of each mode is enforced to remain on the $\mathbb{S}^2$ manifold, and the combination of all three modes is each time enforced to remain an element on the $SO(3)$ manifold. In other words, in each iteration we compute the $SO(3)$-consistent update that is closest to the individual mean-shift updates.

3.4   Initialization in the First Frame

We use mean-shift clustering to initialize the algorithm, and thus build on top of our MF tracking scheme. The procedure is summarized in Fig. 3. We simply run the MF tracking procedure for 100 times, each time starting from a random initial rotation. This returns a redundant set of candidate MFs, within which we need to identify the most dominant cluster in order to complete the initialization. In fact, typically only a very small number of trials will not converge to the dominant MF if there is only one MF in the observed scene. However, the MF estimates are not directly comparable since one and the same MF may indeed be found or represented by any permutation or negation of individual basis vectors, as long as the result remains a right-handed matrix. In fact, there are 24 possible representations for one and the same MF. In order to render the results comparable and identify the dominant MF cluster, we convert the matrices into a canonical form based on a set of simple rules. For instance, the number of possible representations can already be reduced to 4 by simply requiring the basis vector with the potentially highest z-coordinate to be the one corresponding to the z-axis. To finally identify the dominant cluster, we simply group them based on a simple distance metric between rotation matrices, as well as a fixed threshold.

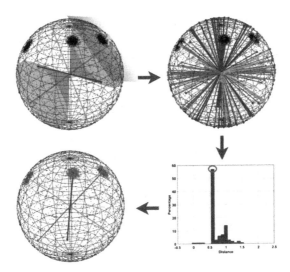

Fig. 3. The mechanism of the initial Manhattan frame seeking. The first figure shows a random initial MF. As indicated by one example, each dominant direction is refined by performing mean-shifts on the corresponding tangential plane. The second figure shows the redundant result obtained after full MF fitting from 100 random starts. The redundancy of the estimated rotation matrices **R** is removed by first converting them into a canonical form, and then performing histogram-based non-maximum suppression. The final result is shown in the fourth figure. For the sake of a clear visualization, the illustrated example is contaminated by a rather significant amount of uniformly distributed noise. Note that the proposed seeking strategy is even able to find multiple MFs in the environment, and thus come up with a mixture of Manhattan frames.

4   Translation Estimation Through Separated 1-D Alignments

In this section, we show that by taking advantage of the MW properties, the translation in each dominant direction can be estimated separately. We then discuss the 1D alignments which rely on kernel density distribution functions. A convergence analysis is given in Sect. 2 of the supplementary material.

4.1   Independence of the Three Translational Degrees of Freedom

Although we are not using an explicit model for representing the density distributions, let us assume for a moment that it is given by a simple Gaussian (i.e. a toy GMM) to see the implications of a Manhattan world and a known absolute orientation of the Manhattan frame. A Gaussian in 3D with mean $\boldsymbol{\mu}$ and covariance $\boldsymbol{\Sigma}$ is simply given by

$$\phi(\mathbf{x}|\boldsymbol{\mu}, \boldsymbol{\Sigma}) = \frac{\exp[-0.5(\mathbf{x} - \boldsymbol{\mu})^T \boldsymbol{\Sigma}^{-1}(\mathbf{x} - \boldsymbol{\mu})]}{\sqrt{(2\pi)^3|\det(\boldsymbol{\Sigma})|}}. \tag{11}$$

There are two Gaussians in two frames and—using the known absolute orientations to unrotate the point clouds—they are separated by a pure translation $\mathbf{t}$. By adding $\mathbf{t}$ to the mean of the Gaussian in the second frame, the kernel correlation between the two Gaussians can be calculated by

$$
\begin{aligned}
\mathrm{D} &= \int \phi(\mathbf{x}|\boldsymbol{\mu}_1, \boldsymbol{\Sigma}_1)\phi(\mathbf{x}|(\boldsymbol{\mu}_2 + \mathbf{t}), \boldsymbol{\Sigma}_2)d\mathbf{x} \\
&= \phi(0|\boldsymbol{\mu}_1 - \boldsymbol{\mu}_2 - \mathbf{t}, \boldsymbol{\Sigma}_1 + \boldsymbol{\Sigma}_2).
\end{aligned} \tag{12}
$$

We now simplify the case by assuming that the unrotated point clouds can be expressed by a 3D Gaussian distribution with a diagonal covariance matrix. This is reasonable since the unrotated point clouds will indeed contain sets of points that are parallel to the basis axes. Let $\boldsymbol{\Sigma}_d = \boldsymbol{\Sigma}_1 + \boldsymbol{\Sigma}_2 = \mathrm{diag}(\sigma_{dx}, \sigma_{dy}, \sigma_{dz})$, and $\boldsymbol{\mu}_d = \boldsymbol{\mu}_1 - \boldsymbol{\mu}_2$. Then the kernel correlation becomes

$$
\begin{aligned}
\mathrm{D} &= \frac{\exp\left[-0.5\left(\frac{(t_x - \mu_{dx})^2}{\sigma_{dx}} + \frac{(t_y - \mu_{dy})^2}{\sigma_{dy}} + \frac{(t_z - \mu_{dz})^2}{\sigma_{dz}}\right)\right]}{\sqrt{(2\pi)^3 \sigma_{dx}\sigma_{dy}\sigma_{dz}}} \\
&= k \cdot e^{\frac{(t_x - \mu_{dx})^2}{-2\sigma_{dx}}} e^{\frac{(t_y - \mu_{dy})^2}{-2\sigma_{dy}}} e^{\frac{(t_z - \mu_{dz})^2}{-2\sigma_{dz}}}.
\end{aligned} \tag{13}
$$

The goal of the alignment in this toy example is to find $\mathbf{t}$ such that D is maximized. It is clear that the above expression involves the product of three independent and positive elements, which means that maximizing each one independently will also maximize the overall distance between the Gaussians. Note that—in practice—the shape of the measured distributions is also influenced by occlusions under motion. However, we confirmed through our experiments that this has a neglible influence on the accuracy of the translation estimation in frame-to-frame motion estimation, as the location of the peaks in the distribution typically remains very stable.

4.2   Alignment of Kernel Density Distributions

Our translation alignment procedure relies on implicit kernel density distribution functions. Assuming that the absolute orientation with respect to the MF is given, each degree of freedom can be solved independently, as in our toy GMM-based example. We therefore compensate for the absolute rotation of the point clouds, and project them onto each basis axis to obtain three independent 1D point sets. Inspired by popular point-set registration works, we then express the 1D point sets via kernel density distribution functions. We sample the function at regular intervals between the minimal and the maximal value. A Gaussian kernel with constant width is used to extract the density at each sampling position. Finally, the alignment between pairs of discretely sampled 1D signals seeks the 1D shift that minimizes the correlation distance between the two signals. It is worth to note that minimizing the correlation distance is equivalent to maximizing the kernel correlation as discussed above. The correlation distance for each pair of 1-D discrete signals is defined as

$$\mathcal{F} = \sum_{i=1}^{n} \left(f(x_i + t) - g(x_i) \right)^2, x_i \in X, \qquad (14)$$

where X denotes a set of sampling positions for which a density is extracted using a Gaussian kernel. The functions f and g record the density at discrete sampling positions. The correlation distance is the sum over the squared differences at each sampling position. t is continuous, and we therefore obtain density values in between the sampled positions by employing linear interpolation. Note that the procedure has linear complexity in the number of points. The convergence analysis of the 1-D alignment is detailed in the supplemental material.

5   Experimental Validation

This section evaluates our algorithm. We start by discussing parameter choices. We then compare our algorithm against two other established state-of-the-art motion estimation solutions on several publicly available datasets. We furthermore provide a reconstruction of a building-scale scene, and conclude by discussing the limitations and failure cases of our method.

Further simulation experiments and analyses are provided in the supplemental material. It contains (1) an evaluation of the robustness of our manifold-constraint mean-shift based MF-seeking strategy and (2) the benefit of aligning the point density distributions along the main axes of the MF.

5.1   Parameter Configuration

In the initial MF seeking (i.e. the initialization of the absolute rotation from scratch), the total number of random starts N_{trial} is set to 100. The apex angle is set to 90° during the initialization and 20° during later tracking. This reduction of the cone apex angle is justified by the assumption that the orientation of the MF does not change too much under smooth motion. Each iterative mean-shift procedure terminates once the angle of the update rotation within one iteration falls below a threshold angle $\theta_{Converge}$, which we set to 1°. The factor c in Eq. (5) is set to 20. Mean-shift updates are furthermore required to have a minimum number N_{min} of surface normal vectors within the dual-cone. The value of N_{min} depends on the resolution of the input depth map. For low resolution sensors (e.g. Kinect v.1, 160×120), $N_{min} = 30$. For high resolution sensors (Kinect v.2, 640×480), $N_{min} = 100$.

The parameters for the translation estimation contain two parts. The first part concerns the extraction of the density distributions. The sampling between the minimum and maximum value along each basis axis is made in constant intervals of $\delta_s = 0.01$ m. The standard deviation σ of the Gaussian kernel for the KDEs is set to 0.03 m. The second part concerns the actual minimization of the correlation distance between each pair of 1D distributions. We simply employ gradient descent with an initial step size of 0.001 m. The search range is furthermore restricted to ±0.1 m.

Table 1. Performance comparison on several indoor datasets.

Dataset	DVO				ICP				Our method			
	$\hat{e}_R$	$\hat{e}_t$	$\tilde{e}_R$	$\tilde{e}_t$	$\hat{e}_R$	$\hat{e}_t$	$\tilde{e}_R$	$\tilde{e}_t$	$\hat{e}_R$	$\hat{e}_t$	$\tilde{e}_R$	$\tilde{e}_t$
TUM 1	4.91	0.15	4.46	0.13	6.64	0.17	6.01	0.15	**1.02**	**0.02**	**0.82**	**0.01**
TUM 2	2.21	0.10	1.59	0.06	9.07	0.27	7.57	0.26	**0.76**	**0.03**	**0.55**	**0.02**
TUM 3	10.90	0.20	3.89	0.07	12.80	0.17	10.17	0.16	**0.94**	**0.04**	**0.70**	**0.02**
TUM 4	**0.57**	**0.02**	**0.47**	**0.02**	8.66	0.29	7.17	0.27	1.01	0.03	0.80	0.03
TUM 5	**0.94**	**0.02**	**0.74**	**0.02**	16.80	0.24	14.19	0.22	1.12	0.04	0.87	**0.02**
IC 1	10.91	1.36	9.37	0.88	6.78	0.15	5.42	0.10	**1.55**	**0.13**	**1.12**	**0.09**
IC 2	6.97	0.70	6.58	0.45	6.31	0.16	5.28	0.10	**1.53**	**0.10**	1.07	**0.08**

5.2 Evaluation on Real Data

We compare the performance of our method against two state-of-the-art, open-source motion estimation implementations for 3D sensors, namely DVO [14] and KinectFusion's ICP [6,7]. DVO uses both RGB images and depth maps while ICP and our algorithm use only depth information. We evaluate the methods on several recently published and challenging benchmark datasets from the TUM RGB-D [32] and IC-NUIM [33] series. The datasets we picked for evaluation are listed below and the results are summarized in Table 1. The selection of the datasets is based on the existence of sufficient MW structure in the observed scenes.

- TUM 1, 2, 3, 4, 5: fr3 (cabinet, structure_notexture/_texture _far/_near)
- IC 1,2: Living Room kt3, Office Room kt3.

Note that for TUM 4, IC 1 and IC 2, our algorithm cannot process the entire sequence due to algorithm limitations that are discussed in the following section. However, in order to remain fair, we evaluate the performance of all algorithms on the same segments of each sequence. A detailed result of the TUM 1 dataset is shown in Fig. 4. We also evaluate each method using the tool given by [32] and provide root-mean-square errors $\hat{e}$ and median errors $\tilde{e}$ per second for both rotation (degree) and translation (meter) estimation in Table 1. The best performing method's error is each time indicated in bold.

It can be seen that in most cases, once the MW assumption is sufficiently met, our result provides very low drift in both rotation and translation. It is outperforming both ICP and DVO in most situations though DVO achieves better performance once there is sufficient texture in the environment. On the other hand, our method remains computationally efficient even on depth images with VGA resolution, and processes frames at about 50 Hz on a CPU. While DVO is real-time capable as well (about 30 Hz), ICP quickly drops in computational efficiency as the number of points increases, and can only work in real-time with the help of a powerful GPU.

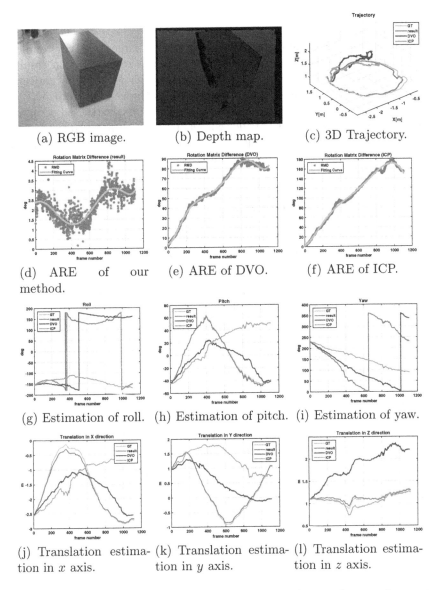

(a) RGB image.   (b) Depth map.   (c) 3D Trajectory.

(d) ARE of our (e) ARE of DVO.   (f) ARE of ICP.
method.

(g) Estimation of roll. (h) Estimation of pitch. (i) Estimation of yaw.

(j) Translation estima- (k) Translation estima- (l) Translation estimation in x axis.   tion in y axis.   tion in z axis.

Fig. 4. Evaluation of our method on the TUM dataset *cabinet* and comparison to two alternative odometry solutions (DVO and ICP). We provide the 3D trajectory, the absolution rotation error (ARE), and the translational error in each degree of freedom for each method. Our method (red curve) outperforms both DVO (blue curve) and ICP (magenta curve) in terms of absolute drift in rotation and translation. Relative pose errors can be found in Table 1. Note that only DVO uses RGB images. (Color figure online)

5.3   3D Reconstruction

In order to demonstrate that our algorithm can work in larger scale environments such as corridors and open-space offices, we present a reconstruction result of the TAMU RGB-D dataset (corridor A const) [34] in Fig. 5. The trajectory is about 40 m long. Our algorithm robustly tracks the camera until only one dominant direction of the MW can be observed. The reconstructed structures (walls and ground, walls at the corridor corner) preserve orthogonality very well, which demonstrates the good quality of the motion estimation. Note that only depth information is used for the tracking. Color information is only used for visualization purposes.

5.4   Limitations and Failure Cases

The existence of a MW structure in the environment is key to the proposed method. Therefore, the effectiveness of our work currently has the following limitations:

- Only one mode of a MF is observed.
- If only two orthogonal planes are observed, the tracking can continue. However, due to the loss of structural information, the density distribution in the unobserved direction becomes very homogeneous, and the estimation of the respective translation becomes inaccurate.
- In the case where two MFs are very close to each other (which could happen in so-called Atlanta environments), our mean-shift scheme may converge in between the two modes, which leads to inaccurate rotation estimation and thus also potentially wrong translation estimation.

Fig. 5. Reconstruction of a corridor scene.

6   Discussion

We present an efficient alternative to the iterative closest point algorithm for real-time tracking of modern depth cameras in Manhattan Worlds. We exploit the common orthogonal structure of man-made environments in order to decouple the estimation of the rotation and the three degrees of freedom of the translation. The derived camera orientation is absolute and thus free of long-term drift, which in turn benefits the accuracy of the translation estimation as well. We achieve not only competitive accuracy, but also superior computational efficiency. Our method operates robustly in large-scale environments, even if the Manhattan World assumption is not fully met. In summary, the presented framework has high value in mobile robotics or industrial applications, where computational load or the lack of texture are major concerns. Code will be released as open-source.

Our future work consists of removing the restriction to pure Manhattan worlds. By adding a real-time mode detection and removal module, we can extend our work to the more general case of piece-wise planar environments. Interestingly, the cascaded mean-shift strategy presented in this work will still be applicable, the only difference being that the underlying manifold will no longer be $SO(3)$, but the manifold of all direction bundles with constant inscribed angles.

Acknowledgement. The research leading to these results is supported by Australian Centre for Robotic Vision. The work is furthermore supported by ARC grants DE150101365. Yi Zhou acknowledges the financial support from the China Scholarship Council for his Ph.D. Scholarship No. 201406020098.

References

1. Bachrach, A., Prentice, S., He, R., Henry, P., Huang, A.S., Krainin, M., Maturana, D., Fox, D., Roy, N.: Estimation, planning, and mapping for autonomous flight using an RGB-D camera in GPS-denied environments. Int. J. Rob. Res. **31**, 1320–1343 (2012)
2. Bohren, J., Rusu, R.B., Jones, E.G., Marder-Eppstein, E., Pantofaru, C., Wise, M., Mösenlechner, L., Meeussen, W., Holzer, S.: Towards autonomous robotic butlers: lessons learned with the PR2. In: 2011 IEEE International Conference on Robotics and Automation (ICRA), pp. 5568–5575. IEEE (2011)
3. Besl, P.J., McKay, N.D.: Method for registration of 3-D shapes. In: Robotics-DL Tentative, International Society for Optics and Photonics, pp. 586–606 (1992)
4. Pomerleau, F., Magnenat, S., Colas, F., Liu, M., Siegwart, R.: Tracking a depth camera: parameter exploration for fast ICP. In: 2011 IEEE/RSJ International Conference on Intelligent Robots and Systems (IROS), pp. 3824–3829. IEEE (2011)
5. Pomerleau, F., Colas, F., Siegwart, R., Magnenat, S.: Comparing ICP variants on real-world data sets. Auton. Rob. **34**, 133–148 (2013)
6. Newcombe, R.A., Izadi, S., Hilliges, O., Molyneaux, D., Kim, D., Davison, A.J., Kohi, P., Shotton, J., Hodges, S., Fitzgibbon, A.: KinectFusion: real-time dense surface mapping and tracking. In: 2011 10th IEEE International Symposium on Mixed and Augmented Reality (ISMAR), pp. 127–136. IEEE (2011)
7. Whelan, T., Kaess, M., Fallon, M., Johannsson, H., Leonard, J., McDonald, J.: Kintinuous: spatially extended KinectFusion. In: RSS Workshop on RGB-D: Advanced Reasoning with Depth Cameras, Sydney, Australia (2012)
8. Jian, B., Vemuri, B.C.: Robust point set registration using gaussian mixture models. IEEE Trans. Pattern Anal. Mach. Intell. **33**, 1633–1645 (2011)
9. Chui, H., Rangarajan, A.: A new algorithm for non-rigid point matching. In: Proceedings of the IEEE Conference on Computer Vision and Pattern Recognition, vol. 2, pp. 44–51. IEEE (2000)
10. Chui, H., Rangarajan, A.: A feature registration framework using mixture models. In: Proceedings of the IEEE Workshop on Mathematical Methods in Biomedical Image Analysis, pp. 190–197. IEEE (2000)
11. Tsin, Y., Kanade, T.: A correlation-based approach to robust point set registration. In: Pajdla, T., Matas, J. (eds.) ECCV 2004. LNCS, vol. 3023, pp. 558–569. Springer, Heidelberg (2004). doi:10.1007/978-3-540-24672-5_44

12. Fitzgibbon, A.W.: Robust registration of 2D and 3D point sets. Image Vis. Comput. **21**, 1145–1153 (2003)
13. Yang, J., Li, H., Jia, Y.: Go-ICP: solving 3D registration efficiently and globally optimally. In: 2013 IEEE International Conference on Computer Vision (ICCV), pp. 1457–1464. IEEE (2013)
14. Kerl, C., Sturm, J., Cremers, D.: Robust odometry estimation for RGB-D cameras. In: 2013 IEEE International Conference on Robotics and Automation (ICRA), pp. 3748–3754. IEEE (2013)
15. Newcombe, R.A., Lovegrove, S.J., Davison, A.J.: DTAM: dense tracking and mapping in real-time. In: 2011 IEEE International Conference on Computer Vision (ICCV), pp. 2320–2327. IEEE (2011)
16. Engel, J., Sturm, J., Cremers, D.: Semi-dense visual odometry for a monocular camera. In: 2013 IEEE International Conference on Computer Vision (ICCV), pp. 1449–1456. IEEE (2013)
17. Engel, J., Schöps, T., Cremers, D.: LSD-SLAM: large-scale direct monocular SLAM. In: Fleet, D., Pajdla, T., Schiele, B., Tuytelaars, T. (eds.) ECCV 2014. LNCS, vol. 8690, pp. 834–849. Springer, Heidelberg (2014). doi:10.1007/978-3-319-10605-2_54
18. Schöps, T., Engel, J., Cremers, D.: Semi-dense visual odometry for ar on a smartphone. In: 2014 IEEE International Symposium on Mixed and Augmented Reality (ISMAR), pp. 145–150. IEEE (2014)
19. Kneip, L., Zhou, Y., Li, H.: SDICP: semi-dense tracking based on iterative closest points. In: Xianghua Xie, M.W.J., Tam, G.K.L. (eds.) Proceedings of the British Machine Vision Conference (BMVC), pp. 100.1–100.12. BMVA Press, Guildford (2015)
20. Weingarten, J., Siegwart, R.: 3D slam using planar segments. In: 2006 IEEE/RSJ International Conference on Intelligent Robots and Systems, pp. 3062–3067. IEEE (2006)
21. Trevor, A.J., Rogers III, J.G., Christensen, H., et al.: Planar surface slam with 3D and 2D sensors. In: 2012 IEEE International Conference on Robotics and Automation (ICRA), pp. 3041–3048. IEEE (2012)
22. Taguchi, Y., Jian, Y.D., Ramalingam, S., Feng, C.: Point-plane slam for hand-held 3D sensors. In: 2013 IEEE International Conference on Robotics and Automation (ICRA), pp. 5182–5189. IEEE (2013)
23. Coughlan, J.M., Yuille, A.L.: Manhattan world: compass direction from a single image by Bayesian inference. In: The Proceedings of the Seventh IEEE International Conference on Computer Vision, vol. 2, pp. 941–947. IEEE (1999)
24. Košecká, J., Zhang, W.: Video compass. In: Heyden, A., Sparr, G., Nielsen, M., Johansen, P. (eds.) ECCV 2002. LNCS, vol. 2353, pp. 476–490. Springer, Heidelberg (2002). doi:10.1007/3-540-47979-1_32
25. Silberman, N., Hoiem, D., Kohli, P., Fergus, R.: Indoor segmentation and support inference from RGBD images. In: Fitzgibbon, A., Lazebnik, S., Perona, P., Sato, Y., Schmid, C. (eds.) ECCV 2012. LNCS, vol. 7576, pp. 746–760. Springer, Heidelberg (2012). doi:10.1007/978-3-642-33715-4_54
26. Straub, J., Rosman, G., Freifeld, O., Leonard, J.J., Fisher, J.W.: A mixture of manhattan frames: beyond the manhattan world. In: 2014 IEEE Conference on Computer Vision and Pattern Recognition (CVPR), pp. 3770–3777. IEEE (2014)
27. Straub, J., Bhandari, N., Leonard, J.J., Fisher III, J.W.: Real-time manhattan world rotation estimation in 3D. In: IROS (2015)
28. Fukunaga, K., Hostetler, L.D.: The estimation of the gradient of a density function, with applications in pattern recognition. IEEE Trans. Inf. Theory **21**, 32–40 (1975)

29. Holz, D., Holzer, S., Rusu, R.B., Behnke, S.: Real-time plane segmentation using RGB-D cameras. In: Röfer, T., Mayer, N.M., Savage, J., Saranlı, U. (eds.) RoboCup 2011. LNCS (LNAI), vol. 7416, pp. 306–317. Springer, Heidelberg (2012). doi:10. 1007/978-3-642-32060-6_26

30. He, K., Sun, J., Tang, X.: Guided image filtering. IEEE Trans. Pattern Anal. Mach. Intell. **35**, 1397–1409 (2013)

31. Carreira-Perpiñán, M.: A review of mean-shift algorithms for clustering. arXiv preprint (2015). arXiv:1503.00687

32. Sturm, J., Engelhard, N., Endres, F., Burgard, W., Cremers, D.: A benchmark for the evaluation of RGB-D slam systems. In: IEEE/RSJ International Conference on Intelligent Robots and Systems (IROS) (2012)

33. Handa, A., Whelan, T., McDonald, J., Davison, A.J.: A benchmark for RGB-D visual odometry, 3D reconstruction and slam. In: IEEE International Conference on Robotics and Automation (ICRA) (2014)

34. Lu, Y., Song, D.: Robustness to lighting variations: an RGB-D indoor visual odometry using line segments. In: 2015 IEEE/RSJ International Conference on Intelligent Robots and Systems (IROS), pp. 688–694. IEEE (2015)

Visual Saliency Detection for RGB-D Images with Generative Model

Song-Tao Wang[1,2], Zhen Zhou[1(✉)], Han-Bing Qu[2], and Bin Li[2]

[1] The Higher Educational Laboratory for Measuring & Control Technology
and Instrumentations of Heilongjiang Province,
Harbin University of Science and Technology, Harbin, China
zhzh49@126.com
[2] Key Laboratory of Pattern Recognition,
Beijing Academy of Science and Technology, Beijing, China

Abstract. In this paper, we propose a saliency detection model for RGB-D images based on the contrasting features of colour and depth with a generative mixture model. The depth feature map is extracted based on superpixel contrast computation with spatial priors. We model the depth saliency map by approximating the density of depth-based contrast features using a Gaussian distribution. Similar to the depth saliency computation, the colour saliency map is computed using a Gaussian distribution based on multi-scale contrasts in superpixels by exploiting low-level cues. By assuming that colour- and depth-based contrast features are conditionally independent, given the classes, a discriminative mixed-membership naive Bayes (DMNB) model is used to calculate the final saliency map from the depth saliency and colour saliency probabilities by applying Bayes' theorem. The Gaussian distribution parameter can be estimated in the DMNB model by using a variational inference-based expectation maximization algorithm. The experimental results on a recent eye tracking database show that the proposed model performs better than other existing models.

1 Introduction

Saliency detection is the problem of identifying the points that attract the visual attention of human beings. Le Callet and Niebur introduced the concepts of overt and covert visual attention and the concepts of bottom-up and top-down processing [11]. Visual attention selectively processes important visual information by filtering out less important information and is an important characteristic of the human visual system (HVS) for visual information processing. Visual attention is one of the most important mechanisms that are deployed in the HVS to cope with large amounts of visual information and reduce the complexity of scene analysis. Visual attention models have been successfully applied in many domains, including multimedia delivery, visual retargeting, quality assessment of images and videos, medical imaging, and 3D image applications [11].

Borji and Itti provided an excellent overview of the current state-of-the-art 2D visual attention modelling and included a taxonomy of models (cognitive,

© Springer International Publishing AG 2017
S.-H. Lai et al. (Eds.): ACCV 2016, Part V, LNCS 10115, pp. 20–35, 2017.
DOI: 10.1007/978-3-319-54193-8_2

Bayesian, decision theoretic, information theoretical, graphical, spectral analysis, pattern classification, and more) [3]. Many saliency measures have emerged that simulate the HVS, which tends to find the most informative regions in 2D scenes [4,13,18]. However, most saliency models disregard the fact that the HVS operates in 3D environments and these models can thus investigate only from 2D images. Eye fixation data are captured while looking at 2D scenes, but depth cues provide additional important information about content in the visual field and therefore can also be considered relevant features for saliency detection. The stereoscopic content carries important additional binocular cues for enhancing human depth perception [5,10]. Today, with the development of 3D display technologies and devices, there are various emerging applications for 3D multimedia, such as 3D video retargeting [16], 3D video quality assessment [9,19] and so forth. Overall, the emerging demand for visual attention-based applications for 3D multimedia has increased the need for computational saliency detection models for 3D multimedia content. In contrast to saliency detection for 2D images, the depth factor must be considered when performing saliency detection for RGB-D images. Therefore, two important challenges when designing 3D saliency models are how to estimate the saliency from depth cues and how to combine the saliency from depth features with those of other 2D low-level features.

In this paper, we propose a new computational saliency detection model for RGB-D images that considers both colour- and depth-based contrast features with a generative mixture model. The main contributions of our approach consist of two aspects: (1) to estimate saliency from depth cues, we propose the creation of depth feature maps based on superpixel contrast computation with spatial priors and model the depth saliency map by approximating the density of depth-based contrast features using a Gaussian distribution, and (2) by assuming that colour-based and depth-based features are conditionally independent given the classes, the discriminative mixed-membership naive Bayes (DMNB) model is used to calculate the final saliency map by applying Bayes' theorem.

2 Related Work

As introduced in the Sect. 1, many computational models of visual attention have been proposed for various 2D multimedia processing applications. However, compared with the set of 2D visual attention models, only a few computational models of 3D visual attention have been proposed [6–8,12,14,17,20]. These models all contain a stage in which 2D saliency features are extracted and used to compute 2D saliency maps. However, depending on the way in which they use depth information in terms of the development of computational models, these models can be classified into three different categories:

(1) Depth-weighting models—This type of model adopts depth information to weight a 2D saliency map to calculate the final saliency map for RGB-D images with feature map fusion [6,17]. Fang et al. proposed a novel 3D saliency detection framework based on colour, luminance, texture and depth

contrast features, which designed a new fusion method to combine the feature maps to obtain the final saliency map for RGB-D images [6]. In [17], colour contrast features and depth contrast features are calculated to construct an effective multi-feature fusion to generate saliency maps, and multiscale enhancement is performed on the saliency map to further improve the detection precision focused on the 3D salient object detection. The models in this category combine 2D features with a depth feature to calculate the final saliency map, but they do not include the depth saliency map in their computation processes.

(2) Depth-saliency models—This type of model combines depth saliency maps and traditional 2D saliency maps simply to obtain saliency maps for RGB-D images [8,12,14]. Ren et al. presented a two-stage 3D salient object detection framework, which first integrates the contrast region with the background, depth and orientation priors to achieve a saliency map and then reconstructs the saliency map globally [14]. Peng et al. proved a simple fusion framework that combines existing RGB-produced saliency with new depth-induced saliency: the former one is estimated from existing RGB models while the latter one is based on the multi-contextual contrast model [12]. Furthermore, Ju et al. proposed a novel saliency method that worked on depth images based on anisotropic centre-surround difference [8]. The models in this category rely on the existence of "depth saliency maps." Depth features are extracted from the depth map to create additional feature maps, which are then used to generate the depth saliency maps (DSM). These depth saliency maps are finally combined with 2D saliency maps using a saliency map pooling strategy to obtain a final 3D saliency map.

(3) Learning-based models—Instead of using a depth saliency map directly, this type of model uses machine learning techniques to build a 3D saliency detection model for RGB-D images based on extracted 2D features and depth features [7,20]. Inspired by the recent success of machine learning techniques in building 2D saliency detection models, Fang et al. proposed a learning-based model for RGB-D images using linear SVM [7]. Zhu et al. proposed a learning-based approach for extracting saliency from RGB-D images, in which discriminative features can be automatically selected by learning several decision trees based on the ground truth, and those features are further utilized to search the saliency regions via the predictions of the trees [20].

From the above description, the key to 3D saliency detection models is determining how to integrate the depth cues with traditional 2D low-level features. In this paper, we propose a learning-based 3D saliency detection model with a generative mixture model that considers both colour- and depth-based contrast features. Instead of simply combining a depth map with 2D saliency maps as in previous studies, we propose a computational saliency detection model for RGB-D images based on the DMNB model [15]. Experimental results from a public eye tracking database demonstrate the improved performance of the proposed model over other strategies.

3   The Proposed Approach

In this section, we introduce a method that integrates the colour saliency probability with the depth saliency probability computed from Gaussian distributions based on multi-scale superpixel contrast features and yields a prediction of the final 3D saliency map using the DMNB model within a Bayesian framework. First, the input RGB-D images are represented by superpixels using multi-scale segmentation. Then, we compute the colour and depth map using the weighted summation and normalization of the colour- and depth-based contrast features, respectively, at different scales. Second, the probability distributions of both the colour and depth saliency are modelled using the Gaussian distribution based on the colour and depth feature maps, respectively. The parameters of the Gaussian distribution can be estimated in the DMNB model using a variational inference-based expectation maximization (EM) algorithm. The general architecture of the proposed framework is presented in Fig. 1.

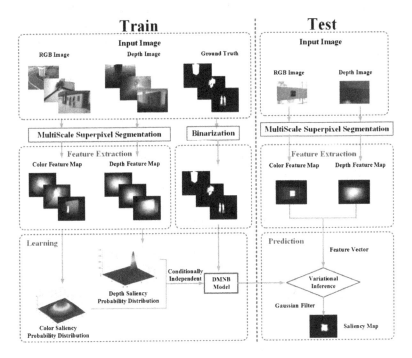

Fig. 1. The flowchart of the proposed model. The framework of our model consists of two stages: the training stage shown in the left part of the figure and the testing stage shown in the right part of the figure. In this work, we perform experiments based on the NLPR dataset in [12].

3.1 Feature Extraction Using Multi-scale Superpixels

We introduce a colour-based contrast feature and a depth-based contrast feature to capture the contrast information of salient regions with spatial priors based on multi-scale superpixels, which are generated at various grid interval parameters $\mathcal{S}$, similar to simple linear iterative clustering (SLIC) [1]. We further impose a spatial prior term on each of the contrast measures holistically, which constrains the pixels that were rendered as salient to be compact as well as centred in the image domain. This spatial prior can also be generalized to consider the spatial distribution of different saliency cues such as the centre prior and background prior [18]. We also observe that the background often presents local or global appearance connectivity with each of four image boundaries. These two features complement each other in detecting 3D saliency cues from different perspectives and, when combined, yield the final 3D saliency value.

RGB-D Images Multi-scale Superpixel Segmentation. For an RGB-D image pair, superpixels are segmented according to both colour and depth cues. We notice that when applying the SLIC algorithm directly to the RGB image and depth map, the segmentation result is unsatisfactory due to the lack of a mutual context relationship. We redefine the distance measurement incorporating depth as shown in Eq. 1:

$$D_s = \sqrt{d_{lab}^2 + \omega_d d_d^2 + \frac{m}{\mathcal{S}} d_{xy}^2} \tag{1}$$

where $d_d = \sqrt{(d_j - d_i)^2}$ denotes the depth distance weighted by ω_d between pixel i and j in the depth map, d_{lab} and d_{xy} are the original distance measurements of colour and spatiality normalized with $\frac{m}{\mathcal{S}}$ in [1], and D_s is the final distance between two pixels in the RGB-D image pair.

We obtain more accurate segmentation results as shown in Fig. 2 by considering the colour and depth cues simultaneously. The boundary between the foreground and the background is segmented more accurately.

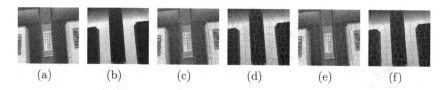

(a)   (b)   (c)   (d)   (e)   (f)

Fig. 2. Visual samples for superpixel segmentation of RGB-D images with $\mathcal{S} = 40$. (a) RGB image, (b) Depth image, (c) Colour-based segmentation, (d) Depth-based segmentation, (e) Colour- and depth-based segmentation result on colour image and (f) Colour- and depth-based segmentation result on depth image. (Color figure online)

Colour-Based Contrast Feature. An input image is oversegmented at L scales, and the colour feature map is formulated as

$$f(p_c^l) = \omega_c^l SC_{GMR}^l \tag{2}$$

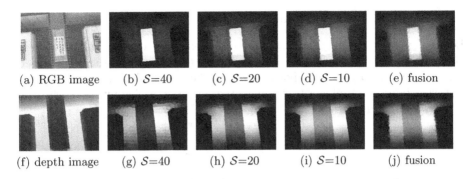

(a) RGB image   (b) $\mathcal{S}$=40   (c) $\mathcal{S}$=20   (d) $\mathcal{S}$=10   (e) fusion

(f) depth image   (g) $\mathcal{S}$=40   (h) $\mathcal{S}$=20   (i) $\mathcal{S}$=10   (j) fusion

Fig. 3. Visual samples of different colour and depth feature maps. Row 1: colour feature maps of the NLPR dataset. Row 2: depth feature maps of the NLPR dataset.

where p_c^l is a quantified histogram in the CIE Lab colour space for each superpixel at any scale l, and SC_{GMR}^l is the colour saliency map generated by graph-based manifold ranking only with background cues similar to [18], in which the RGB image is represented as a single-layer graph with surperpixels as nodes at any l scale. In contrast to [18], the definition of the background priors is inspired by the observation that the patches from the corners of images are more likely to be background and contain considerable scene information that helps distinguish salient objects. With multi-scale fusion, the colour feature map is constructed by weighted summation of $f(p_c^l)$, where the weights are determined by $\sum_{l=1}^{L} \omega_c^l = 1$. The final pixel-wise colour feature map is obtained by assigning the feature value of each superpixel to every pixel belonging to it, as shown in the first row of Fig. 3.

Depth-Based Contrast Feature. Similar to the construction of the colour feature maps, we formulate the depth feature maps based on multi-scale super-pixels in the depth maps:

$$f(p_d^l) = \omega_d^l SD_{GMR}^l \tag{3}$$

where p_d^l is the depth value of the centroid calculated as the mean depth value within the superpixel and SD_{GMR}^l is the depth saliency map generated via graph-based manifold ranking only with background cues. In this work, the weight of the affinity matrix between two nodes in a depth map at any l scales is defined by

$$\omega_{ij}^l = e^{-\frac{(\overline{d_j^l} - \overline{d_i^l})^2}{\sigma^2}} \tag{4}$$

where $\overline{d_j^l}$ and $\overline{d_i^l}$ denote the mean of the superpixel i and superpixel j corresponding to two nodes, respectively, and σ is a constant that controls the strength of the weight in [18]. With multi-scale fusion, the depth feature map is constructed by weighted summation of $f(p_d^l)$, where the weights are determined by $\sum_{l=1}^{L} \omega_d^l = 1$. Visual samples for different depth feature maps are shown in the second row of Fig. 3.

3.2   Bayesian Framework for Saliency Detection

Let the binary random variable z_s denote whether a point belongs to a salient class. Given the observed colour-based contrast feature x_c and the depth-based contrast feature x_d of that point, we formulate the saliency detection as a Bayesian inference problem to estimate the posterior probability at each pixel of the RGB-D image:

$$p(z_s|x_c, x_d) = \frac{p(z_s, x_c, x_d)}{p(x_c, x_d)} \tag{5}$$

where $p(z_s|x_c, x_d)$ is shorthand for the probability of predicting whether a pixel is salient, $p(x_c, x_d)$ is the likelihood of the observed colour-based and depth-based contrast features, and $p(z_s, x_c, x_d)$ is the joint probability of the latent class and observed features, defined as $p(z_s, x_c, x_d) = p(z_s)p(x_c, x_d|z_s)$.

In this paper, the class-conditional mutual information (CMI) is used as a measure of dependence between two features x_c and x_d, which can be defined as $I(x_c, x_d|z_s) = H(x_c|z_s) + H(x_d|z_s) - H(x_c, x_d|z_s)$, where $H(x_c|z_s)$ is the class-conditional entropy of x_c. We employ a CMI threshold τ to discover feature dependencies, as shown in Fig. 4. For CMI between the colour-based contrast feature and depth-based contrast feature less than τ, we assume that x_c and x_d are conditionally independent given the classes z_s, that is, $p(x_c, x_d|z_s) = p(x_c|z_s)p(x_d|z_s)$. This entails the assumption that the distribution of the colour-based contrast features does not change with the depth-based contrast features. Thus, the pixel-wise saliency of the likelihood is given by $p(z_s|x_c, x_d) \propto p(z_s)p(x_c|z_s)p(x_d|z_s)$.

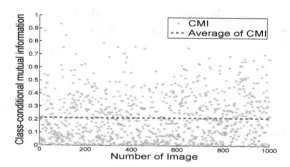

Fig. 4. Visual results for class-conditional mutual information between colour-based contrast features and depth-based contrast features on two RGB-D image datasets.

3.3   Generative Model for Saliency Estimation

Given the graphical model of DMNB for saliency detection shown in Fig. 5, the generative process for $\{x_{1:N}, y\}$ following the DMNB model can be described as follows (Algorithm 1), where $Dir()$ is shorthand for a Dirichlet distribution,

Algorithm 1. Generative process for saliency detection following the DMNB model

1: **Input:** α, η.
2: **Choose a component proportion:** $\theta \sim Dir(\theta|\alpha)$.
3: **For each feature:**
    choose a component $z_j \sim Mult(z_j|\theta)$;
    choose a feature value $x_j \sim p(x_j|z_j, \Omega)$.
4: **Choose the label:** $y \sim p(y|z_j, \eta)$.

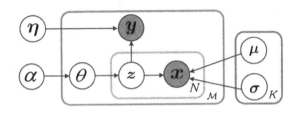

Fig. 5. Graphical models of DMNB for saliency estimation. y and x are the corresponding observed states, and z is the hidden variable.

$Mult()$ is shorthand for a Multinomial distribution, $x_{1:N} = (x_c, x_d)$, $z_{1:N} = z_s = (z_c, z_d)$, N is the number of features, and y is the label that indicates whether the pixel is salient or not.

In this work, both the colour- and depth-based contrast features are assumed to have been generated from a Gaussian distribution with a mean of $\{\mu_{jk}, [j]_1^N\}$ and a variance of $\{\sigma_{jk}^2, [j]_1^N\}$. The marginal distribution of $(x_{1:N}, y)$ is

$$p(x_{1:N}, y|\alpha, \Omega, \eta) = \int p(\theta|\alpha)(\prod_{j=1}^{N} \sum_{z_j} p(z_j|\theta)p(x_j|z_j, \Omega)p(y|z_j, \eta))d\theta \qquad (6)$$

where θ is the prior distribution over K components, $\Omega = \{(\mu_{jk}, \sigma_{jk}^2), [j]_1^N, [k]_1^K\}$ are the parameters for the distributions of N features respectively, $p(x_j|z_j, \Omega) \triangleq \mathcal{N}(x_j|\mu_{jk}, \sigma_{jk}^2)$. In two-class classification, y is either 0 or 1 generated from $Bern(y|\eta)$. Because the DMNB model assumes a generative process for both the labels and features, we use both $\mathcal{X} = \{(x_{ij}), [i]_1^{\mathcal{M}}, [j]_1^N\}$ and $\mathcal{Y} = \{y_i, [i]_1^{\mathcal{M}}\}$ as a collection of $\mathcal{M}$ superpixels in trained images from the generative process to estimate the parameters of the DMNB model such that the likelihood of observing $(\mathcal{X}, \mathcal{Y})$ is maximized. In practice, we may find a proper K using the Dirichlet process mixture model (DPMM) [2]. The DPMM thus provides a nonparametric prior for the parameters of a mixture model that allows the number of mixture components to grow as the training set grows, as shown in Fig. 6.

Due to the latent variables, the computation of the likelihood in Eq. 6 is intractable. In this paper, we use a variational inference method, which alternates between obtaining a tractable lower bound to the true log-likelihood and choosing the model parameters to maximize the lower bound. By a direct application

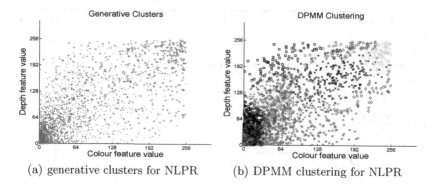

(a) generative clusters for NLPR   (b) DPMM clustering for NLPR

Fig. 6. Visual result for the number of components K in the DMNB model: generative clusters vs DPMM clustering. We find $K = 28$ using DPMM on the NLPR dataset.

of Jensen's inequality [15], the lower bound to $\log p(\boldsymbol{y}, \boldsymbol{x}_{1:N}|\alpha, \Omega, \eta)$ is given by

$$\log p(\boldsymbol{y}, \boldsymbol{x}_{1:N}|\alpha, \Omega, \eta) \geq E_q(\log p(\boldsymbol{y}, \boldsymbol{x}_{1:N}, \boldsymbol{z}_{1:N}|\alpha, \Omega, \eta)) + H(q(\boldsymbol{z}_{1:N}, \theta|\gamma, \phi)) \tag{7}$$

Noticing that $\boldsymbol{x}_{1:N}$ and $\boldsymbol{y}$ are conditionally independent given $\boldsymbol{z}_{1:N}$, we use a variational distribution:

$$q(\boldsymbol{z}_{1:N}, \theta|\gamma, \phi) = q(\theta|\gamma) \prod_{j=1}^{N} q(\boldsymbol{z}_j|\phi) \tag{8}$$

where $q(\theta, \gamma)$ is a K-dimensional Dirichlet distribution for θ, $q(\boldsymbol{z}_j|\phi)$ is Discrete distribution for $\boldsymbol{z}_j$. We use $\mathcal{L}$ to denote the lower bound:

$$\mathcal{L} = E_q[\log p(\theta|\alpha)] + E_q[\log p(\boldsymbol{z}_{1:N}|\theta)] + E_q[\log p(\boldsymbol{x}_{1:N}|\boldsymbol{z}_{1:N}, \gamma)]$$
$$- E_q[\log q(\theta)] - E_q[\log q(\boldsymbol{z}_{1:N})] + E_q[\log p(\boldsymbol{y}|\boldsymbol{z}_{1:N}, \eta)] \tag{9}$$

where $E_q[\log p(\boldsymbol{y}|\boldsymbol{z}_{1:N}, \eta)] \geq \sum_{k=1}^{K} \phi_k(\eta_k \boldsymbol{y} - \frac{e^{\eta_k}}{\xi}) - (\frac{1}{\xi} + \log \xi)$ and $\xi > 0$ is a newly introduced variational parameter. Maximizing the lower-bound function $\mathcal{L}(\gamma_k, \phi_k, \xi; \alpha, \Omega, \eta)$ with respect to the variational parameters yields updated equations for γ_k, ϕ_k and ξ as follows:

$$\phi_k \propto e^{(\Psi(\gamma_k) - \Psi(\sum_{l=1}^{K} \gamma_l) + \frac{1}{N}(\eta_k \boldsymbol{y}_i - \frac{e^{\eta_k}}{\xi} - \sum_{j=1}^{N} \frac{(x_{ij} - \mu_{jk})^2}{2\sigma_{jk}^2}))} \tag{10}$$

$$\gamma_k = \alpha + N\phi_k \tag{11}$$

$$\xi = 1 + \sum_{k=1}^{K} \phi_k e^{\eta_k} \tag{12}$$

Variational parameters $(\gamma^*, \phi^*, \xi^*)$ from the inference step gives the optimal lower bound to the log-likelihood of $(\boldsymbol{x}_i, \boldsymbol{y}_i)$, and maximizing the aggregate lower bound $\sum_{i=1}^{\mathcal{M}} \mathcal{L}(\gamma^*, \phi^*, \xi^*, \alpha, \Omega, \eta)$ over all data points with respect to α, Ω and

Algorithm 2. Variational EM algorithm for DMNB

1: **repeat**
2: **E-step:** Given $(\alpha^{m-1}, \Omega^{m-1}, \eta^{m-1})$, for each feature value and label, find the optimal variational parameters
$(\gamma_i^m, \phi_i^m, \xi_i^m) = \arg\max \mathcal{L}(\gamma_i, \phi_i, \xi_i; \alpha^{m-1}, \Omega^{m-1}, \eta^{m-1})$.
Then, $\mathcal{L}(\gamma_i^m, \phi_i^m, \xi_i^m; \alpha, \Omega, \eta)$ gives a lower bound to $\log p(y_i, x_{1:N}|\alpha, \Omega, \eta)$.
3: **M-step:** Improved estimate of the model parameters (α, Ω, η) are obtained by maximizing the aggregate lower bound:
$(\alpha^m, \Omega^m, \eta^m) = \arg\max_{(\alpha, \Omega, \eta)} \sum_{i=1}^{N} \mathcal{L}(\gamma_i^m, \phi_i^m, \xi_i^m; \alpha, \Omega, \eta)$.
4: **until** $\sum \mathcal{L}(\gamma_i^m, \phi_i^m, \xi_i^m; \alpha^m, \Omega^m, \eta^m) - \sum \mathcal{L}(\gamma_i^{m+1}, \phi_i^{m+1}, \xi_i^{m+1}; \alpha^{m+1}, \Omega^{m+1}, \eta^{m+1})$
$\leq$ threshold

η, respectively, yields the estimated parameters. As for μ, σ and η, we have
$\mu_{jk} = \frac{\sum_{i=1}^{\mathcal{M}} \phi_{ik} x_{ij}}{\sum_{i=1}^{\mathcal{M}} \phi_{ik}}$, $\sigma_{jk} = \frac{\sum_{i=1}^{\mathcal{M}} \phi_{ik}(x_{ij} - \mu_{jk})^2}{\sum_{i=1}^{\mathcal{M}} \phi_{ik}}$, $\eta_k = \log(\frac{\sum_{i=1}^{\mathcal{M}} \phi_{ik} y_i}{\sum_{i=1}^{\mathcal{M}} \frac{\phi_{ik}}{\xi_i}})$.
Based on the variational inference and parameter estimation updates, it is straightforward to construct a variant EM algorithm to estimate (α, Ω, η). Starting with an initial guess $(\alpha^0, \Omega^0, \eta^0)$, the variational EM algorithm alternates between two steps, as follows (Algorithm 2).

After obtaining the DMNB model parameters from the EM algorithm, we can use η to perform saliency prediction. Given the feature $x_{1:N}$, we have

$$E[\log p(y|x_{1:N}, \alpha, \Omega, \eta)] = \begin{cases} \eta^T E[\bar{z}] - E[\log(1 + e^{\eta^T \bar{z}})] & y = 1 \\ 0 - E[\log(1 + e^{\eta^T \bar{z}})] & y = 0 \end{cases} \quad (13)$$

where $\bar{z}$ is an average of $z_{1:N}$ over all of the observed features. The computation for $E[\bar{z}]$ is intractable; therefore, we again introduce the distribution $q(z_{1:N}, \theta)$ and calculate $E_q[\bar{z}]$ as an approximation of $E[\bar{z}]$. In particular, $E_q[\bar{z}] = \phi$; therefore, we only need to compare $\eta^T \phi$ with 0.

4 Experimental Evaluation

4.1 Experimental Setup

Dataset. In this section, we conduct some experiments to demonstrate the performance of our method. We use NLPR dataset[1] to evaluate the performance of the proposed model. The NLPR dataset includes 1000 images of diverse scenes in real 3D environments, where the ground-truth was obtained by requiring five participants to select regions where objects are presented, i.e., the salient regions were marked by hand.

Evaluation Metrics. We introduce two types of measures to evaluate algorithm performance on the benchmark. The first one is the gold standard: F-measure [13]. The second is the receiver operating characteristic (ROC) curve and the

---

[1] http://sites.google.com/site/rgbdsaliency.

area under the ROC curve (AUC). A continuous saliency map can be converted into a binary mask using a threshold, resulting in a pair of precision and recall values when the binary mask is compared against the ground truth. A ROC curve is then obtained by varying the threshold from 0 to 1.

Parameter Setting. To evaluate the quality of the proposed approach, we divided the datasets into two subsets accroding to their CMI values, and we held out 10% of the data for testing purpose and trained on the remaining 90% whose CMI values are less than CMI threshold τ. As shown in Fig. 4, we compute the CMI for all of the RGB-D images, and the parameter τ is set to 0.35, which is a heuristically determined value. We set the $m = 20$ and $\omega_d = 1.0$ in Eq. 1. We set the $L = 3$, $\omega_c^l = 0.2, 0.3, 0.5$, $\omega_d^l = 0.3, 0.3, 0.4$ and $\sigma^2 = 0.1$ in Eqs. 2, 3 and 4 respectively. We initialize the model parameters using all data points and their labels in the training set in Algorithm 1. In particular, we use the mean and standard deviation of the data points in each class to initialize Ω and $\frac{D_c}{D}$ to initialize α_i, where D_c is the number of data points in class c and D is the total number of data points. For the η in the DMNB model, we run a cross validation by holding out 10% of the training data as the validation set and use the parameters generating the best results on the validation set. We find the initial number of components K using the DPMM.

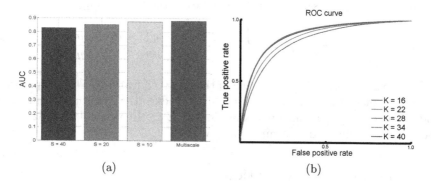

(a)   (b)

Fig. 7. (a) The effects of the number of scales $\mathcal{S}$ on the NLPR dataset. A single scale produces inferior results. (b) The ROC curves for different K components in the DMNB model in terms of the NLPR dataset. The K found using DPMM was adjusted over a wide range to compare the performance. The ROC curves show that changing the parameter K has only a slight effect on the performance.

The Effect of the Parameters. In particular, we performed the experiments while varying $\mathcal{S}$ from Eq. 1 and K from Algorithm 1. Figure 7(a) shows typical results when varying $\mathcal{S}$ from Eq. 1, which illustrates the AUC obtained from the different numbers of superpixels. If only one scale is used, the results are inferior. This justifies our multi-scale approach.

The parameter K in Algorithm 1 is set according to the training set based on DPMM, as shown in Fig. 6. Figure 7(b) shows the ROC curve from changing the number of components K in Algorithm 1. Finally, for all the experiments described below, the parameter K was fixed at 28 - no user fine-tuning was done.

4.2 Qualitative Experiment

During the experiments, we compare our algorithm with five state-of-the-art saliency detection methods, among which three are developed for RGB-D images and two for traditional 2D image analysis. One RGB-D method performs saliency detection at Low-level, Mid-level, and High-level stages and is therefore referred to as LMH [12]. One RGB-D method is based on anisotropic centre-surround difference and is therefore denoted ACSD [8]. The other RGB-D method exploits global priors, which include the background, depth, and orientation priors to achieve a saliency map and is therefore denote GP [14]. The two 2D methods are Hemami's frequency-tuned method [13], which is denoted FT, and the approach from the graph-based manifold ranking [18], which is denoted GMR. For the two 2D saliency approaches, we also add and multiple their results with the DSM produced by our proposed depth feature map; these results are denoted FT+DSM, FT×DSM, GMR+DSM and GMR×DSM. All of the results are produced using the public codes that are offered by the authors of the previously mentioned literature reports.

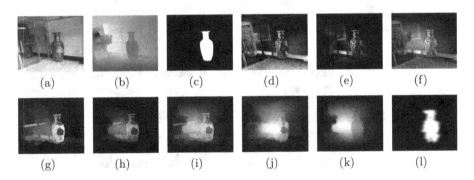

Fig. 8. Visual comparison of the saliency estimations of the different 2D methods with DSM. (a) RGB image, (b) depth image, (c) ground truth, (d) FT, (e) FT×DSM, (f) FT+DSM, (g) GMR, (h) GMR×DSM, (i) GMR+DSM, (j) CSM, (k) DSM, (l) Ours. + indicates a linear combination strategy, and × indicates a weighting method based on multiplication. DSM means depth saliency map, which is produced by our proposed depth feature map. CSM means colour saliency map, which is produced by our proposed colour feature map.

Figure 8 compares our results with FT, FT+DSM, FT×DSM, GMR, GMR+DSM and GMR×DSM. FT detects many uninteresting background pixels as salient because it does not consider any global features. The experiments show that both FT+DSM and FT×DSM are highly improved when incorporated with the DSM. GMR fails to detect many pixels on the prominent objects because it does not define the pseudo-background accurately. Although the simple late fusion strategy achieves improvements, it still suffers from inconsistency in the homogeneous foreground regions and lacks precision around object boundaries, which may be ascribed to treating the appearance and depth correspondence cues in an independent manner. Our approach consistently detects the pixels on the dominant objects within a Bayesian framework with higher accuracy to resolve the issue.

The comparison of the ACSD, LMH and GP RGB-D approaches is presented in Fig. 9. ACSD works on depth images on the assumption that salient objects tend to stand out from the surrounding background, which takes relative depth into consideration. ACSD generates unsatisfying results without colour cues. LMH uses a simple fusion framework that takes advantage of both depth and appearance cues from the low-, mid-, and high-levels. In [12], the background is nicely excluded; however, many pixels on the salient object are not detected as salient. Ren et al. proposed two priors, which are the normalized depth prior and the global-context surface orientation prior [14]. Because their approach uses the two priors, it has problems when such priors are invalid. We can see that the proposed method can accurately locate the salient objects, and produce nearly equal saliency values for the pixels within the target objects.

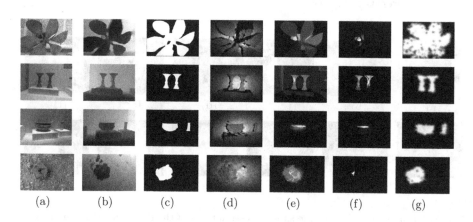

| (a)   (b)   (c)   (d)   (e)   (f)   (g) |

Fig. 9. Visual comparison of the saliency estimations of different 3D methods based on the NLPR dataset. (a) RGB image, (b) Depth image, (c) Ground truth, (d) ACSD, (e) GP, (f) LMH, (g) Ours.

4.3 Quantitative Evaluation

Comparison of the 2D Models Combined with DSM. In this experiment, we first compare the performances of existing 2D saliency models before and after DSM fusing. Figure 10 presents the experimental results, where + and × denote a linear combination strategy and a weighting method, respectively. From Fig. 10(a), we can see the strong influence of using the DSM on the distribution

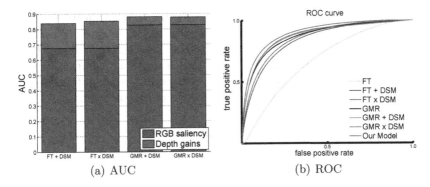

(a) AUC   (b) ROC

Fig. 10. The quantitative comparisons of the performance of depth cues. + means a linear combination strategy and × means a weighting method based on multiplication.

of visual attention in terms of the viewing of 3D content. Although the simple late fusion strategy achieves improvements, it still suffers from inconsistency in the homogeneous foreground regions, which may be ascribed to treating the appearance and depth correspondence cues in an independent manner, as shown in Fig. 8. We also provide the ROC curves for several compared methods in Fig. 10(b). The ROC curves demonstrate that the proposed 3D saliency detection model performs better than the compared methods do.

Comparison of 3D Models. In this paper, the GP model, LMH model and ACSD model are classified as depth-saliency models. Figure 11 shows the quantitative comparisons among these method on the constructed RGBD datasets in terms of ROC curves and F-measures. Interestingly, the LMH method, which uses Bayesian fusion to fuse depth and RGB saliency by simple multiplication, has lower performance compared to the GP method, which uses the Markov Random Field model as a fusion strategy, as shown in Fig. 11(a). However, LMH

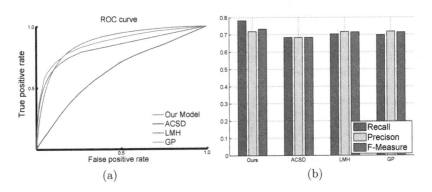

(a)   (b)

Fig. 11. (a) The ROC curves of different 3D saliency detection models in terms of the NLPR dataset. (b) The F-measures of different 3D saliency detection models when used on the NLPR dataset.

and GP achieve better performances than ACSD by using fusion strategies. The proposed RGBD method is superior to the baselines in terms of all the evaluation metrics. Although the ROC curves are very similar, Fig. 11(b) shows that the proposed method improves the recall and F-measure when compared to LMH and GP. This is mainly because the feature extraction using multi-scale super-pixels enhances the consistency and compactness of salient patches.

5 Conclusion

In this study, we proposed a saliency detection model for RGB-D images that considers both colour- and depth-based contrast features with a generative mixture model. The experiments verify that the proposed model's depth-produced saliency can serve as a helpful complement to the existing colour-based saliency models. Compared with other competing 3D models, the experimental results based on a recent eye tracking databases show that the performance of the proposed saliency detection model is promising. We hope that our work is helpful in stimulating further research in the area of 3D saliency detection.

Acknowledgement. This work was supported in part by the Beijing Academy of Science and Technology Youth Backbone Training Plan (2015–16) and Innovation Group Plan of Beijing Academy of Science and Technology (IG201506N).

References

1. Achanta, R., Shaji, A., Smith, K., Lucchi, A., Fua, P., Süsstrunk, S.: Slic superpixels compared to state-of-the-art superpixel methods. IEEE Trans. Pattern Anal. Mach. Intell. **34**, 2274–2282 (2012)
2. Blei, D.M., Jordan, M.I.: Variational inference for Dirichlet process mixtures. Bayesian Anal. **1**, 121–143 (2006)
3. Borji, A., Itti, L.: State-of-the-art in visual attention modelling. IEEE Trans. Pattern Anal. Mach. Intell. **35**, 185–207 (2013)
4. Borji, A., Cheng, M., Jiang, H., Li, J.: Salient object detection: a benchmark. IEEE Trans. Image Process. **24**, 5706–5722 (2015)
5. Desingh, K., Madhava, K.K., Rajan, D., Jawahar, C.V.: Depth really matters: improving visual salient region detection with depth. In: BMVC (2013)
6. Fang, Y., Wang, J., Narwaria, M., Le Callet, L., Lin, W.: Saliency detection for stereoscopic images. IEEE Trans. Image Process. **23**, 2625–2636 (2014)
7. Fang, Y., Lin, W., Fang, Z., Lei, J., Le Callet, P., Yuan, F.: Learning visual saliency for stereoscopic images. In: ICME (2014)
8. Ju, R., Ge, L., Geng, W., Ren, T., Wu, G.: Depth saliency based on anisotropic centre-surround difference. In: ICIP (2014)
9. Kim, H., Lee, S., Bovik, C.A.: Saliency prediction on stereoscopic videos. IEEE Trans. Image Process. **23**, 1476–1490 (2014)
10. Lang, C., Ngugen, T.V., Katti, H., Yadati, K., Kankanhalli, M., Yan, S.: Depth matters: influence of depth cues on visual saliency. In: Fitzgibbon, A., Lazebnik, S., Perona, P., Sato, Y., Schmid, C. (eds.) ECCV 2012. LNCS, vol. 7573, pp. 101–115. Springer, Heidelberg (2012). doi:10.1007/978-3-642-33709-3_8

11. Le Callet, P., Niebur, E.: Visual attention and applications in multimedia technology. Proc. IEEE **101**, 2058–2067 (2013)
12. Peng, H., Li, B., Xiong, W., Hu, W., Ji, R.: RGBD salient object detection: a benchmark and algorithms. In: Fleet, D., Pajdla, T., Schiele, B., Tuytelaars, T. (eds.) ECCV 2014. LNCS, vol. 8691, pp. 92–109. Springer, Heidelberg (2014). doi:10.1007/978-3-319-10578-9_7
13. Radhakrishna, A., Sheila, H., Francisco, E., Sabine, S.: Frequency-tuned salient region detection. In: CVPR (2009)
14. Ren, J., Gong, X., Yu, L., Zhou, W.: Exploiting global priors for RGB-D saliency detection. In: CVPRW (2015)
15. Shan, H., Banerjee, A., Oza, N.C.: Discriminative mixed-membership models. In: ICDM (2009)
16. Wang, J., Fang, Y., Narwaria, M., Lin, W., Le Callet, P.: Stereoscopic image retargeting based on 3D saliency detection. In: ICASSP (2014)
17. Wu, P., Duan, L., Kong, L.: RGB-D salient object detection via feature fusion and multi-scale enhancement. In: Zha, H., Chen, X., Wang, L., Miao, Q. (eds.) CCCV 2015. CCIS, vol. 547, pp. 359–368. Springer, Heidelberg (2015). doi:10. 1007/978-3-662-48570-5_35
18. Yang, C., Zhang, L., Lu, H., Ruan, X., Yang, M.H.: Saliency detection via graph based manifold ranking. In: CVPR (2013)
19. Zhang, Y., Jiang, G., Yu, M., Chen, K.: Stereoscopic visual attention model for 3D video. In: Boll, S., Tian, Q., Zhang, L., Zhang, Z., Chen, Y.-P.P. (eds.) MMM 2010. LNCS, vol. 5916, pp. 314–324. Springer, Heidelberg (2010). doi:10.1007/ 978-3-642-11301-7_33
20. Zhu, L., Cao, Z., Fang, Z., Xiao, Y., Wu, J., Deng, H., Liu, J.: Selective features for RGB-D saliency. In: CAC (2015)

A Coarse-to-Fine Indoor Layout Estimation (CFILE) Method

Yuzhuo Ren[✉], Shangwen Li, Chen Chen, and C.-C. Jay Kuo

University of Southern California, Los Angeles, CA 90089, USA
{yuzhuore,shangwel,chen80}@usc.edu, cckuo@sipi.usc.edu

Abstract. The task of estimating the spatial layout of cluttered indoor scenes from a single RGB image is addressed in this work. Existing solutions to this problem largely rely on hand-crafted features and vanishing lines, and they often fail in highly cluttered indoor scenes. The proposed coarse-to-fine indoor layout estimation (CFILE) method consists of two stages: (1) coarse layout estimation; and (2) fine layout localization. In the first stage, we adopt a fully convolutional neural network (FCN) to obtain a coarse-scale room layout estimate that is close to the ground truth globally. The proposed FCN combines the layout contour property and the surface property so as to provide a robust estimation in the presence of cluttered objects. In the second stage, we formulate an optimization framework that enforces several constraints such as layout contour straightness, surface smoothness and geometric constraints for layout detail refinement. Our proposed system offers the state-of-the-art performance on two commonly used benchmark datasets.

1 Introduction

The task of spatial layout estimation of indoor scenes is to locate the boundaries of the floor, walls and ceiling. The room layout can be represented by either surface boundaries or surfaces themselves, which are two equivalent representations for a room layout. The segmented boundaries and surfaces are valuable for a wide range of computer vision applications such as indoor navigation [1], object detection [2] and augmented reality [1,3–5]. However, there are many challenges in estimating the room layout from a single RGB image, especially in highly cluttered rooms where the ground and wall boundaries are occluded by various objects. Furthermore, indoor scene images may be shot at different viewpoints with large intra-class variations. As a result, high-level reasoning is often required to accurately estimate the spatial layout. For example, the global room model and its associated geometric reasoning can be exploited for this purpose.

The indoor room layout estimation problem has been actively studied in recent years. Hedau *et al.* [6] formulated it as a structured learning problem. It

Electronic supplementary material The online version of this chapter (doi:10.1007/978-3-319-54193-8_3) contains supplementary material, which is available to authorized users.

© Springer International Publishing AG 2017
S.-H. Lai et al. (Eds.): ACCV 2016, Part V, LNCS 10115, pp. 36–51, 2017.
DOI: 10.1007/978-3-319-54193-8_3

first generates hundreds of layout proposals based on inference from vanishing lines. Then, it uses the line membership and the geometric context features to rank the obtained proposals and chooses the one with the highest score as the desired final result.

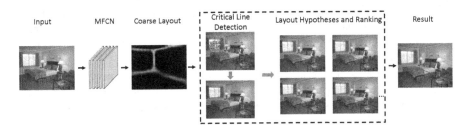

Fig. 1. The pipeline of the proposed coarse-to-fine indoor layout estimation (CFILE) method. For an input indoor image, a coarse layout estimate that contains large surfaces and their boundaries are obtained by a multi-task fully convolutional neural network (MFCN) in the first stage. Then, occluded lines and missing lines are filled in and possible layout choices are ranked according to a pre-defined score function in the second stage. The one with the highest score is chosen as the final output.

In this work, we propose a coarse-to-fine indoor layout estimation (CFILE) method whose pipeline is shown in Fig. 1. The system takes an RGB image as the input and provides a box layout as the output. The CFILE method consists of two stages: (1) coarse layout estimation; and (2) fine layout localization. In the first stage, we adopt a multi-task fully convolutional neural network (MFCN) [7] to obtain a coarse-scale room layout estimate. This is motivated by the strength of the FCN in semantic segmentation [8] and contour detection [9]. The FCN has a strong discriminant power in handling a large variety of indoor scenes using the surface property and the layout contour property. It can provide robust estimation in the presence of cluttered objects, which is close to the ground truth globally. In the second stage, being motivated by structured learning, we formulate an optimization framework that enforces several constraints such as layout contour straightness, surface smoothness and geometric constraints for layout detail refinement.

It is worthwhile to emphasize that the spatial layout estimation problem is different from semantic object segmentation problem in two aspects. First, the aim of the spatial layout problem is to label the semantic surface of an indoor room rather than objects in the room. Second, we have to label occluded surfaces while semantic segmentation does not deal with the occlusion problem at all. Also, unlike in the contour detection problem, occluded layout contours have to be detected.

The major contributions of this work are three folds. First, we use the FCN to learn the labeling of key contours and main surfaces jointly, which are critical to robust estimation of indoor scene layout. The FCN training is elaborated and it is shown that the coarse-scale layout estimate obtained by the FCN is robust and close to the ground truth. Second, we formulate an optimization framework that enforces three constraints (i.e. surface smoothness, contour straightness and proper geometrical structure) to refine the coarse-scale layout estimate. Third, we conduct extensive performance evaluation by comparing the proposed CFILE method and several benchmarking methods on the dataset of Hedau *et al.* [6], and the LSUN validation dataset [10]. It is shown by experimental results that the proposed CFILE method offers the state-of-the-art performance. It outperforms the second best method by 1.16% and 1.32% in Hedau dataset and the LSUN dataset, respectively.

The rest of this paper is organized as follows. Related previous work is reviewed in Sect. 2. The proposed CFILE method is described in detail in Sect. 3. Experimental results are shown in Sect. 4. Concluding remarks are drawn in Sect. 5.

2   Related Work

Structured Learning. The structured learning methodology [11] has been widely used in the context of indoor room layout estimation. The aim of this methodology is to learn the structure of an environment in the presence of imperfect low-level features. It consists of the following two stages [11]. First, a set of layout hypothesis are generated. Second, a score function is defined to evaluate the structure in hypotheses set. The first stage is guided by low-level features such as vanishing lines under the Manhattan assumption. The number of layout hypotheses in the first stage is typically large, and the majority of the hypotheses are of low accuracy due to the presence of clutters. If the quality of hypotheses is low in the first stage, there is no easy way to fix it in the second stage. In the second stage of layout ranking, the score function contains various features such as line membership [6,12], geometric context [6,12], object location [13], etc. The score function cannot handle objects well since they overlap with more than one surface (e.g., between the floor and walls). The occluding objects in turn make the surface appearance quite similar along their boundaries.

Classical Methods for Indoor Layout Estimation. Research on indoor room layout estimation has been active in recent years. Hedau *et al.* [6] formulated it as a structured learning problem. There are many follow-up efforts after this milestone work. They focus on either developing new criteria to reject invalid layout hypotheses or introducing new features to improve the score function in layout ranking.

Different hypothesis evaluation methods were considered in [6,13–18]. Hedau *et al.* [6] reduced noisy lines by first removing clutters. Specifically, they used the line membership together with semantic labeling to evaluate hypotheses. Gupta *et al.* [13] proposed an orientation map that labels three orthogonal surface directions based on line segments, and then, used the orientation map to re-evaluate layout proposals. They detected objects and fit them into 3D boxes. Since an object cannot penetrate the wall, they used the box location as a constraint to reject invalid layout proposals. The work in [2,19] attempted to model objects and spatial layout simultaneously. Hedau *et al.* [20] improved their earlier work in [2,6] by localizing the box more precisely using several cues such as edge- and corner-based features. Ramalingam *et al.* [18] proposed an algorithm to detect Manhattan Junctions and selected the best layout by optimizing a conditional random field whose corners are well aligned with pre-detected Manhattan Junctions. Del Pero *et al.* [17] integrated the camera model, an enclosing room box, frames (windows, doors, pictures), and objects (beds, tables, couches, cabinets) to generate layout hypotheses. Lampert *et al.* [21] improved object detection by maximizing a score function through the branch and bound algorithm.

3D- and Video-based Indoor Layout Estimation. Zhao and Zhu [16] exploited the location information and 3D spatial rules to obtain as many 3D boxes as possible. For example, if a bed is detected, the algorithm will search its neighborhood to look for a side table. Then, they rejected impossible layout hypotheses. Choi *et al.* [22] trained several 3D scene graph models to learn the relation among the scene type, the object type, the object location and layout jointly. Guo *et al.* [14] recovered a 3D model from a single RGBD image by transferring the exemplar layout in the training set to the test image. Fidler *et al.* [23] and Xiang and Savarese [24] represented objects by a deformable 3D cuboid model for improved object detection and then used in layout estimation. Fouhey *et al.* [25] exploited human action and location in time-lapse video to infer functional room geometry. Jiang *et al.* [26] proposed a novel linear method to match cuboids in indoor scenes using RGBD images which effectively gave room layout estimation. Khan *et al.* [27] improved the cuboid representation by generating two types of cuboid hypotheses, one corresponding to regular objects inside a scene, and the other corresponding to the main structures, such as floor and walls.

CNN- and FCN-based Indoor Layout Estimation. The convolutional neural networks (CNN) have had a great impact on various computer vision research topics, such as object detection, scene classification, semantic segmentation, etc. Mallya and Lazebnik [12] used the fully convolutional neural networks (FCN) to learn the informative edge from an RGB image to provide a rough layout. The FCN shares features in convolutional layers and optimize edges detection and geometric context labeling [6,28,29] jointly. The learned contours are used as a new feature in sampling vanishing lines for layout hypotheses generation. Dasgupta and Kuan Fang [30] used an FCN to learn semantic surface labels. Instead of learning edges, their solution adopted the heat map of semantic surfaces obtained by the FCN as the belief map and further optimized it by vanishing lines. Generally speaking, a good layout should satisfy several constraints such as boundary straightness, surface smoothness and proper geometrical structure. However, the CNN is weak in imposing spatial constraints and performing spatial inference. As a result, an inference model was appended in both [12,30] to refine the layout result obtained by CNN.

3 Coarse-to-Fine Indoor Layout Estimation (CFILE)

3.1 System Overview

Most research on indoor layout estimation [6,13–18] is based on the "Manhattan World" assumption. That is, a room contains three orthogonal directions indicated by three groups of vanishing lines. Hedau *et al.* [6] presented a layout model based on 4 rays and a vanishing point. The model can written as

$$\text{Layout} = (l_1, l_2, l_3, l_4, v), \tag{1}$$

where l_i is the ith line and v is the vanishing point. If (l_1, l_2, l_3, l_4, v) can be easily detected without any ambiguity, the layout problem is straightforward. One example is given in Fig. 2(a), where five surfaces are visible in the image without occlusion. However, there exist more challenging cases, where vertices p_i and e_i lie outside the image. One example is shown in Fig. 2(b) where vertices p_2 and p_3 are floor corners and they are likely occluded by objects. Furthermore, line l_2 may be entirely or partially occluded as shown in Fig. 2(c), where lines l_3 and l_4 are wall boundaries, that can be partially (but not fully) occluded, and line l_1 is the ceiling boundary which is likely to be visible.

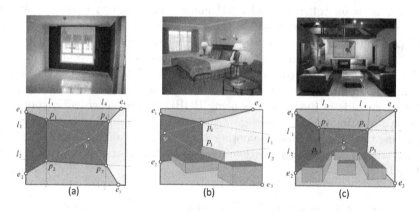

Fig. 2. Illustration of a layout model Layout $= (l_1, l_2, l_3, l_4, v)$ that is parameterized by four lines and a vanishing point: (a) An easy setting where all five surfaces are present; (b) A setting where some surfaces are outside the image; (c) A setting where key boundaries are occluded.

The proposed CFILE system consists of two stages as illustrated in Fig. 1. In the first stage, we propose a multi-task fully convolutional neural network (MFCN) to offer a coarse yet robust layout estimation. Since the CNN is weak in imposing spatial smoothness and conducting geometric reasoning, it cannot provide a fine-scale layout result. In the second stage, we first use the coarse layout from the MFCN as a guidance to detect a set of critical lines. Then, we generate a small set of high quality layout hypotheses based on these critical lines. Finally, we define a score function to select the best layout as the desired output. Detailed tasks in these two stages are elaborated below.

3.2   Coarse Layout Estimation via MFCN

We adopt a multi-task fully convolutional neural network (MFCN) [7,8,12] to learn the coarse layout of indoor scenes. The MFCN [7] shares features in the convolutional layers with those in the fully connected layers and builds different branches for multi-task learning. The total loss of the MFCN is the sum of the losses of different tasks. The proposed two-task network structure is shown in Fig. 3. We use the VGG-16 architecture for fully convolutional layers and train the MFCN for two tasks jointly, i.e. one for layout learning while the other for semantic surface learning (including the floor, left-, right-, center-walls and the ceiling). Our work is different from that in [12], where layout is trained together with geometric context labels [28,29] which contains object labels. Here, we train the layout and semantic surface labels jointly. By removing objects from the concern, the boundaries of semantic surfaces and layout contours can be matched even in occluded regions, leading to a clearer layout. Compared to the work in [30], which adopts the fully convolutional neural network to learn semantic surfaces with a single task network, our network has two branches for coarse layout learning and semantic surface learning, where their learned results can help each other.

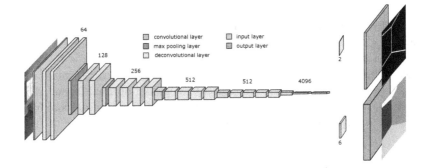

Fig. 3. Illustration of the FCN-VGG16 with two output branches. We use one branch for the coarse layout learning and the other branch for semantic surface learning. The input image size is re-sized to 404 × 404 to match the receptive field size of the filter at the fully convolutional layer.

The receptive field of the filter at the fully connected layer of the FCN-VGG16 is 404 × 404, which is independent of the input image size [8,31]. Xu *et al.* [31] attempted to vary the FCN training image size so as to capture different level of details in image content. If the input image size is larger than the receptive field size, the filter of the fully connected layer looks only at a part of the image. If the input image size is smaller than the receptive field size, it is padded with zeros and spatial resolution is lost in this case. The layout describes the whole image's global structure. We resize the input image to 404 × 404 so that the filter examines the whole image.

3.3   Layout Refinement

There are two steps in structured learning: (1) to generate a hypotheses set; and (2) to define a score function and search for a hypothesis in the hypotheses set that maximizes the score function. Our objective is to improve performance in both steps.

Given an input image $\mathbf{I}$ of size $w \times h \times 3$, the output of the coarse layout from the proposed MFCN in Fig. 3 is a probability function in the form of

$$\mathbf{P}^{(k)} = Pr(\mathbf{L}_{ij} = k|\mathbf{I}), \quad \forall k \in \{0,1\}, \; i \in [1, ..., h], \; j \in [1, ..., w], \quad (2)$$

where $\mathbf{L}$ is an image of size $w \times h$ that maps each pixel $\mathbf{I}_{ij}$ in the original image to a label $\mathbf{L}_{ij} \in \{0,1\}$, in the output image, where 0 denotes a background pixel and 1 denotes a layout pixel. One way to estimate the final layout from the MFCN output is to select the label with the highest score; namely,

$$\hat{\mathbf{L}}_{ij} = \operatorname*{argmax}_{k} \mathbf{P}^{(k)}_{ij} \quad \forall i \in [1, ..., h], \; j \in [1, ..., w]. \quad (3)$$

It is worthwhile to point out that $\hat{\mathbf{L}}_{ij}$ generated from the MFCN output is noisy for the following two reasons. First, the contour from the MFCN is thick and not straight since the convolution and the pooling operations lose the spatial resolution gradually along stages. Second, the occluded floor boundary (e.g., the l_2 line in Fig. 2) is more difficult to detect since it is less visible than other contours (e.g., the l_1, l_3 and l_4 lines in Fig. 2). We need to address these two challenges in defining a score function.

The optimal solution for Eq. (3) is difficult to obtain directly. Instead, we first generate layout hypotheses that are close to the global optimal layout, denoted by $\mathbf{L}^*$, in the layout refinement algorithm. Then, we define a novel score function to rank layout hypotheses and select the one with the highest score as the final result.

Generation of High-Quality Layout Hypotheses. Our objective is to find a set of layout hypotheses that contains fewer yet more robust proposals in the presence of occluders. Then, the best layout with the smallest error can be selected.

Vanishing Line Sampling. We first threshold the layout contour obtained by the MFCN, convert it into a binary mask, and dilate it by 4 pixels to get a binary mask image denoted by C. Then, we apply a vanishing lines detection algorithm [13] to the original image and select those inside the binary mask as critical lines $l_{i(\text{original})}$, shown as solid lines in Fig. 4(c)–(e) for ceiling, wall and floor respectively. Candidate vanishing point v is generated by grid search around the initial v from [13].

Handling Undetected Lines. There exists a case when no vanishing lines are detected inside C because of low contrast, such as wall boundaries, l_3(or l_4). If ceiling corners are available, l_3 (or l_4) are filled in by connecting ceiling corners and vertical vanishing point. If ceiling corners do not present in the image, the missing l_3(or l_4) is estimated by logistic regression using the layout points in $\mathbf{L}$.

Handling Occluded Lines. As discussed earlier, the floor line, l_2, can be entirely or partially occluded. One illustrative example is shown in Fig. 4 where l_2 is partially occluded. If l_2 is partially occluded, the occluded part of l_2 can be recovered by line extension. For entirely occluded l_2, if we simply search lines inside C or uniformly sample lines [12], the layout proposal will not be accurate as the occluded boundary line cannot be recovered. Instead, we automatically fill in occluded lines based on geometric rules. If p_2 (or p_3) is detectable by connecting detected l_3 (or l_4) to e_2v (or e_3v), l_2 is computed as the line passing through the available p_2 or p_3 and the vanishing point l_2 associated with. If neither p_2 nor p_3 is detectable, l_2 is estimated by logistic regression use the layout points in $\mathbf{L}$.

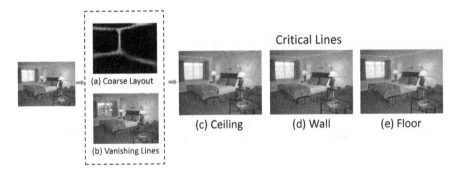

Fig. 4. Illustration of critical lines detection for better layout hypotheses generation. For a given input image, the coarse layout offers a mask that guides vanishing lines selection and critical lines inference. The solid lines indicate detected vanishing lines in C. The dashed wall lines indicate those wall lines that are not detected but inferred inside mask C from ceiling corners. The dashed floor lines indicate those floor lines that are not detected but inferred inside mask C.

In summary, the final l_{critial} used in generating layout hypotheses is the union of three parts as given below:

$$l_{\text{critical}} = l_{i(\text{original})} \cup l_{i(\text{occluded})} \cup l_{i(\text{undetected})}, \qquad (4)$$

where $l_{i(\text{original})}$ denotes detected vanishing lines inside C, $l_{i(\text{occluded})}$ denotes the recovered occluded boundary, and $l_{i(\text{undetected})}$ denotes undetected vanishing lines because of low contrast but recovered from geometric reasoning. These three types of lines are shown in Fig. 4. With $l_{i(\text{original})}$ and vanishing point v, we generate all possible layouts $\mathbf{L}$ using the model described in Sect. 3.1.

Layout Ranking. We use the coarse layout probability map $\mathbf{P}$ as a weight mask to evaluate the layout. The score function is defined as

$$S(\mathbf{L}|\mathbf{P}) = \frac{1}{N} \sum_{i,j} \mathbf{P}_{i,j}, \quad \forall \mathbf{L}_{i,j} = 1, \tag{5}$$

where $\mathbf{P}$ is the output from the MFCN, $\mathbf{L}$ is a layout from the hypotheses set, and N is a normalization factor that is equal to the total number of layout pixels in $\mathbf{L}$. Then, the optimal layout is selected by

$$\mathbf{L}^* = \underset{\mathbf{L}}{\operatorname{argmax}}\, S(\mathbf{L}|\mathbf{P}). \tag{6}$$

The score function is in favor of the layout that is aligned well with the coarse layout. Figure 5 shows one example where the layout hypotheses are ranked using the score function in Eq. (6). The layout with the highest score is chosen to be the final result.

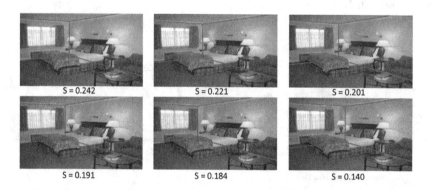

Fig. 5. Example of layout ranking using the proposed score function.

4 Experiments

4.1 Experimental Setup

We evaluate the proposed CFILE method on two popular datasets; namely, Hedau dataset [6] and the LSUN dataset [12]. Hedau dataset contains 209 training images, 53 validation images and 105 test images. Mallya and Lazebnik [12] expanded the Hedau dataset by adding 75 new images to the training set. This expanded dataset is referred to as the Hedau+ dataset. We conduct data augmentation for Hedau+ dataset as done in [12] by cropping, rotation, scaling and luminance adjustment in the training of the MFCN. The LSUN dataset [10] contains 4000 training images, 394 validation images and 1000 test images. Since no ground truth is released for the 1000 test images, we evaluate the proposed method on the validation set only. We resize all images to 404 × 404 by bicubic interpolation, and train two coarse layout models for the two datasets separately. Hedau+ dataset provides both the layout and the geometric context labels but it does not provide semantic surface labels. Thus, we use the layout polygon provided in the dataset to generate semantic surface labels. The LSUN dataset provides surface segmentation but not the layout boundary and semantic surfaces. We relabel the surface segmentation to make the segmentation with the same semantic to have the same label. We detect edges on semantic surface labels and dilate them to a width of 7 pixels. By following [12], we use the NYUDv2 RGBD dataset [32] for semantic segmentation to initialize the MFCN. Also, we set the base learning rate to 10^{-4} with momentum 0.99.

We adopt two performance metrics: the pixel-wise error and the corner error. To compute the pixel-wise error, the obtained layout segmentation is mapped to the ground truth layout segmentation. Then, the pixel-wise error is computed as the percentage of pixels that are wrongly matched. To compute the corner error, we sum up all Euclidean distances between obtained corners and their associated ground truth corners.

4.2 Experimental Results and Discussion

The coarse layout scheme described in Sect. 3.2 is first evaluated using the methodology in [33]. We compare our results, denoted by $MFCN_1$ and $MFCN_2$, against the informative edge method [12], denoted by FCN, in Table 1. Our proposed two coarse layout schemes have higher ODS (fixed contour threshold) and OIS (per-image best threshold) scores. This indicates that they provide more accurate regions for vanishing line samples in layout hypotheses generation.

We use several exemplary images to demonstrate that the proposed coarse layout results are robust and close to the ground truth. That is, we compare visual results of the FCN in [12] and the proposed $MFCN_2$ in Fig. 6. As compared to the layout results of the FCN in [12], the proposed $MFCN_2$ method provides robust and clearer layout results in occluded regions, which are not significantly affected by object boundaries.

Next, we evaluate the performance of the proposed full layout algorithm, CFILE, including the coarse layout estimation and the layout optimization and ranking. The performance of several methods for Hedau dataset and the LSUN

Table 1. Performance comparison of coarse layout results for Hedau test dataset, where the performance metrics are the fixed contour threshold (ODS) and the per-image best threshold (OIS) [33]. We use FCN to indicate the informative edge method in [12]. Both $MFCN_1$ and $MFCN_2$ are proposed in our work. They correspond to the two settings where the layout and semantic surfaces are jointly trained on the original image size ($MFCN_1$) and the downsampled image size 404 × 404 ($MFCN_2$).

	FCN [12]		$MFCN_1$(our)		$MFCN_2$(our)	
Metrics	ODS	OIS	ODS	OIS	ODS	OIS
Hedau dataset	0.255	0.263	0.265	0.284	0.265	0.291

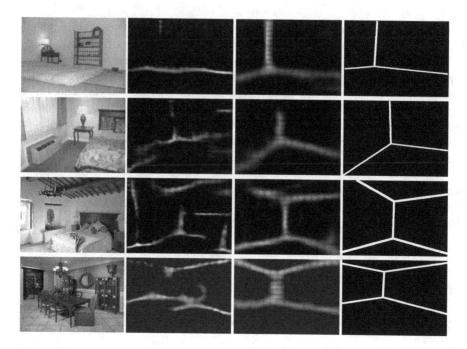

Fig. 6. Comparison of coarse layout results (from left to right): the input image, the coarse layout result of the FCN in [12], the coarse layout results of the proposed MFCN$_2$ and the ground truth. The results of the MFCN$_2$ are more robust, and it provides clearer contours in occluded regions. The first two examples are from the Hedau dataset and the last two examples are from the LSUN dataset.

dataset is compared in Tables 2 and 3, respectively. The proposed CFILE method achieves state-of-the-art performance. It outperforms the second best algorithm by 1.16% for Hedau dataset and 1.32% for the LSUN dataset.

The best six results of the proposed CFILE method for Hedau test images are visualized in Fig. 7. It can be observed from these six examples that the coarse layout estimation algorithm is robust in highly cluttered rooms (see the second row and the fourth row). The layout refinement algorithm can recover occluded boundaries accurately in Fig. 7(a)–(e). It can also select the best layout among several possible choices. The three worst results of the proposed CFILE method for Hedau test images are visualized in Fig. 8. Figure 8(a) shows one example where the fine layout result is misled by the wrong coarse layout estimate. Figure 8(b) is a difficult case. The left wall and right wall have the same appearance and there are several confusing wall boundaries. Figure 8(c) gives the worst example of the CFILE method with accuracy 79.4%. However, it is still higher than the worst example reported in [12] with accuracy 61.05%. The ceiling boundary is confusing in Fig. 8(c). The proposed CFILE method selects the ceiling line overlapping with the coarse layout. More visual results from the LSUN dataset are shown in Fig. 9.

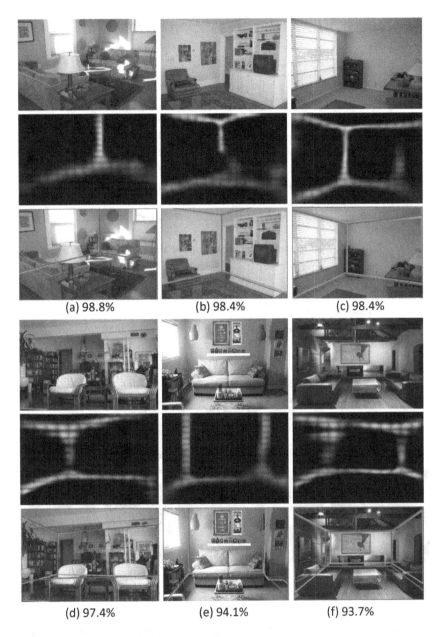

(a) 98.8%   (b) 98.4%   (c) 98.4%

(d) 97.4%   (e) 94.1%   (f) 93.7%

Fig. 7. Visualization of the six best results of the CFILE method in Hedau test dataset (from top to bottom): original images, the coarse layout estimates from MFCN, our results with pixel-wise accuracy (where the ground truth is shown in green and our result is shown in red) (Color figure online).

Table 2. Performance benchmarking for Hedau dataset.

Method	Pixel error (%)
Hedau et al. (2009) [6]	21.20
Del Pero et al. (2012) [17]	16.30
Gupta et al. (2010) [13]	16.20
Zhao and Zho (2013) [16]	14.50
Ramalingam et al. (2013) [18]	13.34
Mallya and Lazebnik (2015) [12]	12.83
Schwing and Urtasun (2012) [34]	12.80
Del Pero et al. (2013) [35]	12.70
Dasgupta and Kuan Fang (2016) [30]	9.73
Proposed CFILE	**8.67**

Table 3. Performance benchmarking for the LSUN dataset.

Method	Corner error (%)	Pixel error (%)
Hedau et al. (2009) [6]	15.48	24.23
Mallya and Lazebnik (2015) [12]	11.02	16.71
Dasgupta and Kuan Fang (2016) [30]	8.20	10.63
Proposed CFILE	**7.95**	**9.31**

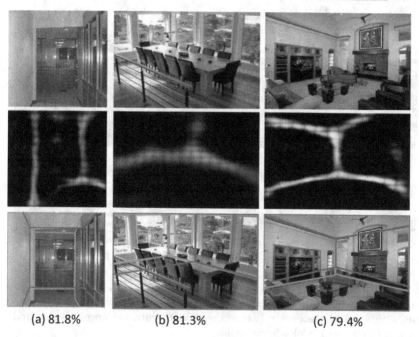

(a) 81.8%   (b) 81.3%   (c) 79.4%

Fig. 8. Visualization of the three worst results of the CFILE method in Hedau test dataset (from top to bottom): original images, the coarse layout estimates from MFCN, our results with pixel-wise accuracy (where the ground truth is shown in green and our result is shown in red) (Color figure online).

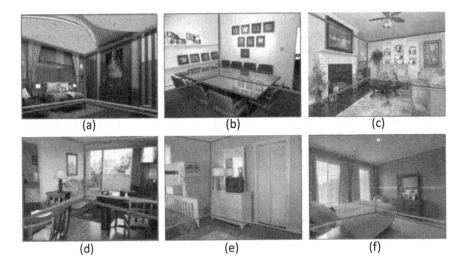

Fig. 9. Visualization of layout results of the CFILE method in the LSUN validation set. Ground truth is shown in green and our result is shown in red (Color figure online).

5   Conclusion and Future Work

A coarse-to-fine indoor layout estimation (CFILE) method was proposed to estimate the room layout from an RGB image. We adopted a multi-task fully convolutional neural network (MFCN) to provide a robust coarse layout estimate for a variety of indoor scenes with joint layout and semantic surface training. However, CNN is weak in enforcing spatial constraints. To address this problem, we formulated an optimization framework that enforces several constraints such as layout contour straightness, surface smoothness and geometric constraints for layout detail refinement. It was demonstrated by experimental results that the proposed CFILE system yields the best performance on two commonly used benchmark datasets. It is an interesting topic to investigate how the improved scene layout estimation can help in achieving a better performance for geometry estimation, clutter identification, and semantic segmentation.

Acknowledgement. Computation for the work described in this paper was supported by the University of Southern California's Center for High-Performance Computing (hpc.usc.edu).

References

1. Karsch, K., Hedau, V., Forsyth, D., Hoiem, D.: Rendering synthetic objects into legacy photographs. ACM Trans. Graph. (TOG) **30**, Article no. 157 (2011)
2. Hedau, V., Hoiem, D., Forsyth, D.: Thinking inside the box: using appearance models and context based on room geometry. In: Daniilidis, K., Maragos, P., Paragios, N. (eds.) ECCV 2010. LNCS, vol. 6316, pp. 224–237. Springer, Heidelberg (2010). doi:10.1007/978-3-642-15567-3_17

3. Liu, C., Schwing, A.G., Kundu, K., Urtasun, R., Fidler, S.: Rent3d: floor-plan priors for monocular layout estimation. In: 2015 IEEE Conference on Computer Vision and Pattern Recognition (CVPR), pp. 3413–3421. IEEE (2015)
4. Xiao, J., Furukawa, Y.: Reconstructing the worlds museums. Int. J. Comput. Vis. **110**, 243–258 (2014)
5. Martin-Brualla, R., He, Y., Russell, B.C., Seitz, S.M.: The 3D jigsaw puzzle: mapping large indoor spaces. In: Fleet, D., Pajdla, T., Schiele, B., Tuytelaars, T. (eds.) ECCV 2014. LNCS, vol. 8691, pp. 1–16. Springer, Heidelberg (2014). doi:10.1007/978-3-319-10578-9_1
6. Hedau, V., Hoiem, D., Forsyth, D.: Recovering the spatial layout of cluttered rooms. In: 2009 IEEE 12th International Conference on Computer vision, pp. 1849–1856. IEEE (2009)
7. Dai, J., He, K., Sun, J.: Instance-aware semantic segmentation via multi-task network cascades. arXiv preprint (2015). arXiv:1512.04412
8. Long, J., Shelhamer, E., Darrell, T.: Fully convolutional networks for semantic segmentation. In: Proceedings of the IEEE Conference on Computer Vision and Pattern Recognition, pp. 3431–3440 (2015)
9. Xie, S., Tu, Z.: Holistically-nested edge detection. In: Proceedings of the IEEE International Conference on Computer Vision, pp. 1395–1403 (2015)
10. http://lsun.cs.princeton.edu/2015.html
11. Nowozin, S., Lampert, C.H.: Structured learning and prediction in computer vision. Found. Trends Comput. Graph. Vis. **6**, 185–365 (2011)
12. Mallya, A., Lazebnik, S.: Learning informative edge maps for indoor scene layout prediction. In: Proceedings of the IEEE International Conference on Computer Vision, pp. 936–944 (2015)
13. Gupta, A., Hebert, M., Kanade, T., Blei, D.M.: Estimating spatial layout of rooms using volumetric reasoning about objects and surfaces. In: Advances in Neural Information Processing Systems, pp. 1288–1296 (2010)
14. Guo, R., Zou, C., Hoiem, D.: Predicting complete 3d models of indoor scenes. arXiv preprint (2015). arXiv:1504.02437
15. Schwing, A., Fidler, S., Pollefeys, M., Urtasun, R.: Box in the box: joint 3d layout and object reasoning from single images. In: Proceedings of the IEEE International Conference on Computer Vision, pp. 353–360 (2013)
16. Zhao, Y., Zhu, S.C.: Scene parsing by integrating function, geometry and appearance models. In: Proceedings of the IEEE Conference on Computer Vision and Pattern Recognition, pp. 3119–3126 (2013)
17. Del Pero, L., Bowdish, J., Fried, D., Kermgard, B., Hartley, E., Barnard, K.: Bayesian geometric modeling of indoor scenes. In: 2012 IEEE Conference on Computer Vision and Pattern Recognition (CVPR), pp. 2719–2726. IEEE (2012)
18. Ramalingam, S., Pillai, J., Jain, A., Taguchi, Y.: Manhattan junction catalogue for spatial reasoning of indoor scenes. In: Proceedings of the IEEE Conference on Computer Vision and Pattern Recognition, pp. 3065–3072 (2013)
19. Wang, H., Gould, S., Roller, D.: Discriminative learning with latent variables for cluttered indoor scene understanding. Commun. ACM **56**, 92–99 (2013)
20. Hedau, V., Hoiem, D., Forsyth, D.: Recovering free space of indoor scenes from a single image. In: 2012 IEEE Conference on Computer Vision and Pattern Recognition (CVPR), pp. 2807–2814. IEEE (2012)
21. Lampert, C.H., Blaschko, M.B., Hofmann, T.: Efficient subwindow search: a branch and bound framework for object localization. IEEE Trans. Pattern Anal. Mach. Intell. **31**, 2129–2142 (2009)

22. Choi, W., Chao, Y.W., Pantofaru, C., Savarese, S.: understanding indoor scenes using 3d geometric phrases. In: Proceedings of the IEEE Conference on Computer Vision and Pattern Recognition, pp. 33–40 (2013)

23. Fidler, S., Dickinson, S., Urtasun, R.: 3d object detection and viewpoint estimation with a deformable 3d cuboid model. In: Advances in Neural Information Processing Systems, pp. 611–619 (2012)

24. Xiang, Y., Savarese, S.: Estimating the aspect layout of object categories. In: 2012 IEEE Conference on Computer Vision and Pattern Recognition (CVPR), pp. 3410–3417. IEEE (2012)

25. Fouhey, D.F., Delaitre, V., Gupta, A., Efros, A.A., Laptev, I., Sivic, J.: People watching: human actions as a cue for single view geometry. Int. J. Comput. Vis. **110**, 259–274 (2014)

26. Jiang, H., Xiao, J.: A linear approach to matching cuboids in RGBD images. In: Proceedings of the IEEE Conference on Computer Vision and Pattern Recognition, pp. 2171–2178 (2013)

27. Khan, S.H., He, X., Bennamoun, M., Sohel, F., Togneri, R.: Separating objects and clutter in indoor scenes. In: Proceedings of the IEEE Conference on Computer Vision and Pattern Recognition, pp. 4603–4611 (2015)

28. Hoiem, D., Efros, A.A., Hebert, M.: Geometric context from a single image. In: Tenth IEEE International Conference on Computer Vision, ICCV 2005, vol.1, pp. 654–661. IEEE (2005)

29. Hoiem, D., Efros, A.A., Hebert, M.: Recovering surface layout from an image. Int. J. Comput. Vis. **75**, 151–172 (2007)

30. Dasgupta, S., Kuan Fang, K.: Delay: robust spatial layout estimation for cluttered indoor scenes. In: 2016 IEEE Conference on Computer Vision and Pattern Recognition (CVPR), IEEE (2016)

31. Xu, H., Venugopalan, S., Ramanishka, V., Rohrbach, M., Saenko, K.: A multi-scale multiple instance video description network. arXiv preprint (2015). arXiv:1505.05914

32. Gupta, S., Arbelaez, P., Malik, J.: Perceptual organization and recognition of indoor scenes from RGB-D images. In: Proceedings of the IEEE Conference on Computer Vision and Pattern Recognition, pp. 564–571 (2013)

33. Arbelaez, P., Maire, M., Fowlkes, C., Malik, J.: Contour detection and hierarchical image segmentation. IEEE Trans. Pattern Anal. Mach. Intell. **33**, 898–916 (2011)

34. Schwing, A.G., Urtasun, R.: Efficient exact inference for 3d indoor scene understanding. In: Fitzgibbon, A., Lazebnik, S., Perona, P., Sato, Y., Schmid, C. (eds.) ECCV 2012. LNCS, vol. 7577, pp. 299–313. Springer, Heidelberg (2012). doi:10.1007/978-3-642-33783-3_22

35. Del Pero, L., Bowdish, J., Kermgard, B., Hartley, E., Barnard, K.: Understanding Bayesian rooms using composite 3d object models. In: Proceedings of the IEEE Conference on Computer Vision and Pattern Recognition, pp. 153–160 (2013)

Unifying Algebraic Solvers for Scaled Euclidean Registration from Point, Line and Plane Constraints

Folker Wientapper[1]([⊠]) and Arjan Kuijper[2]

[1] VRAR, Fraunhofer IGD, Darmstadt, Germany
`folker.wientapper@igd.fraunhofer.de`
[2] MAVC, TU Darmstadt, Darmstadt, Germany

Abstract. We investigate recent state-of-the-art algorithms for absolute pose problems (PnP and GPnP) and analyse their applicability to a more general type, namely the scaled Euclidean registration from point-to-point, point-to-line and point-to plane correspondences. Similar to previous formulations we first compress the original set of equations to a least squares error function that only depends on the non-linear rotation parameters and a small symmetric coefficient matrix of fixed size. Then, in a second step the rotation is solved with algorithms which are derived using methods from algebraic geometry such as the Gröbner basis method. In previous approaches the first compression step was usually tailored to a specific correspondence types and problem instances. Here, we propose a unified formulation based on a representation with orthogonal complements which allows to combine different types of constraints elegantly in one single framework. We show that with our unified formulation existing polynomial solvers can be interchangeably applied to problem instances other than those they were originally proposed for. It becomes possible to compare them on various registrations problems with respect to accuracy, numerical stability, and computational speed. Our compression procedure not only preserves linear complexity, it is even faster than previous formulations. For the second step we also derive an own algebraic equation solver, which can additionally handle the registration from 3D point-to-point correspondences, where other solvers surprisingly fail.

1 Introduction

We consider the problem of finding optimal similarity transformations which relate a set of 3D points to other corresponding points, lines or planes in a different coordinate system. The registration from point-to-point correspondences is a fundamental problem in computer vision and has applications in many fields, such as the correct alignment of independent Structure-from-Motion

Electronic supplementary material The online version of this chapter (doi:10. 1007/978-3-319-54193-8_4) contains supplementary material, which is available to authorized users.

© Springer International Publishing AG 2017
S.-H. Lai et al. (Eds.): ACCV 2016, Part V, LNCS 10115, pp. 52–66, 2017.
DOI: 10.1007/978-3-319-54193-8_4

reconstructions, handling of drift in loop-closure, hand-eye calibration, and many more. The two most prominent algorithms for solving this registration problem were proposed by Umeyama [1] and Horn [2]. The latter is also an integral part of the well-known ICP-algorithm for aligning point clouds. However, these algorithms are not applicable to point-to-line or point-to-plane correspondences. Usually, iterative methods which only converge locally are used for these cases [3,4].

Olsson et al. [5] were the first to propose an algorithm to find the global optimum for registration problems of this kind. The algorithm is based on an iterative Branch-and-Bound procedure using convex under-estimators to solve for the rotation. Although it guarantees to find the global optimum, it is also computationally demanding. The same authors showed later [6] that their algorithm is also applicable to the Perspective-n-Point problem (PnP).

Regarding the related field of perspective registration problems (PnP and GPnP) substantial progress has been made in the last years. Hesch and Roumeliotis [7] proposed an algorithm for the central PnP problem, where the original problem is reduced to a polynomial equation system of fixed size irrespective of the number of used correspondences. This approach has been extended to pose estimation from generalized cameras (Generalized PnP) with and without scale [8,9]. In this paper, we propose a formulation which extends them even further to various 3d registration problems considered by Olsson et al. At the same time, their closed-form character and all of their desirable properties are preserved: they remain non-iterative, applicable to minimal as well as to overconstrained problem instances, and capable of providing all minima at once.

2 Unified Mathematical Framework for Registration Problems

2.1 Objective Function and Vector-Matrix Representation

Suppose we have a set of K points, $\mathbf{x}_k \in \mathbb{R}^3$. Our goal is to find a transformation consisting of a rotation, $\mathbf{R} \in SO(3)$, a translation, $\mathbf{t} \in \mathbb{R}^3$, and an (inverse) scaling $s^{-1} \in \mathbb{R}_+$, so that the transformed points,

$$\mathbf{x}_k' = s^{-1}(\mathbf{R}\mathbf{x}_k + \mathbf{t}), \tag{1}$$

are as close as possible to their corresponding geometric entities, which may either be a plane, π_k, a line, l_k, or another point, p_k. For point correspondences we measure the Euclidean distance between the transformed point $\mathbf{x}_k'$ and the reference point p_k. For lines and planes we use the orthogonal distance, i.e. the length of the shortest vector that connects the transformed point with some other point on the line or the plane. We completely describe a geometric entity with an *offset point*, $\mathbf{y}_k$, and an *orthogonal complement matrix*, $\mathbf{N}_k$, whose columns are orthonormal vectors which are perpendicular to the affine subspace of the geometric entity. Then, the squared error can be written for all correspondence types in a uniform way

$$d_k^2 = \|\mathbf{N}_k^{\mathsf{T}}(\mathbf{x}_k' - \mathbf{y}_k)\|_2^2. \tag{2}$$

The registration problem can then be formulated as the least-squares minimization of the total error for all correspondences,

$$\underset{R \in SO(3), t \in \mathbb{R}^3, s \in \mathbb{R}_+}{\arg \min} \sum_{k=1}^{K} \| \mathbf{N}_k^\mathsf{T}(\mathbf{R}\mathbf{x}_k + \mathbf{t} - s\mathbf{y}_k) \|_2^2. \tag{3}$$

By multiplying the error function with the squared scale, i.e. $s^2 \sum d_k^2$, we have decoupled it from the other unknowns, and it now scales the offset points, $\mathbf{y}_k$. This only affects the absolute value of the error, but not the location of the minima. By stacking the rows of the rotation matrix, the scaling and the translation vector into a parameter vector, $\mathbf{s}^\mathsf{T} = [\mathbf{R}_{11}, \mathbf{R}_{12}, \ldots, \mathbf{R}_{33}, s, \mathbf{t}^\mathsf{T}]$, Eq. 3 can be re-factored and written in matrix-vector form,

$$\underset{\mathbf{R}, \mathbf{t}, s}{\arg \min} \| \mathbf{A}\mathbf{s} \|_2^2. \tag{4}$$

The corresponding rows, $\mathbf{A}_k$, belonging to the correspondence k can be written compactly using the kronecker product ('⊗'):

$$\mathbf{A}_k \in \mathbb{R}^{n_k \times 13} = \left[\mathbf{N}_k^\mathsf{T} \otimes \mathbf{x}_k^\mathsf{T}, \; -\mathbf{N}_k^\mathsf{T}\mathbf{y}_k, \; \mathbf{N}_k^\mathsf{T} \right], \qquad n_k \in \{1, 2, 3\}. \tag{5}$$

Each row of $\mathbf{A}$ corresponds to an equation in a (possibly overconstrained) homogeneous system of equations representing the least-squares problem of Eq. 4. Intuitively, the different types of correspondences should also impose different numbers of effective constraints on the registration problem. In our formulation this is achieved naturally by the size of the orthogonal complement matrix. For *point-to-plane correspondences*, $\mathbf{N}_k^{(\pi)} \in \mathbb{R}^{3 \times 1}$ is given by the normal vector of the plane which results in one equation per correspondence. *Points* can be interpreted as entities that span a zero-dimensional subspace, so any vector in $\mathbb{R}^3$ belongs to the orthogonal complement of a point. Thus, we are free to choose an arbitrary orthogonal matrix for $\mathbf{N}_k^{(p)} \in \mathbb{R}^{3 \times 3}$ yielding three equations per correspondence. In particular, the identity matrix, $\mathbf{I}^{3 \times 3}$, is a convenient choice. In case of *Point-to-line correspondences* often there is no orthogonal complement matrix at hand, but a bearing vector, $\mathbf{v}_k$, instead. For example, in the PnP problem and its generalized variant the bearing vectors are the vectors connecting the homogenized image points and the camera centers. One can easily obtain a matrix $\mathbf{N}_k^{(l)} \in \mathbb{R}^{3 \times 2}$ by means of an orthogonalization method like the Gram-Schmidt Algorithm or a QR-decomposition of $\mathbf{v}_k$. Alternatively, it is also possible to construct 3×3 matrices $\tilde{\mathbf{N}}_k^{(l)}$ with rank two directly from the bearing vectors. Two possible options are:

$$\tilde{\mathbf{N}}_k^{(l)} \in \mathbb{R}^{3 \times 3} = \begin{cases} [\mathbf{v}_k]_\times & \text{(cross product form)} \\ \mathbf{I} - \mathbf{v}_k \mathbf{v}_k^\mathsf{T} & \text{(annihilator form)} \end{cases}, \tag{6}$$

where $[\mathbf{v}_k]_\times$ is the skew-symmetric cross product matrix of $\mathbf{v}_k$. Using the *cross product form* results in a similar system of equations than in the well-known DLT-Algorithm [10]. Since both variants of $\tilde{\mathbf{N}}_k^{(l)}$ have only rank two, there are also only two rows of $\mathbf{A}_k$ which are linearly independent, and one might be tempted to leave one of them out. We emphasize, however, that if $\mathbf{v}_k$ is normalized ($\mathbf{v}_k^\mathsf{T}\mathbf{v}_k = 1$) then we have $\tilde{\mathbf{N}}_k\tilde{\mathbf{N}}_k^\mathsf{T} = \mathbf{N}_k\mathbf{N}_k^\mathsf{T}$, so the squared error, $d_k^2 = (\mathbf{x}_k' - \mathbf{y}_k)^\mathsf{T}\mathbf{N}_k\mathbf{N}_k^\mathsf{T}(\mathbf{x}_k' - \mathbf{y}_k)$, of the objective function (Eq. 2) will remain unchanged regardless of which matrix is used to represent the orthogonal complement. This will be different if one decides to leave out one of the equations, and thus, it should be avoided in order to not obtain biased solutions.

2.2   Thin-SVD-Based Linear Parameter Elimination

Based on the matrix-vector notation of the objective function, the linear parameters, s and $\mathbf{t}$, can now be eliminated by representing them in terms of the nonlinear parameters, $\mathbf{r}^\mathsf{T} = [\mathbf{R}_{11}, \mathbf{R}_{12}, \ldots, \mathbf{R}_{33}]$, using the pseudo-inverse and by back-substituting the resulting expression. Specifically, taking the derivative of Eq. 4 with respect to s and $\mathbf{t}$ and setting it zero yields the first order optimality conditions for $[s, \mathbf{t}^\mathsf{T}]$

$$\mathbf{A}_{st}^\mathsf{T}\mathbf{A}_r\mathbf{r} + \mathbf{A}_{st}^\mathsf{T}\mathbf{A}_{st}\begin{bmatrix} s \\ \mathbf{t} \end{bmatrix} = \mathbf{0}. \tag{7}$$

Here, we have partitioned $\mathbf{A} = [\mathbf{A}_r\ \mathbf{A}_{st}]$ column-wise into the submatrices $\mathbf{A}_r$ and $\mathbf{A}_{st}$ which belong to the non-linear parameters, $\mathbf{r}$, and linear parameters, s and $\mathbf{t}$, respectively. Hence, we can express $[s, \mathbf{t}^\mathsf{T}]$ as a function of $\mathbf{r}$ using the *Moore-Penrose pseudoinverse*, $\mathbf{A}_{st}^\dagger = (\mathbf{A}_{st}^\mathsf{T}\mathbf{A}_{st})^{-1}\mathbf{A}_{st}^\mathsf{T}$ which in turn can be computed efficiently and numerically stably with the Singular-Value-decomposition, $\mathbf{A}_{st} = \mathbf{U}\mathbf{\Sigma}\mathbf{V}^\mathsf{T}$ [11],

$$\begin{bmatrix} s \\ \mathbf{t} \end{bmatrix} = -\underbrace{\mathbf{V}\mathbf{\Sigma}^\dagger\mathbf{U}^\mathsf{T}}_{\mathbf{A}_{st}^\dagger}\mathbf{A}_r\mathbf{r}. \tag{8}$$

By plugging this expression back into Eq. 4 and by defining the fixed size matrix

$$\mathbf{M}_h \in \mathbb{R}^{9\times 9} = \mathbf{A}_r^\mathsf{T}\mathbf{A}_r - \mathbf{A}_r^\mathsf{T}\mathbf{U}(\mathbf{\Sigma}^\dagger\mathbf{\Sigma})\mathbf{U}^\mathsf{T}\mathbf{A}_r, \tag{9}$$

we can describe the registration problem by the following constrained minimization problem which now only depends on the rotation parameters $\mathbf{r}$,

$$\underset{\mathbf{R}\in SO(3)}{\arg\min}\ \{\mathbf{r}^\mathsf{T}\mathbf{M}_h\mathbf{r}\}, \tag{10}$$

The expression for $\mathbf{M}_h$ has been greatly simplified thanks to the orthogonality of the singular vector matrices, $\mathbf{V}^\mathsf{T}\mathbf{V} = \mathbf{I}$ and $\mathbf{U}^\mathsf{T}\mathbf{U} = \mathbf{I}$. The diagonal matrix ($\mathbf{\Sigma}^\dagger\mathbf{\Sigma}$) usually is a 4×4 identity matrix. Only in certain degenerate configurations it may also have zeros on its diagonal whenever some singular values of $\mathbf{A}_{st}$ are zero (or smaller than a machine precision dependent threshold). It then acts as a column selector for $\mathbf{U}$, so that only those singular vectors are chosen effectively span the column space of $\mathbf{A}_{st}$. Rewriting $\mathbf{M}_h$ in Eq. 9 as $\mathbf{M}_h = \mathbf{A}_r^\mathsf{T}(\mathbf{I} - \mathbf{U}(\mathbf{\Sigma}^\dagger\mathbf{\Sigma})\mathbf{U}^\mathsf{T})\mathbf{A}_r$ one can see that $\mathbf{A}_r$ is again projected onto the orthogonal complement $(\mathbf{I} - \mathbf{U}(\boldsymbol{\Sigma}^\dagger \boldsymbol{\Sigma})\mathbf{U}^\mathsf{T})$ of the column space of $\mathbf{A}_{st}$. This mechanism allows to elegantly cope with some degenerate configurations which we will discuss later in this paper.

Considering the computational effort, one might still ask, whether using the Singular-Value-Decomposition for computing the pseudoinverse is a good choice, because for general $n \times m$ matrices the complexity for its computation is $\mathcal{O}(n^2 m + m^3)$ (see [11]). In our case $\mathbf{A}_{st} \in \mathbb{R}^{n \times 4}$, so even though $m = 4$ is constant, we are still left with quadratic complexity with respect to the number of equations or constraints, n. It is important to note, however, that only the first four columns of $\mathbf{U}$ are needed. In this case - which is often referred to as *thin SVD* or *reduced SVD* - the number of computations can be reduced to $\mathcal{O}(nm^2 + m^3)$. This is an important part of our formulation because it still allows to compute $\mathbf{M}_h$ in linear time and thus also the effort for the whole registration problem remains linear. Most matrix libraries offer appropriate routine options.[1] As we show in our simulations it is even much faster than any of the'closed-form' derivations proposed in earlier papers.

2.3   Relation to Existing Approaches

We would like to point out commonalities and differences on how the registration problem is formulated in previous approaches for the PnP and Generalized PnP case [7–9]. The typical procedure is to describe the problem by a system of (noise-free) equations which in their most general form are as follows:

$$\underbrace{\begin{bmatrix} \mathbf{v}_1 & \mathbf{y}_1 & -\mathbf{I}^{3\times3} \\ & \ddots & \vdots & \vdots \\ & & \mathbf{v}_K & \mathbf{y}_K & -\mathbf{I}^{3\times3} \end{bmatrix}}_{=:\tilde{\mathbf{A}}} \underbrace{\begin{bmatrix} \lambda_1 \\ \vdots \\ \lambda_K \\ s \\ t \end{bmatrix}}_{=:\tilde{\mathbf{u}}} = \underbrace{\begin{bmatrix} \mathbf{R} \\ & \ddots \\ & & \mathbf{R} \end{bmatrix}}_{=:\tilde{\mathbf{W}}} \underbrace{\begin{bmatrix} \mathbf{x}_1 \\ \vdots \\ \mathbf{x}_K \end{bmatrix}}_{=:\tilde{\mathbf{x}}}, \tag{11}$$

where additional virtual depth parameters $\lambda_k \in \mathbb{R}$ for the points are introduced. These ensure that in the noiseless case the scaled image points, $\mathbf{y}_k + \lambda_k \mathbf{v}_k$, coincide with their corresponding transformed points $\mathbf{x}'_k$. As in our case, the next step consists in eliminating the linear parameters $\tilde{\mathbf{u}}$ by means of the pseudoinverse, $\tilde{\mathbf{A}}^\dagger = (\tilde{\mathbf{A}}^\mathsf{T}\tilde{\mathbf{A}})^{-1}\tilde{\mathbf{A}}^\mathsf{T}$, and expressing them in terms of the rotation parameters

$$\tilde{\mathbf{u}} = \tilde{\mathbf{A}}^\dagger \tilde{\mathbf{W}} \tilde{\mathbf{x}}. \tag{12}$$

After inserting this expression back into Eq. 11 the resulting system of equations depends only on the rotation $\mathbf{R} \in SO(3)$ (or the vectorized rotation matrix $\mathbf{r}$).

---

[1] Matlab provides the option `economy` to the SVD-routine. In LAPACK the routine DGESVD comes with the option JOBU=`S` or JOBU=`O`. In Eigen one can set ComputeThinU in the constructor of the template JacobiSVD<.>.

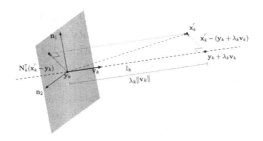

Fig. 1. Visualization of the equivalence of the geometric errors minimized in previous approaches (red) and in our approach (blue). (Color figure online)

The final minimization problem to determine the rotation has again the following form:

$$\arg\min_{\mathbf{R}\in SO(3)} \{\mathbf{r}^\mathsf{T}\tilde{\mathbf{M}}\mathbf{r}\}, \tag{13}$$

with

$$\tilde{\mathbf{M}} = \begin{bmatrix} \mathbf{I}\otimes\mathbf{x}_1, \cdots, \mathbf{I}\otimes\mathbf{x}_K \end{bmatrix} (\mathbf{I} - \tilde{\mathbf{A}}\tilde{\mathbf{A}}^\dagger) \begin{bmatrix} \mathbf{I}\otimes\mathbf{x}_1^\mathsf{T} \\ \vdots \\ \mathbf{I}\otimes\mathbf{x}_K^\mathsf{T} \end{bmatrix}. \tag{14}$$

It is important to note that this formulation for the PnP problem and its generalization minimizes the same error as in our case with point-to-line correspondences. The previous approaches [7–9] minimize the Euclidean distance between the transformed point $\mathbf{x}_k'$ and the point $\mathbf{y}_k + \lambda\mathbf{v}_k$ which represents the line l_k parameterized by the depth value λ_k. Since the depth parameter is included as optimization variable inside the whole minimization problem, it will attain its optimal value when the vector $\mathbf{x}_k' - \mathbf{y}_k + \lambda\mathbf{v}_k$ is exactly orthogonal to the line l_k or the vector $\mathbf{v}_k$ (see Fig. 1). Otherwise the error could still be reduced by changing λ_k while leaving the other parameters fixed. In our formulation the vector $\mathbf{x}_k' - \mathbf{y}_k$ is directly projected onto the orthogonal complement of the line which is spanned by the columns of the matrix $\mathbf{N}_k = [\mathbf{n}_1, \mathbf{n}_2]$. As a consequence, the lengths of both vectors, $\|\mathbf{x}_k' - \mathbf{y}_k + \lambda\mathbf{v}_k\|$ and $\|\mathbf{N}_k^\mathsf{T}(\mathbf{x}_k' - \mathbf{y}_k)\|$, are equal at the minimum of their respective objective functions. We also note that for non-degenerate configurations the resulting matrices, $\tilde{\mathbf{M}}$ and $\mathbf{M}$, are identical up to small numerical differences when computed with the previous approaches and with ours.

The major difference of previous formulations is that the geometric entities are described by their *affine subspaces* (represented by the bearing vector $\mathbf{v}_k$) and not by their *orthogonal complement* as in our case. This makes it necessary to introduce the virtual depth parameters λ_k. The downside is that the involved matrices $\tilde{\mathbf{A}}$ and $\tilde{\mathbf{W}}$ become very sparse and much larger than in our case. In particular, computing the pseudoinverse of $\tilde{\mathbf{A}} \in \mathbb{R}^{3K\times K+4}$ is prohibitively costly if one resorts to standard techniques for dense matrices. For this reason an important aspect in the aforementioned papers is the presentation of a custom-made computation of the pseudoinverse of $\tilde{\mathbf{A}}$ and the final composition of $\tilde{\mathbf{M}}$. Special care was taken to exploit the sparsity of the matrices and thus to preserve the linear complexity of the whole algorithm. Yet still, the computation of $\tilde{\mathbf{M}}$ remains up to one order of magnitude slower than in our proposed method, as we show in our simulations.

Furthermore, the derivations of $\tilde{\mathbf{M}}$ make specific assumptions on the targeted problem instance, so they are only applicable to PnP-type problems (or to 3d point-to-line registrations). It would be possible to extend these subspace-based parameterizations to the registration of point-to-plane correspondences. One would then introduce two "depth" parameters per plane and a 3×2 matrix $\mathbf{V}_k$ whose orthonormal column vectors span the subspace of the plane instead of a single bearing vector $\mathbf{v}_k$. However, computations would only get more complicated as one would have to track down the type of correspondence along the whole process of generating $\tilde{\mathbf{M}}$. In our case, once the matrix $\mathbf{A}$ is set up (Eq. 5), all information on the correspondence type is essentially hidden. In order to compute the pseudoinverse of the dense matrix $\mathbf{A}_u$ and finally $\mathbf{M}$ one can always use the same technique, no matter if $\mathbf{A}_u$ was composed from point-to-point, point-to-line, point-to-plane correspondences or any mixture of them.

2.4   Minimal Number of Constraints and the Inhomogeneous Case

So far, we have restricted our discussion on the full seven DoF problem, i.e. the Euclidean registration *with scale*. Intuitively, one will also need seven constraining equations for the problem in Eq. 4 to be solvable in general. It does not matter from which kind of correspondence types these constraints are obtained, the important part is that the minimal number of seven effective constraints are reached in total and that each 3D point-to-plane, point-to-line or point-to-point yields one, two or three constraints, respectively. For example one can compute the registration parameters from seven point-to-plane correspondences only, where each correspondence gives rise to one equation. In previous approaches to pose-and-scale estimation [9,12] at least 'three-and-a-half' 2D-3D correspondences are needed, which is in accordance with our formulation, where constraints arising from 2d image point measurements are translated into 3D point-to-line correspondences. For general configurations, i.e. when the image measurements are distributed in more than one camera (also referred to as the *non-central case* [8]), the sub-matrix $\mathbf{A}_{st}$ has full rank four. So in the process of eliminating the linear parameters by means of the pseudoinverse, $\mathbf{A}_{st}^{\dagger}$, the final matrix $\mathbf{M}_h \in \mathbb{R}^{9 \times 9}$ will have at least rank three, which is a necessary requirement for solving for the remaining three DoF of the rotation. Since the seven DoF problem forms a homogeneous system of equations we also refer to this as the *homogeneous case*.

By contrast, if the scale parameter is already known or only the rotation and translation is to be estimated, then one would not eliminate the scale parameter. The correct procedure is then to compute the SVD only on the matrix $\mathbf{A}_t$, i.e.

the columns of $\mathbf{A} = [\mathbf{A}_{rs}\ \mathbf{A}_t]$ which belong to the translation part, and finally solve the slightly modified problem

$$\underset{\mathbf{R}\in SO(3)}{\arg\min}\left\{[\mathbf{r}^\mathsf{T}, 1]\,\mathbf{M}_i\begin{bmatrix}\mathbf{r}\\1\end{bmatrix}\right\}, \quad \text{with } \mathbf{M}_i \in \mathbb{R}^{10\times 10}, \tag{15}$$

which we call the *inhomogeneous case*.

There is an important connection between the homogeneous and the inhomogeneous version, which happens when all lines and planes have one common intersection point. This corresponds to the situation, when the camera pose is estimated from measurements in a single camera only as in the *central PnP or PnL* case. Clearly, for single-view pose estimation the scale parameter is meaningless and cannot be computed. In the homogeneous case the problem is therefore ill-conditioned which manifests itself in the matrix $\mathbf{A}_{st} \in \mathbb{R}^{n\times 4}$ having only rank three. As a consequence, the pseudoinverse cannot be computed using the explicit formula $\mathbf{A}_{st}^\dagger = (\mathbf{A}_{st}^\mathsf{T}\mathbf{A}_{st})^{-1}\mathbf{A}_{st}^\mathsf{T}$ because $(\mathbf{A}_{st}^\mathsf{T}\mathbf{A}_{st})$ is singular. However, by using the SVD instead, as proposed in Sect. 2.2, this degeneracy is automatically handled correctly by means of the matrix $\mathbf{\Sigma}^\dagger\mathbf{\Sigma}$. This leads to the important property that the solutions to the remaining parameters, $\mathbf{R}$ and $\mathbf{t}$, can still be computed even if only six effective constraints are provided (e.g. three 2D-3D correspondences for the PnP problem), because the resulting matrix $\mathbf{M}_h$ still has rank three. As for the inhomogeneous case we note, that the column $\mathbf{a}_s$ of $\mathbf{A} = [\mathbf{A}_r, \mathbf{a}_s, \mathbf{A}_t]$ belonging to the scale parameter is a linear combination of the columns of $\mathbf{A}_t$ and therefore its projection onto the orthogonal complement of $\mathbf{A}_t$, i.e. $(\mathbf{I} - \mathbf{U}(\mathbf{\Sigma}^\dagger\mathbf{\Sigma})\mathbf{U}^\mathsf{T})$, is zero. Consequently, the last row and the column of the resulting matrix $\mathbf{M}_i$ will then also be zero and the upper-left 9×9 sub-matrix in $\mathbf{M}_i$ is identical to $\mathbf{M}_h$.

To summarize, in the central case both matrices $\mathbf{M}_i$ and $\mathbf{M}_h$ carry the same information for the solution of the rotation. And as both, the central case (like [7,13]) and the homogeneous non-central case [9], can be represented by 9×9 matrices $\mathbf{M}_h$, we expect that the corresponding algebraic solvers for the nonlinear rotation can be used interchangeably for both types of problems. Further, we expect that an algebraic solver working on a 10×10 matrix $\mathbf{M}_i$ capable of solving both the central case and the inhomogeneous non-central case, such as the one inside the approach of Kneip *et al.* [8], can be applied to all problems considered here.

2.5 Efficiently Pre-rotating Reference Points

Often it is advantageous to work with a modified $\widehat{\mathbf{M}}_h$ that is derived by simply pre-rotating the reference points $\mathbf{x}_k$ with some rotation matrix $\mathbf{R}_0$, i.e. $\widehat{\mathbf{x}}_k = \mathbf{R}_0\mathbf{x}_k$. Any solution, $\widehat{\mathbf{R}}$, obtained on the basis of $\widehat{\mathbf{M}}_h$ is then also a rotated version of the original solution $\mathbf{R}$, i.e. $\widehat{\mathbf{R}} = \mathbf{R}\mathbf{R}_0^\mathsf{T}$. The algebraic solvers which will be discussed in the next Section may fail to determine the correct solutions in all cases. In particular, solvers based on the Cayley parameterization for rotation matrices will not succeed whenever the correct solution for the rotation has an angle of π. In this case one can re-evaluate the problem for different $\widehat{\mathbf{M}}_h$ and collect all solutions. Although not published by Hesch and Roumeliotis [7], the same authors implemented this strategy as an improved version of the DLS-algorithm[2], which was later also adopted by others [14]. Another use case is the post-refinement of solutions for $\mathbf{R}$ with a second order Newton minimization applied on a matrix $\widehat{\mathbf{M}}_h$ with the rotation $\widehat{\mathbf{R}}$ being optimized is close to the identity. This was done e.g. by Zheng et al. and Kneip et al. [8,13].

Instead of re-evaluating $\widehat{\mathbf{M}}_h$ each time for the rotated points $\widehat{\mathbf{x}}_k$ from scratch, we observe that it is also possible to manipulate the matrix $\mathbf{M}_h$, directly. This has the same effect but it can be computed in constant time, whereas a full re-evaluation requires a linear effort with respect to the number of correspondences. Recomputing $\widehat{\mathbf{M}}_h$ for some $\mathbf{R}_0$ thus becomes a negligible operation compared to the actual solving step of the algebraic solver.

As can be seen from 5, rotating the reference points $\mathbf{x}_k$ leaves $\mathbf{A}_{st}$ unchanged and $\mathbf{A}_r$ changes as follows,

$$\widehat{\mathbf{A}}_r = \mathbf{A}_r \mathcal{R}_0, \quad \text{with } \mathcal{R}_0 = \begin{bmatrix} \mathbf{R}_0 & & \\ & \mathbf{R}_0 & \\ & & \mathbf{R}_0 \end{bmatrix}, \tag{16}$$

which together with Eq. 9 yields

$$\widehat{\mathbf{M}}_h = \mathcal{R}_0^\mathsf{T} \mathbf{M}_h \mathcal{R}_0. \tag{17}$$

One can partition $\mathbf{M}_h$ into nine 3×3 submatrices $\mathbf{M}_h^{(i,j)}$ and transform each of them individually, i.e. $\widehat{\mathbf{M}}_h^{(i,j)} = \mathbf{R}_0^\mathsf{T} \mathbf{M}_h^{(i,j)} \mathbf{R}_0$. Thus, it is possible to exploit the sparsity of $\mathcal{R}_0$ and to avoid explicitly constructing it as a matrix.

Considering again the Cayley-parameterization, the traditional procedure consists of re-evaluating the problem for two extra *randomly gerenated* $\mathbf{R}_0$. We note, that the set of Cayley singularities actually forms a two-dimensional manifold. So, in order to guarantee that $\widehat{\mathbf{R}}$ never is near the set of these singularities for all evaluations, one actually has to perform four evaluations in total. This is because for three arbitrary pre-rotating matrices, $\mathbf{R}_i$, $i \in 1, 2, 3$, one can always find a forth rotation, $\mathbf{R}_4$, so that $\mathbf{R}_4^\mathsf{T} \mathbf{R}_i$ has an rotation angle of π. Instead of generating the pre-rotation matrix $\mathbf{R}_0$ randomly, we propose to select it from the canonical set of rotations,

$$\mathbf{R}_0 \in \left\{ \begin{bmatrix} 1 & & \\ & 1 & \\ & & 1 \end{bmatrix}, \begin{bmatrix} 1 & & \\ & -1 & \\ & & -1 \end{bmatrix}, \begin{bmatrix} -1 & & \\ & 1 & \\ & & -1 \end{bmatrix}, \begin{bmatrix} -1 & & \\ & -1 & \\ & & 1 \end{bmatrix} \right\}, \tag{18}$$

where the relative rotation between any two of these elements has the angle π. In this case, recomputing $\widehat{\mathbf{M}}_h$ simply amounts to changing signs of some of the entries in $\mathbf{M}_h$. This can be achieved almost instantly and it alleviates some of the common objections against the use of Cayley parameterization.

---

[2] The implementation of DLS is available at http://www-users.cs.umn.edu/~joel/.

3   Algebraic Solvers for the Rotation

We will now turn our attention towards solving for the nonlinear rotation, i.e. finding all solutions of Eq. 10 or 15. To this end the rotation matrix is parameterized either by quaternions or via Cayley parameters. For a quaternion, $\mathbf{q} = [q_0, q_1, q_2, q_3]^\mathsf{T}$, with real part q_0, the rotation matrix is given by

$$\mathbf{R}(q) = \frac{1}{\|\mathbf{q}\|} \begin{bmatrix} q_0^2 + q_1^2 - q_2^2 - q_3^2, & 2(q_1 q_2 - q_0 q_3), & 2(q_1 q_2 - q_0 q_3) \\ 2(q_1 q_2 + q_0 q_3), & q_0^2 - q_1^2 + q_2^2 - q_3^2, & 2(q_2 q_3 - q_0 q_1) \\ 2(q_1 q_3 - q_0 q_2), & 2(q_2 q_3 + q_0 q_1), & q_0^2 - q_1^2 - q_2^2 + q_3^2 \end{bmatrix}. \quad (19)$$

The Cayley parameterization is given by simply fixing $q_0 = 1$.

The first order optimality conditions are obtained by taking the derivative of the error function with respect to the four quaternion parameters, which leads to a system of four equations with monomials in q_i of degree three (three equations for Cayley parametrization).

$$2\frac{\partial}{\partial q_i}\mathbf{r}(\mathbf{q})^T \cdot \mathbf{M} \cdot \mathbf{r}(\mathbf{q}) = 0 \quad (20)$$

3.1   UPnP Solver

For the UPnP-solver [8] four additional equations were added, which are the derivatives of the squared unit norm constraint of the quaternion. The solver is derived for the generalized PnP problem without scale for the minimal case of three 2d–3d correspondences, but in practice it can be applied to any number. The derivation of the solver by means of an automatic Gröbner Basis solver generator requires that the two-fold symmetry of quaternions is considered [15]. It is also necessary to model the input data in a consistent way in Z_p. A C++ Version of the final algorithm can be found inside the OpenGV framework[3]. We separated the linear parameter elimination step from the actual rotation solver, so we are able to evaluate them separately.

3.2   DLS/gDLS Solver

The DLS-solver [7] uses the Macaulay-Resultant-Matrix method to solve the algebraic equations. It uses the Cayley parametrization. The solver returns at most 27 real solutions.

The solver of gDLS [9] is a transcription of the DLS solver from Matlab code to C++ using Eigen as math library. Apart from that they are absolutely identical and we use the gDLS-version for efficiency reasons. It can be found inside the Theia-Library[4]. For the evaluations we again separate everything related to setting up the matrix $\mathbf{M}_h$ from the actual solver.

---

[3] http://laurentkneip.github.io/opengv/.
[4] http://www.theia-sfm.org/.

3.3   Own Solver for the Homogeneous Case

We also developed an own solver following the main ideas presented by Kneip et al. [8], but with the difference that it works on the 9×9 homogeneous matrix $\mathbf{M}_h$ instead of $\mathbf{M}_i$. We used Kukelova's *Automatic Solver Generator* [16] for its derivation. Several modifications were necessary, including the consideration of the two-fold symmetry of quaternions [15] by working only with polynomials of even degree. Furthermore, we replaced the default random $\mathbb{Z}_p$-instantiation module with an own that generates 'noiseless' integer measurements before a Gröbner Basis is derived.

Our final solver uses an elimination template matrix of size 184×176. As for the UPnP-solver the size of the final action matrix is 8×8, so our algorithms also returns at most eight real solutions[5].

4   Evaluation

4.1   Accuracy

We conducted several synthetic evaluations for measuring the accuracy of the solvers. Our focus is on the applicability of our unifying framework presented in Sect. 2. Therefore, we replace the computation of $\mathbf{M}_h$ by our version and only use the polynomial solvers inside the algorithms (see Sect. 3). The solvers are denoted by **DLS/gDLS (OC)** [7,9], **UPnP (OC)** [8], and **own solver (OC)** (Sect. 3.3), where *'OC'* refers to the substitution with our *orthogonal complement* formulation. We only use the raw polynomial solvers, so no root-polishing is applied on the obtained solutions afterwards (as opposed to the original UPnP algorithm). For the DLS/gDLS solver which uses the cayley parametrization, we employ the strategy outlined in Sect. 2.5, i.e. we solve the problem four times, collect all solutions, and among the duplicates we only keep the ones which have the smaller error according to Eq. 10. We also apply the same strategy for the UPnP solver and our own solver, although they do not suffer from the singularities by the Cayley parameterization. We explain the reason for that in a separate evaluation below in Sect. 4.2.

General Configurations for Point, Line and Plane Registration. In a first experiment we analysed the accuracy of the full Euclidean registration with scale for general geometric configurations.[6] We evaluated point-to-point, point-to-line, and point-to-plane registration separately.

We created the evaluation data by first generating Gaussian distributed transformed points $\mathbf{x}'_k$ with identity covariance. Given random ground-truth rotation $\mathbf{R}_{GT}$, translation $\mathbf{t}_{GT}$ and scale $\mathbf{s}_{GT} \in [0.1, 10]$ the inverse

---

[5] Our algorithm and the evaluation framework is available for academic purposes. Please contact the authors.

[6] More evaluations including the classical PnP problem and various degenerate configurations can be found in the supplemental material corresponding to this paper.

transformation is applied to obtain the reference points $\mathbf{x}_k$. The subspace spanned by a geometric entity and its orthogonal complement are obtained by partitioning the columns of a random orthogonal matrix into $\mathbf{N}_k$ and $\mathbf{V}_k$. On the subspace a point is chosen as offset point $\mathbf{y}_k$. Finally, Gaussian 3d noise is added to $\mathbf{x}_k$, whose covariance was kept fixed to 0.001 times the identity matrix throughout the evaluation run.

In the experiment we varied the number of input correspondences from which the matrix $\mathbf{M}_h$ was constructed which, in turn, was used as input for the algorithms. We evaluated the mean error of the rotation, translation, and scale with respect to ground-truth. Figure 2 summarizes the results. It can be seen that it is possible to successfully estimate the transformation parameters with all three solvers for point-to-line and for point-to-plane correspondences and with similar accuracy. This is an important result in several aspects. While Sweeney *et al.* [9] demonstrated that DLS-Algorithm can be extended from the classical PnP Problem to the Generalized PnP Problem with scale, we show here that it can also be used for registration from point-to-plane constraints. The same is true for the UPnP-solver, which also has never been used for point-to-plane registration. In addition, we note that the UPnP-Algorithm was originally proposed to solve the classical and generalized pose problem with *fixed scale*. Our evaluations demonstrate that it can also be used to solve the 7 DoF-problem including the scale as free parameter.

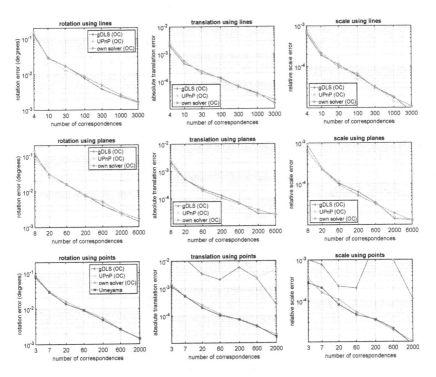

Fig. 2. Mean errors of the estimated rotation, translation, and scale for general geometric configurations with varying numbers of input correspondences.

However, we also observe, that UPnP and gDLS cannot be extended to point-to-point registration as they fail completely in estimating correct results. By contrast, our own solver succeeds in this scenario. We compare it to the algorithm of Umeyama [1] which is the standard algorithm for this case. Both algorithms are almost identical regarding the accuracy.

4.2   Numerical Stability Under Strong Noise

It has already been observed previously that the numerical stability of the UPnP-Algorithm degrades for the central case (homogeneous case) when strong noise is present. Then it may still return very accurate solutions sometimes, but it also happens more frequently that none of the returned solutions is anywhere near the correct rotation. When evaluating the median value of the error instead of the mean value (see [8,14]) the algorithm still showed superior performance.

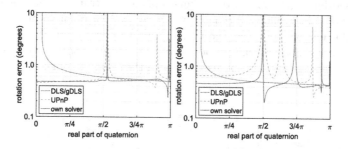

Fig. 3. Two independent evaluations showing the influence of the true rotation on the stability of the rotation solvers under strong noise.

We further investigated this behaviour by analysing the stability as a function of the true rotation. We generated a set of 20 3D point-to-line correspondences using the identity matrix as initial ground truth rotation. We then added a fairly large quantity of noise to the data. The magnitude of the noise corresponded to five pixels standard deviation when projected onto the image plane in the PnP case. Next we rotated the reference points $\mathbf{x}_k$ with a smoothly varying rotation, $\mathbf{R}_0(\alpha)$. The parameter α was used for the real part q_0 and was varied in the range $[0, 1]$. The imaginary values were all set to $q_i = \sqrt{(1 - \alpha^2)}/\sqrt{3}$ for $i \in \{1, 2, 3\}$. The rest of the data, i.e. $\mathbf{y}_k$ and $\mathbf{N}_k$ and the noise, was left unchanged. For each corresponding matrix $\mathbf{M}_h(\alpha)$ we estimate the rotation using the algorithms from Sect. 3 and evaluated the error with respect to the ground-truth rotation $\mathbf{R}_0(\alpha)$. Figure 3 shows two independent evaluations. As expected, the DLS/gDLS algorithm fails to compute the correct solution near $\alpha = 0$, which represents an element in the set of Cayley-degeneracies. The UPnP-solver and our own also exhibit singularities. As opposed to the Cayley-parameterization their location is not known in advance. However, as our analysis shows, they are also dependent on the rotation. This implies that the stability of these algorithms can be significantly improved by re-evaluating the problem for different pre-rotations of the reference points $\mathbf{x}_k$, which can be done very efficiently as shown in Sect. 2.5.

4.3 Runtime Analysis

We evaluated the run-time performance of the algorithms which were all implemented in C++ and executed single threaded with 3.5 GHz clock rate. Figure 4 shows the timings for the linear parameter elimination part inside the gDLS and UPnP algorithm compared to our orthogonal complement based approach. We used 2d-3d correspondences from the PnP problem as input. Both version exhibit linear complexity. For more than 13 input correspondences our approach is faster up to a factor of approximately 2.5. Table 1 shows the mean execution times of the different algebraic solvers. The gDLS-solver is fastest and takes less that 0.9 ms, followed closely by our solver. UPnP is approximately 50% slower than gDLS.

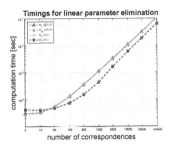

Fig. 4. Computational time of our linear parameter elimination step versus previous approaches.

Table 1. Mean execution times of the different rotation solvers.

UPnP [8] (Sect. 3.1)	gDLS [9] (Sect. 3.2)	our solver (Sect. 3.3)
1.299 ms	0.871 ms	0.908 ms

5 Conclusion

We presented a further generalization for Euclidean transformation problems. We model the point-to-point, point-to-line and point-to-plane constraints using an orthogonal complement representation, which makes it possible that they can be used together in one single framework very elegantly and efficiently. Our formulation also allows to use the different existing algebraic solvers for the rotation interchangeably, so they can be compared against each other on varying problems. We also propose an own solver, which additionally solves the case of point-to-point registration with high precision, where existing solvers failed.

References

1. Umeyama, S.: Least-squares estimation of transformation parameters between two point patterns. IEEE Trans. Pattern Anal. Mach. Intell. (PAMI) **13**, 376–380 (1991)
2. Horn, B.K.P.: Closed-form solution of absolute orientation using unit quaternions. J. Opt. Soc. Am. (JOSA) **4**, 629–642 (1987)
3. Low, K.: Linear least-squares optimization for point-to- plane ICP surface registration. Technical report TR04-004, University of North Carolina (2004)
4. Newcombe, R.A., Izadi, S., Hilliges, O., Molyneaux, D., Kim, D., Davison, A.J., Kohi, P., Shotton, J., Hodges, S., Fitzgibbon, A.: Kinectfusion: real-time dense surface mapping and tracking. In: IEEE Proceedings of International Symposium on Mixed and Augmented Reality (ISMAR), pp. 127–136 (2011)
5. Olsson, C., Kahl, F., Oskarsson, M.: The registration problem revisited: Optimal solutions from points, lines and planes. In: IEEE Proceedings of Conference on Computer Vision and Pattern Recognition (CVPR), vol. 1, pp. 1206–1213 (2006)
6. Olsson, C., Kahl, F., Oskarsson, M.: Branch-and-bound methods for euclidean registration problems. IEEE Tran. Pattern Anal. Mach. Intell. (PAMI) **31**, 783–794 (2009)
7. Hesch, J.A., Roumeliotis, S.I.: A direct least-squares (DLS) method for PnP. In: IEEE Computer Society, Los Alamitos (2011)
8. Kneip, L., Li, H., Seo, Y.: UPnP: an optimal $O(n)$ solution to the absolute pose problem with universal applicability. In: Fleet, D., Pajdla, T., Schiele, B., Tuytelaars, T. (eds.) ECCV 2014. LNCS, vol. 8689, pp. 127–142. Springer, Heidelberg (2014). doi:10.1007/978-3-319-10590-1_9
9. Sweeney, C., Fragoso, V., Höllerer, T., Turk, M.: GDLS: a scalable solution to the generalized pose and scale problem. In: Fleet, D., Pajdla, T., Schiele, B., Tuytelaars, T. (eds.) ECCV 2014. LNCS, vol. 8692, pp. 16–31. Springer, Heidelberg (2014). doi:10.1007/978-3-319-10593-2_2
10. Hartley, R.I., Zisserman, A.: Multiple View Geometry in Computer Vision, 2nd edn. Cambridge University Press, Cambridge (2004). ISBN: 0521540518
11. Golub, G.H., Van Loan, C.F.: Matrix computations: Johns Hopkins Studies in the Mathematical Sciences. The Johns Hopkins University Press, Baltimore (1996)
12. Ventura, J., Arth, C., Reitmayr, G., Schmalstieg, D.: A minimal solution to the generalized pose-and-scale problem. In: IEEE Proceedings of Conference on Computer Vision and Pattern Recognition (CVPR), pp. 422–429. IEEE Computer Society, Los Alamitos (2014)
13. Zheng, Y., Kuang, Y., Sugimoto, S., Åström, K., Okutomi, M.: Revisiting the PnP problem: a fast, general and optimal solution. In: IEEE Proceedings of International Conference on Computer Vision (ICCV), pp. 2344–2351 (2013)
14. Nakano, G.: Globally optimal DLS method for PnP problem with Cayley parameterization. In: Xianghua Xie, M.W.J., Tam, G.K.L. (eds.) Proceedings of the British Machine Vision Conference (BMVC), pp. 78.1–78.11. BMVA Press, London (2015)
15. Ask, E., Kuang, Y., Åström, K.: Exploiting p-fold symmetries for faster polynomial equation solving. In: International Conference on Pattern Recognition (ICPR), pp. 3232–3235 (2012)
16. Kukelova, Z., Bujnak, M., Pajdla, T.: Automatic generator of minimal problem solvers. In: Forsyth, D., Torr, P., Zisserman, A. (eds.) ECCV 2008. LNCS, vol. 5304, pp. 302–315. Springer, Heidelberg (2008). doi:10.1007/978-3-540-88690-7_23

Generalized Fusion Moves for Continuous Label Optimization

Christopher Zach$^{(\boxtimes)}$

Toshiba Research Europe, Cambridge, UK
christopher.m.zach@gmail.com

Abstract. Energy-minimization methods are ubiquitous in computer vision and related fields. Low-level computer vision problems are typically defined on regular pixel lattices and seek to assign discrete or continuous values (or both) to each pixel such that a combined data term and a spatial smoothness prior are minimized. In this work we propose to minimize difficult energies using repeated generalized fusion moves. In contrast to standard fusion moves, the fusion step optimizes over binary and continuous sets of variables representing label ranges. Further, each fusion step can optimize over additional continuous unknowns. We demonstrate the general method on a variational-inspired stereo approach, and optionally optimize over radiometric changes between the images as well.

1 Introduction

Many computer vision applications rely on finding a most-probable label assignment for each pixel as an important subproblem. The dominant formulation as a maximum a-posteriori problem leads to a corresponding energy minimization task, where the energy is typically comprised of per-pixel data terms and smoothness terms defined over small pixel neighborhoods. Often, the admissible set of labels is naturally continuous or very large and therefore "almost continuous." There is a lot of work on approximate discrete inference, which is applicable for finite label sets, and continuous labeling problems are often solved with discrete methods after discretizing the label space. Continuous labeling problems with convex energies are relatively easy to solve by standard convex minimization methods. Therefore, continuous labeling tasks with non-convex energies are more interesting and usually much more relevant in applications.

In this work we consider continuous labeling problems with *piecewise convex* energy, which includes as an important special case truncated convex terms. Determining a minimizer of such problems can be interpreted as first finding which of the convex branches is active and subsequent estimation of the continuous labels. Thus, piece-wise convex energies naturally lead to a discrete-continuous structure for the unknowns, with the discrete state describing the convex branch and the continuous labels defining the desired solution. We build on the convex discrete-continuous (DC-MRF) formulation proposed in [1] for such problem classes. While in principle this method is directly applicable for a

© Springer International Publishing AG 2017
S.-H. Lai et al. (Eds.): ACCV 2016, Part V, LNCS 10115, pp. 67–81, 2017.
DOI: 10.1007/978-3-319-54193-8_5

wide class of labeling problems, the computational cost and the quality of the relaxation can be prohibitive. Therefore we propose to use a generalized fusion move strategy, and employ the DC-MRF formulation only as a subroutine to solve each fusion step. In contrast to existing fusion move approaches for solving continuous labeling problems our generalized fusion move enables (i) the refinement of participating labeling proposals and (ii) allows optimization over additional continuous unknowns. The first advantage reduces the requirements on smart proposal generation and—we believe—also decreases the bias introduced by the exact details of the utilized proposal generation strategy. The second advantage allows more efficient joint optimization over several sets of unknowns (such as joint estimation of disparities and radiometric alignment demonstrated in Fig. 1 and Sect. 7), since (depending on the problem structure) proposals need only to account for a subset of unknowns.

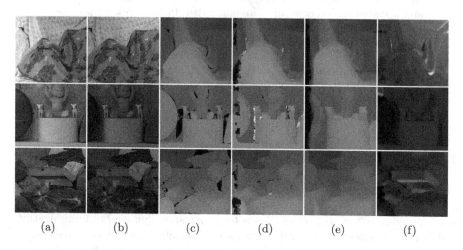

(a)   (b)   (c)   (d)   (e)   (f)

Fig. 1. Simultaneous estimation of continuous-valued disparity map $d(\mathbf{x})$ and per-pixel radiometric gain factor $\gamma(\mathbf{x})$. (a) left image; (b) right image; (c) true disparity; (d) disparity estimated using 5×5 ZNCC and belief propagation using truncated L^1-smoothness prior; (e) estimated disparity d using generalized fusion moves; (f) estimated gain γ. Irregularly shaped shadows and highlights are successfully recovered without "fattening" at occlusions. As a problem in a multidimensional discrete label space, this would be intractably large. This paper's generalized fusion moves allow efficient optimization over non-convex energies in continuous label spaces. Best viewed on screen.

2   Related Work

Move-making algorithms for discrete labeling problems on loopy graphs are an efficient alternative to e.g. belief propagation or message-passing methods for approximate inference. In particular, α-expansion and α-β-swaps [2] are often employed for low-level computer vision tasks auch as segmentation and stereo.

The success of move-making algorithms depends on the "richness" of the allowed moves in each step, and a lot of research is devoted to extending α-expansion and α-β-swaps to enable more powerful moves (e.g. [3–5]). Note that e.g. α-expansion is a very restricted move: for each node (pixel) either the current label is kept, or a node is relabeled to a particular label α. These moves are iterated over all possible labels α until covergence.

Our work shares a lot of motivation with the "range moves" concept originally proposed in [6] and refined later in [7–9]: here each move-making step can keep the current label at a node, or switch to a label out of a contiguous label range. Thus, each move is much more expressive than e.g. pure α-expansion, but the pairwise (smoothness) cost in these works is restricted to truncated convex priors.

For labeling tasks with continuous state spaces (such as computational stereo and optical flow with subpixel accuracy) the algorithms mentioned above can not be directly applied. Very often continuous state spaces allow direct energy minimization to obtain a labeling (one umbrella term is "variational optical flow"), but these methods often do not cope well with the highly non-convex structure of the underlying energy and can return poor local minima. One can expect to escape such local minima by using a suitable move making algorithm allowing larger update steps in the labeling. To our knowledge the first notion of a move-making method for continuous labeling problems is the "optimal splicing" concept introduced in [10], but the general "fusion move" technique was popularized in [11] (for discrete label spaces) and in [12] (for continuously valued problems). The underlying idea is simple: two labeling proposals (with underlying discrete or continuous state spaces) are optimally merged to yield a solution with lower energy. How the two proposals should be optimally merged is subject to a binary segmentation problem, which usually can be efficiently solved. These fusion move steps are repeated to obtain label assignments with decreasing energy. The α-expansion method can now be understood as particular instance of a fusion move method with the current best solution and a constant labeling as proposals to merge. The quality of the result clearly depends on the proposals: it is e.g. demonstrated in [6,13] that the choice of proposals may introduce a particular bias in the returned solution even if the optimized energy has no such bias. If the energy to minimize is differentiable, new proposals can be generated by gradient steps [14].

Our setting explicitly addresses continuous state spaces, but retains a discrete domain, e.g. a regular pixel grid with 4-connected neighborhoods. Thus, our setting is different to move making algorithms for label optimization derived on continuous image domains such as [15,16].

This work is based on the convex relaxation framework for discrete-continuous random fields presented in [1,17], which was subsequently generalized to a larger class of dual objectives [18] and further extended to spatially continuous image domains [19].

3 Notations and Background

Notations: In the following we use the notations $\imath_C(x)$ and $\imath\{x \in C\}$ to write a constraint $x \in C$ in functional form, i.e. $\imath_C(x) = 0$ iff $x \in C$ and ∞ otherwise. Further, we will make extensive use of the *perspective* of a convex function f: $(x, y) \mapsto xf(y/x)$ for $x > 0$ (see e.g. [20]). We denote the lower semi-continuous extension of the perspective to the case $x = 0$ by $f_{\oslash}$, pronounced "persp f". $f_{\oslash}$ can be computed as the biconjugate of the standard perspective, and usually one obtains $f_{\oslash}(0, y) = \imath_{\{0\}}(y)$. In the context of this work the perspective of a (convex) function f can be understood as convex extension of the conditional

$$(x, y) \mapsto \begin{cases} f(y) & \text{if } x = 1 \\ 0 & \text{if } x = 0 \\ \infty & \text{otherwise.} \end{cases} \tag{1}$$

Further, we denote the unit simplex by $\Delta_n \overset{\text{def}}{=} \{\mathbf{x} \in [0,1]^n : \sum_i x_i = 1\}$.

We represent an image domain as finite rectangular lattice over pixels $s \in \mathcal{V}$ with an edge set $\mathcal{E}$ induced by a 4-neighborhood connectivity. Thus, in this setting the degree $\deg(s)$ of a node s, which we are going to use later, is always four.

The DC-MRF Model: In this section we briefly review the DC-MRF formulation for inference proposed in [1], which generalizes approximate discrete inference (discrete state spaces and domains) to continuous-valued label spaces by replacing the standard constant potentials with convex potential functions. For given families of convex functions $\{f_s^i\}_{s \in \mathcal{V}}$ and $\{g_{st}^{ij}\}_{(s,t) \in \mathcal{E}}$ (with $i, j \in \{1, \ldots, L\}$) the discrete-continuous formulation reads as

$$E_{\text{DC-MRF}}(\mathbf{x}, \mathbf{y}) = \sum_{s,i} (f_s^i)_{\oslash}(x_s^i, y_s^i) + \sum_{(s,t) \in \mathcal{E}} \sum_{i,j} (g_{st}^{ij})_{\oslash}(x_{st}^{ij}, y_{st \to s}^{ij}, y_{st \to t}^{ij}) \tag{2}$$

subject to the following marginalization and "decomposition" constraints

$$x_s^i = \sum_j x_{st}^{ij} \quad x_t^j = \sum_i x_{st}^{ij} \quad y_s^i = \sum_j y_{st \to s}^{ij} \quad y_t^j = \sum_i y_{st \to t}^{ij}, \tag{3}$$

and simplex constraints $x_s \in \Delta_L$, $x_{st} \in \Delta_{L^2}$. The unknown vector $\mathbf{x}$ collects the "pseudo-marginals" (i.e. x_{st} indicates a one-hot encoding of the active potential function f_{st}^{ij} state at edge (s, t)). The unknowns $\mathbf{y}$ indirectly represent the assigned continuous labels in the solution, which are actually given by the ratio $\mathbf{y} \div \mathbf{x}$ (element-wise division). The DC-MRF model is an extension of the standard local-polytope relaxation for discrete labeling problems by allowing the unary and pairwise potentials now to be arbitrary piecewise convex functions with continuous label arguments. The formulation Eq. 2 is used in [1] to model convex relaxations of non-convex continuous labeling tasks. In particular, the data term for a continuous labeling problem is allowed to be piecewise convex instead of globally convex, but the same construction applies to piecewise convex higher-order potentials.

3.1   Partial Optimality and Autarkies

In Sect. 5 we will describe an approach to potentially speed up minimization of instances of $E_{\text{DC}-\text{MRF}}$ by first solving a simpler surrogate problem, which allows to fix some (in the ideal case all) x_s^i to either 0 or 1 before fully minimizing the discrete-continuous model. This surrogate problem is a standard (not necessarily submodular) binary labeling task with at most pairwise potentials. The underlying technique in Sect. 5 is heavily inspired by the methods proposed in [21–23] to certify partial optimality of label assignment for certain discrete inference problems. In the following exposition we follow in particular [23] (specializing the notation to the case of binary label spaces $\mathcal{L} = \{0, 1\}$): if we have two label assignments $\mathbf{k}, \mathbf{l} : \mathcal{V} \to \mathcal{L}$, then we introduce the component-wise minimum $\mathbf{k} \wedge \mathbf{l}$ and maximum $\mathbf{k} \vee \mathbf{l}$ via

$$(\mathbf{k} \wedge \mathbf{l})_s = \min(k_s, l_s) \qquad (\mathbf{k} \vee \mathbf{l})_s = \max(k_s, l_s).$$

Note with our binary label set these definitions coincide with a component-wise logical and and logical or. Given two label assignments $\mathbf{l}^{\min}, \mathbf{l}^{\max}$ such that $l_s^{\min} \leq l_s^{\max}$ we define a "clamp" operation for another labeling k

$$\text{clamp}(\mathbf{k}; \mathbf{l}^{\min}, \mathbf{l}^{\max}) \overset{\text{def}}{=} (\mathbf{k} \vee \mathbf{l}^{\min}) \wedge \mathbf{l}^{\max}.$$

A pair of labelings $(\mathbf{l}^{\min}, \mathbf{l}^{\max})$ is called a *weak autarky*, if for all label assignment $\mathbf{k}$ we have

$$\mathbf{f}(\text{clamp}(\mathbf{k}; \mathbf{l}^{\min}, \mathbf{l}^{\max})) \leq \mathbf{f}(\mathbf{k}).$$

If the inequality is strict for all $\mathbf{k}$ such that $\mathbf{k} \neq \text{clamp}(\mathbf{k}; \mathbf{l}^{\min}, \mathbf{l}^{\max})$, then $(\mathbf{l}^{\min}, \mathbf{l}^{\max})$ forms a *strong autarky*. Weak autarkies ensure that there exists at least one optimal solution that is "sandwiched" by $\mathbf{l}^{\min}$ and $\mathbf{l}^{\max}$, and strong autarkies guarantee that every optimal solution lies between $\mathbf{l}^{\min}$ and $\mathbf{l}^{\max}$. If we have a strong autarky available, we can reduce the search space in advance. For binary labeling problems (as ours), a strong autarky $(\mathbf{l}^{\min}, \mathbf{l}^{\max})$ allows to fix the binary state at nodes s whenever $l_s^{\min} = l_s^{\max}$. The following two results are essential for our construction:

Result 1 *(Theorem 1 in [23]). Let* $\mathbf{f} = \mathbf{g} + \mathbf{h}$, *and let* $(\mathbf{l}^{\min}, \mathbf{l}^{\max})$ *be strong autarky for* $\mathbf{g}$ *and a weak autarky for* $\mathbf{h}$. *Then* $(\mathbf{l}^{\min}, \mathbf{l}^{\max})$ *is a strong autarky for* $\mathbf{f}$.

This result is easily verified by checking the strong autarky condition. The following statement provides sufficient conditions for a one-sided autarky to be a weak one for $\mathbf{h}$:

Result 2 [22,23]. *For each* $s \in \mathcal{V}$ *let* $\mathcal{K}_s \subseteq \mathcal{L}$ *be a subset of states. If* $\mathbf{h}$ *satisfies (for* $l_s, l_t \in \mathcal{L}$, $k_s \in \mathcal{K}_s$, $k_t \in \mathcal{K}_t$)*

$$h_s(l_s \vee k_s) \leq h_s(l_s)$$

and

$$h_{st}(l_s \vee k_s, l_t \vee k_t) \leq h_{st}(l_s, l_t),$$

then $(\mathbf{k}^{\min}, 1)$ *is a weak autarky for* $\mathbf{h}$ *for all* $\mathbf{k}^{\min} \in \bigotimes \mathcal{K}_s$.

In a nutshell, $\mathbf{g}$ are submodular potentials (and therefore efficient to solve for exactly) constructed from the original potentials $\mathbf{f}$, that in a carefully designed way favor "smaller" labels (smaller in terms of an arbitrary chosen linear order of labels). If an optimal labeling $\mathbf{k}$ for potentials $\mathbf{g}$ returns a "large" label k_s at node s as its optimal choice, then none of the smaller labels $l_s < k_s$ can appear at s in an optimal solution for $\mathbf{f}$.

Autarkies are a refined (but computationally also more expensive) variant of dead-end elimination theorems (e.g. [24] and we refer to [25–27] for dead-end elimination methods in continuous label spaces).

4   Discrete-Continuous Fusion Moves

Let $\mathcal{G} = (\mathcal{V}, \mathcal{E})$ be an underlying graph (usually a 4-conntected or 8-connected grid), and the task is to solve a continuous label assignment problem w.l.o.g. with at most pairwise terms,

$$E_{\text{Labeling}}(\mathbf{z}) = \sum_{s \in \mathcal{V}} f_s(z_s) + \sum_{(s,t) \in \mathcal{E}} g_{st}(z_s, z_t) \tag{4}$$

for a node-specific data term f_s and an edge-specific smoothness term g_{st}. If f_s and g_{st} can be conveniently written as piecewise convex functions (e.g. $f_s(z) = \min_{i \in \{1,\ldots,N_s\}} \tilde{f}_s^i(z)$ with $\tilde{f}_s^i$ convex), then the DC-MRF relaxation is in principle applicable, but this global relaxation might be weak and very expensive to solve. One method to approximately solve a continuous labeling problem such as E_{Labeling} are fusion moves, which repeatedly merges two proposals with continuous label values assigned to each pixel. Optimal combination of proposals is achieved by solving a binary segmentation task in each iteration. Fusion moves require the exact specification of proposal labelings, and the fusion move itself does not refine the continuous labels.

In many applications the smoothness term has a parametric, piecewise convex shape with a small number of branches (e.g. truncated linear or quadratic pairwise costs). Further, the data term can be highly non-parametric (such as matching costs used in computational stereo and optical flow), but convex surrogate costs valid around a current continuous proposal can often be found (and such approximations are successfully used in the literature, in particular for variational optical flow).

We propose to extend the concept of fusion moves in order allow simultaneous refinement of the continuous labels in addition to the per-node binary decision, which of the two proposals to select. In the simplest setting we assume that g_{st} is convex, i.e. non-convexity of the overall problem is introduced only via the node-specific data term f_s. Further, given two proposal labelings, $\bar{\mathbf{z}}^0$ and $\bar{\mathbf{z}}^1$, our problem under consideration is to determine a combined label assignment $\mathbf{z}$, that is a minimizer of

$$E_{\text{Fusion}}(\mathbf{x}, \mathbf{z}) = \sum_s \sum_{i \in \{0,1\}} x_s^i f_s^i(z_s) + \sum_{(s,t)} \sum_{i,j \in \{0,1\}} g_{st}(z_s, z_t) \tag{5}$$

such that $x_s^i \geq 0$, $x_s^0 + x_s^1 = 1$, and the labels z_s are "near" to either $\bar{z}_s^0$ or $\bar{z}_s^1$,

$$z_s \in \begin{cases} [l_s^0, u_s^0] & \text{if } x_s^0 = 1 \\ [l_s^1, u_s^1] & \text{if } x_s^1 = 1 \end{cases}$$

for appropriate intervals $[l_s^i, u_s^i]$ containing $\bar{z}_s^i$. We define f_s^i as the restriction of f_s to the range $[l_s^i, u_s^i]$. In this context being "near" to either $\bar{z}_s^i$ ($i \in \{0,1\}$) means that f_s^i is convex in $[l_s^i, u_s^i]$ and g_{st} is convex in $[l_s^i, u_s^i] \times [l_s^j, u_s^j]$ for all $i, j \in \{0,1\}$. W.l.o.g. we assume that $[l_s^0, u_s^0]$ and $[l_s^1, u_s^1]$ are non-overlapping. We denote by g_{st}^{ij} the restriction of g_{st} to $[l_s^i, u_s^i] \times [l_s^j, u_s^j]$, and obtain

$$E_{\text{Fusion}}(\mathbf{x}, \mathbf{z}) = \sum_s \sum_{i \in \{0,1\}} x_s^i f_s^i(z_s) + \sum_{(s,t)} \sum_{i,j \in \{0,1\}} x_{st}^{ij} g_{st}^{ij}(z_s, z_t) \tag{6}$$

subject to the marginalization constraints on $\mathbf{x}$, $x_s^i = \sum_j x_{st}^{ij}$ and $x_t^j = \sum_i x_{st}^{ij}$, and simplex constraints $x_s \in \Delta_2$, $x_{st} \in \Delta_4$. This energy is still not convex, and we use the convex relaxation for piece-wise convex labeling problems proposed in [1] to arrive at

$$E_{\text{DC-Fusion}}(\mathbf{x}, \mathbf{y}) = \sum_s \sum_{i \in \{0,1\}} (f_s^i)_\oslash (x_s^i, y_s^i) \tag{7}$$
$$+ \sum_{(s,t)} \sum_{i,j \in \{0,1\}} (g_{st}^{ij})_\oslash (x_{st}^{ij}, y_{st \to s}^{ij}, y_{st \to t}^{ij})$$

subject to the marginalization/decomposition constraints in Eq. 3, and the respective simplex constraints on $\mathbf{x}$. Recall that the continuous labels $\mathbf{z}$ are represented via the ratios $\mathbf{y} \div \mathbf{x}$. The convex relaxation can be made stronger (not necessarily strictly stronger) by moving the unary cost function f_s^i to the pairwise ones [17]. In particular, we evenly distribute f_s^i to the adjacent edges, i.e. we introduce

$$h_{st}^{ij} \overset{\text{def}}{=} g_{st}^{ij} + \frac{1}{\deg(s)} f_s^i + \frac{1}{\deg(t)} f_t^j \tag{8}$$

and rewrite $E_{\text{DC-Fusion}}$ above as

$$\breve{E}_{\text{DC-Fusion}}(\mathbf{x}, \mathbf{y}) = \sum_{(s,t)} \sum_{i,j \in \{0,1\}} (h_{st}^{ij})_\oslash (x_{st}^{ij}, y_{st \to s}^{ij}, y_{st \to t}^{ij}) \tag{9}$$

subject to the same constraints. Note that

$$\min_{\mathbf{x}, \mathbf{y}} \breve{E}_{\text{DC-Fusion}}(\mathbf{x}, \mathbf{y}) \geq \min_{\mathbf{x}, \mathbf{y}} E_{\text{DC-Fusion}}(\mathbf{x}, \mathbf{y}),$$

since $\breve{E}_{\text{DC-Fusion}}$ is a tighter relaxation than $E_{\text{DC-Fusion}}$. Note that the structure of $E_{\text{DC-Fusion}}$ is generally simpler than $\breve{E}_{\text{DC-Fusion}}$ (the former has e.g. fewer constraints). In our examples below the computational advantage of $E_{\text{DC-Fusion}}$ over $\breve{E}_{\text{DC-Fusion}}$ turns out to be minimal, consequently we generally employ the stronger relaxation $\breve{E}_{\text{DC-Fusion}}$ in the following unless explicitly noted. Ultimately, either Eq. 7 or 9 is the convex optimization problem to solve in each discrete-continuous fusion step.

We have described the discrete-continuous fusion moves for a setting where the unknown at each node/pixel is just a continuous label. These fusion moves can be immediately generalized to vector-valued labeling problems (as illustrated in Sect. 7) and even to mixed discrete-continuous state spaces.

Implementation: To our knowledge there is no fast combinatorial algorithm to minimize either Eq. 7 or 9, and one has to revert to generic methods from convex optimization. We utilize a first order method [28] due to its simplicity and relative efficiency to determine a minimizer of the convex programs Eqs. 7 and 9, respectively. The employed method maintains primal and dual variables, which we found beneficial over purely optimizing a (smoothed) dual as proposed in [17,18]. Since optimizing $E_{\text{DC-Fusion}}$ (or $\breve{E}_{\text{DC-Fusion}}$) may lead to fractional values for x_s^i (which can be understood as a per-pixel soft preference for proposal i), we determine a suitable threshold to binarize $\mathbf{x}$ by sweeping over the $[0,1]$ range. The threshold value ρ leading to the smallest original energy is applied. The label at pixel s in the updated proposal is determined as $\bar{z}_s^0 \leftarrow y_s^{i_s^*}/x_s^{i_s^*}$, where $i_s^* = 0$ if $x_s^0 \geq \rho$ and 1 otherwise.

5 Partial Optimality

Neither $E_{\text{DC-Fusion}}$ nor $\breve{E}_{\text{DC-Fusion}}$ can be optimized by a fast combinatorial algorithm, and both energies require to our knowledge a generic optimization approach for non-smooth convex problems. Consequently, it can be beneficial, if the optimal state x_s^i of many nodes/pixels can be determined in advance by a faster method, i.e. before fully optimizing $E_{\text{DC-Fusion}}$. In this section we propose to solve a surrogate discrete problem with only binary labels in order to commit early to either $x_s^0 = 1$ or $x_s^1 = 1$ in $E_{\text{DC-Fusion}}/\breve{E}_{\text{DC-Fusion}}$ without fully minimizing the full optimization problem. Usually, this early committment will allow only a subset of pixels to be labeled in advance. Since our surrogate problem is just a discrete binary segmentation problem with at most pairwise potentials, this labeling can be solved much faster than $E_{\text{DC-Fusion}}$.

Construction of g: In order to construct a surrogate problem over binary labels, which allows us to determine a partial labeling (recall Sect. 3.1), we need to construct submodular potentials $\mathbf{g} = (g_{st})_{st \in \mathcal{E}}$ as follows: if for an $s \in \mathcal{V}$ one has $1 \in \mathcal{K}_s$, then h_{st} has to satisfy the following constrains,

$$h_{st}^{10}(z_s, z_t) \leq h_{st}^{00}(z_s', z_t') \qquad \forall (z_s, z_t) \in R_{st}^{10}, (z_s', z_t') \in R_{st}^{00}$$
$$h_{st}^{11}(z_s, z_t) \leq h_{st}^{10}(z_s', z_t') \qquad \forall (z_s, z_t) \in R_{st}^{11}, (z_s', z_t') \in R_{st}^{10},$$

where $R_{st}^{ij} \stackrel{\text{def}}{=} [l_s^i, u_s^i] \times [l_t^j, u_t^j]$. If $1 \in \mathcal{K}_t$, then the following constraints have to be added,

$$h_{st}^{01}(z_s, z_t) \leq h_{st}^{00}(z_s', z_t') \qquad \forall (z_s, z_t) \in R_{st}^{01}, (z_s', z_t') \in R_{st}^{00}$$
$$h_{st}^{11}(z_s, z_t) \leq h_{st}^{01}(z_s', z_t') \qquad \forall (z_s, z_t) \in R_{st}^{11}, (z_s', z_t') \in R_{st}^{01}.$$

If $\mathcal{K}_s = \{0\}$ (i.e. it is already known that state 0 is not part of any optimal solution at s), then this node does not add any constraints since $l_s \vee 0 = l_s$. We define

$$\underline{h}_{st}^{ij} \stackrel{\text{def}}{=} \min_{(z_s, z_t) \in R_{st}^{ij}} h_{st}^{ij}(z_s, z_t) \qquad \overline{h}_{st}^{ij} \stackrel{\text{def}}{=} \max_{(z_s, z_t) \in R_{st}^{ij}} h_{st}^{ij}(z_s, z_t)$$

(similar for f). Dropping the subscript st for clarity, and using $\mathbf{h} = \mathbf{f} - \mathbf{g}$, the constraints on $\mathbf{h}$ rewritten in terms of $\mathbf{g}$ read as

$$g^{00} \leq \underline{f}^{00} + \min\{\underline{f}^{01} - \overline{f}^{01}, \underline{f}^{10} - \overline{f}^{10}\}$$
$$g^{01} \leq \underline{f}^{01} + g^{11} - \overline{f}^{11}$$
$$g^{10} \leq \underline{f}^{10} + g^{11} - \overline{f}^{11}.$$

Further we have the submodularity constraint, $g^{00} \leq g^{01} + g^{10} - g^{11}$. One particular solution (in analogy to [22,23]) is to assign

$$g^{11} = \overline{f}^{11} \qquad\qquad g^{01} = \underline{f}^{01} \qquad\qquad g^{10} = \underline{f}^{10}$$

and

$$g^{00} = \min\left\{g^{01} + g^{10} - g^{11}, \underline{f}^{00} + \min\left\{\underline{f}^{01} - \overline{f}^{01}, \underline{f}^{10} - \overline{f}^{10}\right\}\right\}.$$

Intuitively, $\mathbf{g}$ is constructed to be submodular and to "favor" label 0 in its solution. Thus, if $\mathbf{l} = (l_s)_{s \in \mathcal{V}}$ is the optimal binary labeling for potentials $\mathbf{g}$, then $l_s = 1$ implies that $x_s^1 = 1$ in the fusion move energy $E_{\text{DC-Fusion}}$ (assuming that $\mathbf{l}$ is the unique optimal solution for $\mathbf{g}$). We solve the submodular problem induced by $\mathbf{g}$ to fix x_s in $E_{\text{DC-Fusion}}$ in advance where possible.

6   Example 1: TV-L^1-Variational Stereo

The first application demonstrates how the proposed generalized fusion moves can be used to improve the results of a variational stereo approach. For a given rectified pair of (grayscale) images I^L and I^R one is interested in computing a dense disparity map d such that $I^L(x) \approx I^R(x+d)$ for each pixel x (where $x+d$ is a shorthand notation for $x + (d, 0)^T$). Variational methods for dense disparity estimation seek a minimizer of

$$E_{\text{stereo}}(\mathbf{d}) = \int_\Omega \phi\left(I^L(x) - I^R(x + d(x))\right) dx + \Psi(\mathbf{d}), \tag{10}$$

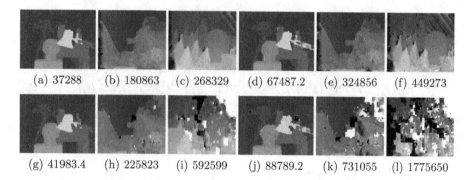

(a) 37288   (b) 180863   (c) 268329   (d) 67487.2   (e) 324856   (f) 449273

(g) 41983.4   (h) 225823   (i) 592599   (j) 88789.2   (k) 731055   (l) 1775650

Fig. 2. Top row: result of generalized fusion moves. Bottom row: disparity maps obtained using a variational multi-scale approach. The left three columns use $\lambda = 2L$ and the three right ones $\lambda = 5L$. We also report the resulting energy values $E_{L^1\text{-stereo}}$ below the images.

where ϕ is a function penalizing intensity differences, and Ψ is the regularization (smoothness) term. The data term above assumes brightness constancy, and can be replaced by different expressions. Even if ϕ and Ψ are convex functions, the energy in Eq. 10 is usually not, since the warped image I^R as a function of $\mathbf{d}$, $\mathbf{d} \mapsto I^R \circ (\mathrm{Id} + \mathbf{d})$, is not convex. Consequently, $I^R(x + d(x))$ is typically linearized around a current linearization point $\bar{d}$, i.e.

$$I^R(x + d) \approx I^R(x + \bar{d}) + (d - \bar{d}) \cdot \nabla_x I^R(x + \bar{d}).$$

In order to cope with the limited validity of the above approximation, typical variational methods for dense disparity (or dense optical flow) estimation build on a multi-scale, coarse-to-fine scheme. If we use a linear interpolation to sample I^R at fractional positions, for disparity estimation the above relation is *exact*, if $\mathbf{d} - \bar{\mathbf{d}}$ is sufficiently bounded. Due to its robustness and simplicity we focus in the following on the L^1 intensity difference as the data term, i.e. $\phi(\cdot) = |\cdot|$. Further, we employ the total variation regularization for the smoothness term Ψ, which allows discontinuities in the solution and is still globally convex.

Since we are operating on a discrete domain (a regular pixel grid), the continuous energy Eq. 10 has a discrete counterpart (with our choice of ϕ and Ψ),

$$E_{L^1\text{-stereo}}(\mathbf{d}) = \sum_s \left| I_s^L - I^R(s + d_s) \right| + \|\nabla \mathbf{d}\|_1 , \qquad (11)$$

where ∇ is a discrete gradient operator (e.g. computed via finite differences). If we add respective bounds constraints on d_s for all s (which depend on the current linearization point $\bar{d}$) the energy in Eq. 11 is convex (it is even a linear program with our choice of the data and smoothness terms). If we knew a linearization point $\bar{\mathbf{d}}$ close to an optimal solution in advance, then minimzing E_{stereo} would just return a refined (and optimal) disparity map $\mathbf{d}$. We do not have a good disparity map $\bar{\mathbf{d}}$ available, but we can hypothesize any $\bar{\mathbf{d}}^1$ and try to merge good aspects of $\bar{\mathbf{d}}^1$ into our current best solution $\bar{\mathbf{d}}^0$.

Let δ be the radius of the "trust region", where the linearization of image intensities holds. If linear interpolation is used to sample from I^R, then $\delta = 0.5$ pixels. One DC fusion move amounts to solve (note that $\mathbf{y} = \mathbf{x} \odot \mathbf{d}$ with $\mathbf{d}$ our desired continuous labeling)

$$E_{L^1\text{-stereo-fusion}}(\mathbf{x}, \mathbf{y}) = \sum_{s,t} \sum_{i,j \in \{0,1\}} (h_{st}^{ij})_{\oslash}(x_{st}^{ij}, y_{st\to s}^{ij}, y_{st\to t}^{ij}) \qquad (12)$$

$$\text{s.t. } x_s^i = x_{st}^{i0} + x_{st}^{i1} \qquad\qquad x_t^j = x_{st}^{0j} + x_{st}^{1j}$$
$$y_s^i = y_{st\to s}^{i0} + y_{st\to s}^{i1} \qquad\qquad y_t^j = y_{st\to t}^{0j} + y_{st\to t}^{1j}$$

and $x_s^0 + x_s^1 = 1$, $\mathbf{x} \geq 0$, where

$$h_{st}^{ij}(d_s, d_t) = \frac{\lambda}{\deg(s)} \left| I_s^R + (d_s - \bar{d}_s^i)\nabla_x I_s^R - I_s^L \right|$$
$$+ \frac{\lambda}{\deg(t)} \left| I_t^R + (d_t - \bar{d}_t^j)\nabla_x I_t^R - I_t^L \right|$$
$$+ |d_s - d_t| + \imath_{[-\delta,\delta]^2} \left(d_s - \bar{d}_s^i, d_t - \bar{d}_t^j \right). \qquad (13)$$

The perspective of the above function actually appearing in Eq. 12 is

$$(h_{st}^{ij})_{\oslash}(x, y_s, y_t) = \frac{\lambda}{\deg(s)} \left| x(I_s^R - \bar{d}_s^i \nabla_x I_s^R - I_s^L) + y_s \nabla_x I_s^R \right|$$
$$+ \frac{\lambda}{\deg(t)} \left| x(I_t^R - \bar{d}_t^j \nabla_x I_t^R - I_t^L) + y_t \nabla_x I_t^R \right|$$
$$+ |y_s - y_t| + \imath_{\geq 0}(x)$$
$$+ \imath\{y_s \in x[\bar{d}_s^i - \delta, \bar{d}_s^i + \delta], y_t \in x[\bar{d}_t^j - \delta, \bar{d}_t^j + \delta]\}.$$

Each fusion step minimizes Eq. 12. We initialize one proposal as local best-cost solution using absolute intensity differences, and the merged proposals are constant but integral disparity hypotheses in a random order. The results shown in the numerical experiments are obtained after one full round of fusion moves, i.e. after L fusion steps. L is the maximum disparity. Figure 2 compares the results of optimizing $E_{L^1\text{-stereo}}$ via the proposed generalized fusion moves with the results obtained by direct variational minimization using a coarse-to-fine framework and frequent relinearization (warping) steps (20 per image pyramid level in our test). We chose $\lambda = 2L$ and $\lambda = 5L$ in $E_{L^1\text{-stereo}}$ (in order to compensate for varying number of disparities). Direct variational methods optimizing the non-convex energy $E_{L^1\text{-stereo}}$ work well in some cases (especially with strong smoothness terms), but have difficulties in recovering from mistakes at coarser levels and are generally prone to miss finer details.

7 Example 2: Towards a Generative Model for Stereo

In this section we consider a stereo problem similar to the formulation in the previous section, but we explicitly allow radiometric differences between the images. Radiometric changes are usually addressed in computational stereo by using an appropriately invariant similarity measure such as zero-mean NCC, the census transform or mutual information (see e.g. [30]). In this section we take a similar path as e.g. [31] by *jointly* determining a disparity map and radiometric alignment between images. Consequently, we still employ a local, pixel-wise similarity cost,

$$\left|\gamma_s I_s^L - I^R(s + d_s)\right|, \tag{14}$$

where γ_s is a spatially varying radiometric gain to compensate illumination and exposure differences between I^L and I^R. Note that a spatial prior on γ_s is needed to avoid a nontrivial solution. The advantage of retaining a pixel-wise matching cost is e.g. that the typical "foreground fattening" effect [32] of radiometrically robust but patch-based matching costs is avoided. In order to keeps matters simple, we do not aim for a fully generative model and consequently do not optimize over an additional latent "clean" image I^*. As with the disparity map $\mathbf{d}$ our prior assumption is that γ is bounded from above and below, and that γ is piecewise constant. Hence, we extend Eq. 12 such that there are two continuous unknowns per pixel, the disparity d_s and the gain compensation γ_s:

$$E_{L^1\text{-stereo++-fusion}}(\mathbf{x}, \mathbf{y}, \mathbf{g}) = \sum_{s,t} \sum_{i,j \in \{0,1\}} (h_{st}^{ij})_\oslash (x_{st}^{ij}, y_{st \to s}^{ij}, y_{st \to t}^{ij}, g_{st \to s}^{ij}, g_{st \to t}^{ij}) \tag{15}$$

such that

$$x_s^i = x_{st}^{i0} + x_{st}^{i1} \qquad x_t^j = x_{st}^{0j} + x_{st}^{1j}$$
$$y_s^i = y_{st \to s}^{i0} + y_{st \to s}^{i1} \qquad y_t^j = y_{st \to t}^{0j} + y_{st \to t}^{1j}$$
$$g_s^i = g_{st \to s}^{i0} + g_{st \to s}^{i1} \qquad g_t^j = g_{st \to t}^{0j} + g_{st \to t}^{1j}$$

and $x_s^0 + x_s^1 = 1$, $\mathbf{x} \geq 0$, where $(h_{st}^{ij})_\oslash$ is the perspective of

$$h_{st}^{ij}(d_s, d_t, \gamma_s, \gamma_t) = \frac{\lambda}{\deg(s)} \left| I_s^R + (d_s - \bar{d}_s^i) \nabla_x I_s^R - \gamma_s I_s^L \right|$$
$$+ \frac{\lambda}{\deg(t)} \left| I_t^R + (d_t - \bar{d}_t^j) \nabla_x I_t^R - \gamma_t I_t^L \right|$$
$$+ |d_s - d_t| + \alpha |\gamma_s - \gamma_t|$$
$$+ \imath_{[-\delta,\delta]^2} \left(d_s - \bar{d}_s^i, d_t - \bar{d}_t^j \right) + \imath_{[\gamma^{\min}, \gamma^{\max}]^2}(\gamma_s, \gamma_t). \tag{16}$$

Observe that we do not prefer a particular value of γ_s since we use a uniform prior in the range $[\gamma^{\min}, \gamma^{\max}]$. In our experiments we set $\gamma^{\min} = 1/4$ and $\gamma^{\max} = 4$.

In Fig. 3 we show estimated depth maps for radiometrically varying benchmark data [30] using the same low resolution setup as in [29]. Our approach is able to optimize the standard resolution of the benchmark data as displayed in Fig. 1. All results are generated with fixed values $\lambda = 2L$ (where L is the maximum disparity) and $\alpha = 50$. We use $L = 80$ in Fig. 1 and $L = 40$ in Fig. 3.

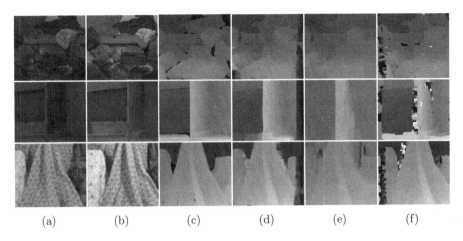

(a) (b) (c) (d) (e) (f)

Fig. 3. Joint estimation of disparities and brightness changes and a comparison to [29]. (a) left image; (b) right image; (c) true disparity; (d) estimated disparity d using generalized fusion moves; (e) result from [29]; (f) disparity estimated using 5×5 ZNCC and belief propagation using truncated L^1-smoothness prior. Best viewed on screen.

8 Conclusion and Future Work

In this work we generalize standard fusion moves—which optimally merge two given proposals—to fusion moves that may refine the proposals and which can optimize over additional continuous latent variables. Consequently, the proposal labelings provided in each fusion step can be inexact, which reduces the burden on smart proposal generation techniques. Additionally, the generalized fusion moves allow inclusion of extra continuous unknowns into the energy to be minimized without the need of including these into the proposal labelings (Fig. 4).

(a) (b) (c) (d)

Fig. 4. Comparison between the weaker relaxation $E_{\text{DC-Fusion}}$ (a, b) and the stronger one $\check{E}_{\text{DC-Fusion}}$ (c, d) for TV-L^1 stereo. (b, d) illustrate the solution $\{x_s^1\}_{s \in \mathcal{V}}$ for a particular fusion move, which ideally should be binary. (b) is less binary than (d), but in this case the returned label assignments (a, c) are very similar (in their visual appearances and final energies).

The proposed discrete-continuous fusion moves are very efficient in terms of memory consumption, but the optimization task is expensive compared to a combinatorial discrete fusion move (the run-times range from minutes to hours depending on the problem instance). On the other hand, each move can do much more work, so the total number of fusions is expected to be lower. In contrast to discrete labelling solutions, however, the first order methods typically employed to minimize convex problems are trivially data parallel and amenable to GPU implementation. We also conducted initial experiments to utilize an early committment approach based on a variant for partial optimality [22,23], but unfortunately most pixels remained unlabeled. Investigating into refined formulations of partial optimality in a discrete-continuous context is left for future work. Another interesting direction for future research is a quantitative analysis of how the proposal generation influences the effective labeling prior.

References

1. Zach, C., Kohli, P.: A convex discrete-continuous approach for Markov random fields. In: Fitzgibbon, A., Lazebnik, S., Perona, P., Sato, Y., Schmid, C. (eds.) ECCV 2012. LNCS, vol. 7577, pp. 386–399. Springer, Heidelberg (2012). doi:10.1007/978-3-642-33783-3_28
2. Boykov, Y., Veksler, O., Zabih, R.: Fast approximate energy minimization via graph cuts. IEEE Trans. Pattern Anal. Mach. Intell. **23**, 1222–1239 (2001)
3. Gould, S., Amat, F., Koller, D.: Alphabet SOUP: a framework for approximate energy minimization. In: Proceedings of CVPR, pp. 903–910 (2009)
4. Carr, P., Hartley, R.: Solving multilabel graph cut problems with multilabel swap. In: 2009 Digital Image Computing: Techniques and Applications, DICTA 2009, pp. 532–539 (2009)
5. Schmidt, M., Alahari, K.: Generalized fast approximate energy minimization via graph cuts: alpha-expansion beta-shrink moves. In: Proceedings of UAI, pp. 653–660 (2011)
6. Veksler, O.: Graph cut based optimization for MRFs with truncated convex priors. In: Proceedings of CVPR (2007)
7. Kumar, M.P., Veksler, O., Torr, P.: Improved moves for truncated convex models. J. Mach. Learn. Res. **12**, 31–67 (2011)
8. Veksler, O.: Multi-label moves for MRFs with truncated convex priors. Int. J. Comput. Vis. **98**, 1–14 (2012)
9. Jezierska, A., Talbot, H., Veksler, O., Wesierski, D.: A fast solver for truncated-convex priors: quantized-convex split moves. In: Boykov, Y., Kahl, F., Lempitsky, V., Schmidt, F.R. (eds.) EMMCVPR 2011. LNCS, vol. 6819, pp. 45–58. Springer, Heidelberg (2011). doi:10.1007/978-3-642-23094-3_4
10. Woodford, O., Reid, I., Torr, P., Fitzgibbon, A.: Fields of experts for image-based rendering. In: Proceedings of BMVC (2006)
11. Lempitsky, V., Rother, C., Blake, A.: Logcut–efficient graph cut optimization for Markov random fields. In: Proceedings of ICCV (2007)
12. Lempitsky, V., Rother, C., Roth, S., Blake, A.: Fusion moves for Markov random field optimization. IEEE Trans. Pattern Anal. Mach. Intell. **32**, 1392–1405 (2010)
13. Woodford, O., Torr, P., Reid, I., Fitzgibbon, A.: Global stereo reconstruction under second-order smoothness priors. IEEE Trans. Pattern Anal. Mach. Intell. **31**, 2115–2128 (2009)
14. Ishikawa, H.: Higher-order gradient descent by fusion-move graph cut. In: Proceedings of ICCV (2009)

15. Trobin, W., Pock, T., Cremers, D., Bischof, H.: Continuous energy minimization via repeated binary fusion. In: Forsyth, D., Torr, P., Zisserman, A. (eds.) ECCV 2008. LNCS, vol. 5305, pp. 677–690. Springer, Heidelberg (2008). doi:10.1007/978-3-540-88693-8_50

16. Olsson, C., Byrod, M., Overgaard, N., Kahl, F.: Extending continuous cuts: anisotropic metrics and expansion moves. In: Proceedings of CVPR, pp. 405–412 (2009)

17. Zach, C.: Dual decomposition for joint discrete-continuous optimization. In: Proceedings of AISTATS (2013)

18. Fix, A., Agarwal, S.: Duality and the continuous graphical model. In: Fleet, D., Pajdla, T., Schiele, B., Tuytelaars, T. (eds.) ECCV 2014. LNCS, vol. 8691, pp. 266–281. Springer, Heidelberg (2014). doi:10.1007/978-3-319-10578-9_18

19. Möllenhoff, T., Laude, E., Moeller, M., Lellmann, J., Cremers, D.: Sublabel-accurate relaxation of nonconvex energies. In: Proceedings of CVPR (2016)

20. Dacorogna, B., Maréchal, P.: The role of perspective functions in convexity, poly-convexity, rank-one convexity and separate convexity. J. Convex Anal. **15**, 271–284 (2008)

21. Kovtun, I.: Partial optimal labeling search for a NP-hard subclass of (max, +) problems. In: Michaelis, B., Krell, G. (eds.) DAGM 2003. LNCS, vol. 2781, pp. 402–409. Springer, Heidelberg (2003). doi:10.1007/978-3-540-45243-0_52

22. Kovtun, I.: Sufficient condition for partial optimality for (max, +) labeling problems and its usage. Technical report, International Research and Training Centre for Information Technologies and Systems (2010)

23. Shekhovtsov, A., Hlavac, V.: On partial optimality by auxiliary submodular problems. In: Control Systems and Computers, no. 2 (2011)

24. Desmet, J., Maeyer, M.D., Hazes, B., Lasters, I.: The dead-end elimination theorem and its use in protein side-chain positioning. Nature **356**, 539–542 (1992)

25. Georgiev, I., Lilien, R.H., Donald, B.R.: The minimized dead-end elimination criterion and its application to protein redesign in a hybrid scoring and search algorithm for computing partition functions over molecular ensembles. J. Comput. Chem. **29**, 1527–1542 (2008)

26. Gainza, P., Roberts, K.E., Donald, B.R.: Protein design using continuous rotamers. PLoS Comput. Biol. **8**, e1002335 (2012)

27. Zach, C.: A principled approach for coarse-to-fine map inference. In: Proceedings of CVPR, pp. 1330–1337 (2014)

28. Pock, T., Chambolle, A.: Diagonal preconditioning for first order primal-dual algorithms in convex optimization. In: Proceedings of ICCV, pp. 1762–1769 (2011)

29. Seitz, S., Baker, S.: Filter flow. In: Proceedings of ICCV, pp. 143–150 (2009)

30. Hirschmüller, H., Scharstein, D.: Evaluation of stereo matching costs on images with radiometric differences. IEEE Trans. Pattern Anal. Mach. Intell. **31**, 1582–1599 (2009)

31. Strecha, C., Tuytelaars, T., Van Gool, L.: Dense matching of multiple wide-baseline views. In: Proceedings of ICCV, pp. 1194–1201 (2003)

32. Sizintsev, M., Wildes, R.: Efficient stereo with accurate 3-D boundaries. Proc. BMVC **25**(1–25), 10 (2006)

Image Attributes, Language, and Recognition

Learning to Describe E-Commerce Images from Noisy Online Data

Takuya Yashima$^{(\boxtimes)}$, Naoaki Okazaki, Kentaro Inui,
Kota Yamaguchi, and Takayuki Okatani

Tohoku University, Sendai, Japan
yashima@vision.is.tohoku.ac.jp

Abstract. Recent study shows successful results in generating a proper language description for the given image, where the focus is on detecting and describing the contextual relationship in the image, such as the kind of object, relationship between two objects, or the action. In this paper, we turn our attention to more subjective components of descriptions that contain rich expressions to modify objects – namely attribute expressions. We start by collecting a large amount of product images from the online market site Etsy, and consider learning a language generation model using a popular combination of a convolutional neural network (CNN) and a recurrent neural network (RNN). Our Etsy dataset contains unique noise characteristics often arising in the online market. We first apply natural language processing techniques to extract high-quality, learnable examples in the real-world noisy data. We learn a generation model from product images with associated title descriptions, and examine how e-commerce specific meta-data and fine-tuning improve the generated expression. The experimental results suggest that we are able to learn from the noisy online data and produce a product description that is closer to a man-made description with possibly subjective attribute expressions.

1 Introduction

Imagine you are a shop owner and trying to sell a handmade miniature doll. How would you advertise your product? Probably giving a good description is one of the effective strategies. For example, stating *Enchanting and unique fairy doll with walking stick, the perfect gift for children* would sound more appealing to customers than just stating *miniature doll for sale*. In this paper, we consider automatically generating *good* natural language descriptions for product images which have rich and appealing expressions.

Natural language generation has become a popular topic as the vision community makes a significant progress in deep models to generate a word sequence given an image [12,27]. Existing generation attempts focus mostly on detecting and describing the contextual relationship in the image [18], such as a kind of object in the scene (e.g., *a man in the beach*) or the action of the subject given a scene (e.g., *a man is holding a surfboard*). In this paper, we turn

© Springer International Publishing AG 2017
S.-H. Lai et al. (Eds.): ACCV 2016, Part V, LNCS 10115, pp. 85–100, 2017.
DOI: 10.1007/978-3-319-54193-8_6

our attention to generating proper descriptions for product images with rich attribute expressions.

Attributes have been extensively studied in the community [3,14,23]. Typical assumption is that there are visually recognizable attributes, and we can build a supervised dataset for recognition. However, as we deal with open-world vocabulary on the web, we often face much complex concepts consisting of phrases rather than a single word. The plausible approach would be to model attributes in terms of a language sequence instead of individual words. The challenge is that attribute expressions can be subjective and ambiguous. Attribute-rich expressions, such as *antique copper flower decoration*, or *enchanting and unique fairy doll*, require higher-level judgement on the concept out of lower-level appearance cues. Even humans do not always agree on the meaning of abstract concepts, such as coolness or cuteness. The concept ambiguity brings a major challenge in building a large-scale corpus of conceptually obscure attributes [20,29].

We attempt to learn attribute expressions using large-scale e-commerce data. Product images in e-commerce websites typically depict a single object without much consideration to the contextual relationship within an image, in contrast to natural images [18,21,25]. Product descriptions must convey, specific color, shape, pattern, material, or even subjective and abstract concepts out of the single image with a short title, or with a longer description in the item detail for those interested in buying the product, e.g., *Beautiful hand felted and heathered purple & fuschia wool bowl*. Although e-commerce data look appealing in terms of data availability and scale, descriptions and meta-data such as tags contain web-specific noise due to the nature of online markets, such as fragmented texts for search optimization or imbalance of distribution arising from shop owners. Naively learning a generation model results in poor product description, e.g., *made to order*. In this paper, we apply natural language processing to extract images and texts suitable for learning a generation model.

Our language generation model is based on the popular image-to-sequence architecture consisting of a convolutional neural network (CNN) and a recurrent neural network (RNN) [12,27]. We learn a generation model using the product images and associated titles from the pre-processed dataset, and show that we are able to generate a reasonable description to the given product image. We also examine how e-commerce meta-data (product category) and optimization to the dataset (fine-tuning) affect the generation process. We annotate a handful of images using crowdsourcing and compare the quality of generated attribute expressions using machine translation metrics. The results suggest that e-commerce meta-data together with fine-tuning generate a product description closer to human.

We summarize our contribution in the following.

- We propose to learn attributes in the form of natural language expression, to deal with the exponentially many combination of open-world modifier vocabulary.
- We collect a large-scale dataset of product images from online market Etsy, as well as human annotation of product descriptions using crowdsourcing for evaluation purpose. We release data to the academic community[1].

---

[1] http://vision.is.tohoku.ac.jp/~kyamagu/research/etsy-dataset.

- We propose a simple yet effective data cleansing approach to transform e-commerce data into a corpus suitable for learning.
- Our empirical study shows that our model can generate a description with attribute expressions using noisy e-commerce data. The study also suggests utilizing e-commerce meta-data can further improve the description quality.

2   Related Work

Language Generation. Generating a natural language description from the given image has been an active topic of research in the vision community [6,11,12,21,27,28]. Early attempts have used retrieval-based approach to generate a sentence [11,21], and recently a deep-learning approach becomes a popular choice. For example, Vinyals *et al.* [27] shows that they can generate a high-quality language description of the scene image using a combination of a CNN and a RNN. Karpathy *et al.* [12] also shows that their model can generate partial descriptions of given image regions, as well as a whole image. Antol *et al.* [1] studies a model which is able to generate sentences in answer to various questions about given images.

In this paper, we are interested in generating a natural language expression that is rich in attribute. Previous work mostly focuses on natural images where the major goal is to understand the scene semantics and spatial arrangement, and produce an objective description. The closest to our motivation is perhaps the work by Mathews *et al.* [20] that studies a model to generate more expressive description with sentiment. They build a new dataset by asking crowd workers to re-write description of images contained in MS-COCO, and report successful generation with sentiment, for instance, beautiful, happy, or great. We take a different approach of utilizing e-commerce data to build an attribute-rich corpus of descriptions.

Attribute Recognition. Semantic or expressive attributes have been actively studied in the community as a means of ontological entity [16] or localizable visual elements [3], expecting that these semantic information can be useful for many applications. In this work, we consider attribute expressions as a natural language description that modifies an object (specifically, a product) and conveys details possibly with abstract words. The attribute expressions are from open-world vocabulary in the real-world e-commerce data. In that sense, we have a similar spirit to weakly supervised learning [5,9,24]. We propose to use a sequence generation model rather than attempting to learn a classifier from exponentially many combinations of attribute expressions.

Vision in E-Commerce. Several attempts have been made to apply computer vision in e-commerce applications [7,8,13,14,19], perhaps for the usefulness in a specific scenario such as better user experience in retrieval or product recommendation. The earlier work by Berg *et al.* [3] propose a method of automatic visual attribute discovery using web data, specifically product images from shopping websites. Our work has the same motivation that we wish to learn language description of attributes from the e-commerce data, though we use variety of products and try to capture abstract attributes using language generation model. Very recently, Zakrewsky *et al.* [30] reports an attempt of popularity prediction of products offered in Etsy. The results suggest the potential usefulness of image feature for selling strategies, such as advertisement.

3 Language Generation Model

Our language generation is based on the combination of convolutional neural networks (CNN) to obtain image representation and recurrent neural networks (RNN), using LSTM cells to translate the representation into a sequence of words [12,27,32]. In addition to the image input, we also consider inserting e-commerce meta-data to the RNN. In this paper, we utilize the category of product as extra information available in the e-commerce scenario, and feed into the RNN as a one-hot vector. Note that each product could have more than one category, such as a main category and sub categories, but in this paper we use only the main category for simplicity. Figure 1 illustrates our generation model.

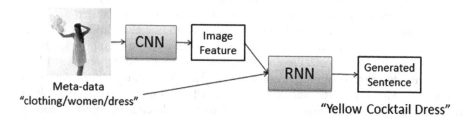

Fig. 1. Our language generation model combining CNN and RNN.

Let us denote the input product image I's feature by $\mathbf{z}_v = \mathrm{CNN}\,(I)$, the one-hot vector of the product category in meta-data by $\mathbf{z}_c$, and the one-hot vector of the currently generated word at description position t by $\mathbf{x}_t$. Our sequence generator is then expressed by:

$$H_{in} = \begin{cases} [1; W_{hi}\,[\mathbf{z}_v; \mathbf{z}_c]\,; 0] & (t=1) \\ [1; W_{hx}\mathbf{x}_t; \mathbf{h}_{t-1}] & (otherwise) \end{cases} \tag{1}$$

$$(i, f, o, g) = W_{LSTM} H_{in}, \tag{2}$$

$$\mathbf{c}_t = f \odot \mathbf{c}_{t-1} + i \odot g, \tag{3}$$

$$\mathbf{h}_t = o \odot \tanh(\mathbf{c}_t), \tag{4}$$

$$\mathbf{y}_t = \mathrm{softmax}(W_{oh}\mathbf{h}_t + b_o), \tag{5}$$

where $W_{hi}, W_{hx}, W_{LSTM}, W_{oh}, b_o$ are weights and biases of the network. We learn these parameters from the dataset. Gates i, f, o, g are controlling whether each input or output is used or not, allowing the model to deal with the vanishing gradient problem. We feed the image and category input only at the beginning. The output y_t represents an unnormalized probability of each word, and has the length equal to the vocabulary size $+ 1$ to represent a special END token to indicate the end of a description.

To learn the network, we use product image, title and category information. The learning procedure starts by setting $h_0 = 0$, y_1 to the desired word in the description ($y_t = y_1$ indicates the first word in the sequence), and x_1 to a special START symbol to indicate the beginning of the sequence. We feed the rest of the words from the ground truth until we reach the special END token at the end. We learn the model to maximize the log probability in the dataset. At test time, we first set $h_0 = 0$, x_1 to the START token, and feed the image representation z_v with the category vector z_c. Once we get an output, we draw a word according to y_t and set the word to x_t, the word predicted at the previous step (so when $t \geq 2$, each x_t is y_{t-1}). We repeat the process until we observe the END token.

When training, we use Stochastic Gradient Descent, set the initial learning rate to `1.0e-3`, and lower as the process iterates. In this paper, we do not back-propagate the gradient to CNN and separately train CNN and RNN. We evaluate how different CNN models perform in Sect. 5.

4   Building Corpus from E-Commerce Data

We collect and build the image-text corpus from the online market site Etsy. We prepare pairs of a product image and title as well as product meta-data suitable for learning attribute expressions. The challenge here is how to choose good descriptions for learning. In this section, we briefly describe the e-commerce data and our approach to extract useful data using syntactic analysis and clustering.

4.1   E-Commerce Data

Etsy is an online market for hand-made products [31]. Figure 2 shows some examples of product images from the website. Each listing contains various information, such as image, title, detailed description, tags, materials, shop owner, price, etc. We crawled product listings from the website and downloaded over two million product images.

We apply various pre-processing steps to transform the crawled raw data into a useful corpus to learn attribute expressions. Note that this semi-automated approach to build a corpus is distinct from the previous language generation efforts where the approach is to start from supervised dataset with clean annno-tations [18]. Our corpus is from the real-world market, and as common in any Web data [21, 25], the raw listing data from Etsy contain a lot of useless data for learning, due to a huge amount of near-duplicates and inappropriate language use for search engine optimization. For example, we observed the following titles:

- *Army of two Airsoft Paintball BB Softair Gun Prop Helmet Salem Costume Cosplay Goggle Mask Maske Masque jason MA102 et*

title   : Twat Finished Framed Subversive Cross Stitch Art
category : Needlecraft, Cross Stitch, Flower
tags   : twat, british, cross stitch framed, twat cross stitch, ...
materials : dmc floss, aida fabric, metal and glass frame
shop id   : 5520172
description : One for the Brits! For a twat, for your twat...

title   : Felted Wool Soft Sculpture Fancy Girl Doll
category : Dolls and Miniatures, Soft Sculpture, Human Figure Doll
tags   : soft sculpture, felted wool, pink, doll, handmade, ...
materials : wool, rhinestone, beads, cotton
shop id   : 5520172
description : This stuffed doll has been handmade with felted wool and...

Fig. 2. Product data in Etsy dataset.

– *teacher notepad set - bird on apple/teacher personalized stationary/ personalized stationery/teacher notepad/teacher gift/notepad.*

Using such raw data to learn a generation model results in poor language quality in the output.

4.2   Syntactic Analysis

One common observation in Etsy is that there are fragments of noun phrases, often considered as a list of keywords targeting at search engine optimization. Although generating search-optimized description could be useful in some application, we are not aiming at learning keyword fragments in this paper. We attempt to identify such fragmented description by syntactic analysis.

We first apply Stanford Parser [4] and estimate part-of-speech (POS) tags, such as noun or verb, for each word in the title. In this paper, we define malformed descriptions by the following criteria:

– more than 5 noun phrases in a row, or
– more than 5 special symbols such as slash, dash, and comma.

Figure 3 shows a few accepted and rejected examples from Etsy data. Note that due to the discrepancy between our Etsy titles and the corpus Stanford Parser is trained on, we found even the syntactic parser frequently failed to assign correct POS tags for each word. We did not apply any special pre-processing for such cases since most of the failed POS tagging resulted in the sequence of nouns, which in turn leads to reject.

4.3   Near-Duplicate Removal

Another e-commerce specific issue is that there is a huge amount of near-duplicates. Near-duplicates are commonly occurring phenomena in web data.

| peppermint, tangerine, and hemp lip balms | coffee bag backpack | green jadeite jade beaded, natural green white color beads size 8mm charm necklace |

(a) Syntactically accepted

| chunky pink and purple butterfly cuddle critter cape set newborn to 3 months photo prop | travel journal diary notebook sketch book - keep calm and go to canada - ivory | texas license plate bird house |

(b) Syntactically rejected

Fig. 3. Accepted and rejected examples after syntactic analysis. Some products have grammatically invalid title due to the optimization to search engine. (Color figure online)

Here, we define near-duplicate items as products whose titles are similar and differ only in a small part of the descriptions such as shown in Table 1. Those near-duplicates add a strong bias towards specific phrasing and affect the quality of the generated description. Without a precaution, the trained model always generates a similar phrase for any kind of images. In Etsy, we observe near-duplicates among the products offered by the same shop and listed in the same kind of categories, such as a series of products under the same category, as shown in Table 1. We find that such textual near-duplicates also exhibit visually similar appearance. Note that near-duplicates can happen for visually similar items but with different description, such as items under the same category but from a different shop. However, we find that such cases are considerably rare compared to the textual near-duplicates in Etsy, perhaps due to the nature of a hand-made market where majority of products are one-of-a-kind.

We automatically identify near-duplicates using shop identity and title description. We apply the following procedure to sub-sample product images from the raw online data.

Table 1. Examples of near-duplicate products.

 CUSTOM iPad SLEEVE 1, 2, New, 3 Black Chevron Lime Green Personalized Monogram

 CUSTOM iPad SLEEVE 1, 2, New, 3 Light Pink Chevron Gray Fancy Script PERSONALIZED Monogram

 CUSTOM iPad SLEEVE 1, 2, New, 3 Black Damask Hot Pink Personalized Monogram

 CUSTOM iPad SLEEVE 1, 2, New, 3 Dark Blue Lattice Lime Green PERSONALIZED Monogram

 CUSTOM iPad SLEEVE 1, 2, New, 3 Blue Diagonal Green PERSONALIZED Monogram

 CUSTOM iPad SLEEVE 1, 2, New, 3 Light Blue Orange Floral Pattern Teal PERSONALIZED Monogram

1. Group products if they are sold by the same shop and belonging to the same category.
2. For each group, measure the similarity of the descriptions between all pairs within the group.
3. Depending on the pairwise similarity, divide the group into sub-groups by DBSCAN clustering.
4. Randomly sub-sample pre-defined number of product images from each subgroup.

Our approach is based on the observation that the same shop tend to sell near-duplicates. We divide products into sub-groups to diversify the variation of descriptions. We divide the group into sub-groups based on thresholding on the Jaccard similarity:

$$J_G = \frac{S_1 \cap S_2 \cdots \cap S_n}{S_1 \cup S_2 \cdots \cup S_n}, \tag{6}$$

where S_i represents a set of words in the title description. Low-similarity within a cluster indicates the group contains variety of descriptions, and consists of subtly different products. We apply Density-Based Spatial Clustering of Applications with Noise (DBSCAN) [10] implemented in `sklearn` to obtain sub-clusters. Figure 4 shows an example of groups. The purpose of the sub-sampling approach is to trim excessive amount of similar products while keeping variety.

After clustering, we randomly pick a certain number of products from each cluster. We determine the number of samples per cluster K_G using the following threshold:

$$K_G = \frac{1}{N} \sum_{k=1}^{m} n_k + \sigma. \tag{7}$$

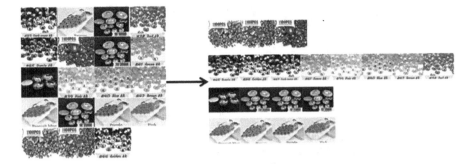

Fig. 4. Near-duplicates clustering. Our clustering only uses meta-data and textual descriptions to identify near-duplicates, but the resulting clusters tend to be visually coherent.

Here, N is the total number of the groups, n_k is the number of the products in the group k and σ is the standard deviation of the distribution of the number of products in the whole groups. We leave out some of the products if the number of products in a certain group or a subgroup is far above the average. After the sub-sampling process, all kinds of products should have been equally distributed in the corpus.

Out of over 2 million products from the raw Etsy products, we first selected 400k image-text pairs by syntactic analysis, and applied near-duplicate removal. We obtained approximately 340k product images for our corpus. We take 75% of the images for training and reserved the rest (25%) for testing in our experiment. We picked up 100 images from testing set for human annotation, and did not use for quantitative evaluation due to the difficulty in obtaining ground truth.

5 Experiment

We evaluate our learned generation model by measuring how close the generated expressions are to human. For that purpose, we collect human annotations to a small number of testing images and measure the performance using machine-translation metrics.

5.1 Human Annotation

We use crowdsourcing to collect human description of the product images. We designed a crowdsourcing task to describe the given product image. Figure 5 depicts our annotation interface. We asked workers to come up with a descriptive and appealing title to sell the given product. During the annotation task, we provide workers the original title and the category information to make sure they understand what kind of products they are trying to sell. We used Amazon Mechanical Turk and asked 5 workers per image to collect reference descriptions for 100 test images. As seen in Fig. 5, our images are quite different from natural

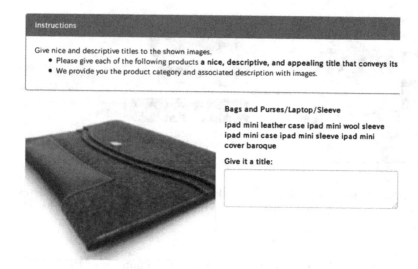

Fig. 5. Crowdsourcing task to collect human description.

scene images. Often we observed phrases rather than a complete sentence in the human annotation. This observation suggests that the essence of attribute expression is indeed in the modifiers to the object rather than the recognition of subject-verb relationships.

5.2 Evaluation Metrics

We use seven metrics to evaluate the quality of our generated descriptions: BLEU-1 [22], BLEU-2, BLEU-3, BLEU-4, ROUGE [17], METEOR [2], and CIDEr [26]. These metrics have been widely used in natural language processing, such as machine translation, automatic text summarization, and recently in language generation. Although our goal is to produce attribute-aware phrases but not necessarily sentences, these metrics can be directly used to evaluate our model using the reference human description. BLEU-N evaluates descriptions based on precision on N-grams, ROUGE is also based on N-grams but intended for recall, METEOR is designed for image descriptions, and CIDEr is also proposed for image descriptions and using N-gram based method. We use the *coco-caption* implementation [26] to calculate the above evaluation metrics.

5.3 Experimental Conditions

We use AlexNet [15] for the CNN architecture of our generation model, and extract a 4,096 dimensional representation from fc7 layer given an input image and its product category. In order to see how domain-knowledge affects the quality of language generation, we compare a CNN model trained on ImageNet, and a model fine-tuned to predict 32 product categories in Etsy dataset. Our recurrent neural network consists of a single hidden layer with 4,096 dimensional image feature and a 32 dimensional one-hot category indicator for an input. We use LSTM implementation of [12]. We compare the following models in the experiment.

- **Category+Fine-tune:** Our proposed model with a fine-tuned CNN and a category vector for RNN input.
- **Category+Pre-train:** Our proposed model with a pre-trained CNN and a category vector for RNN input.
- **Fine-tune:** A fine-tuned CNN with RNN without a category vector.
- **Pre-train:** A pre-trained CNN with RNN without a category vector.
- **MS-COCO:** A reference CNN+RNN model trained on MS-COCO dataset [18] without any training in our corpus.

We include MS-COCO model to evaluate how domain-transfer affects the quality of generated description. Note that MS-COCO dataset contains more objective descriptions for explaining objects, actions, and scene in the given image.

5.4 Quantitative Results

We summarize the performance evaluation in Table 2. Note that all scores are in percentage. Our Category+Fine-tune model achieves the best performance, except for BLEU-3 and BLEU-4, in which Fine-tune model achieves the best. We suspect overfitting happened in the Fine-tune only case where the model learned to predict certain 3- or 4-word phrases such as *Made to order*, some happened to be unexpectedly correct, and resulted in the sudden increase BLEU increase. However, we did not observe a similar improvement in other scores, such as ROUGE or CIDEr. We conjecture this possibly-outlier result could be attributed to BLEU's evaluation method.

From the result, we observe that introducing the category vector has the largest impact on the description quality. We assume that category information supplements semantic knowledge in the image feature even if the category is not apparent from the product appearance, and that results in stabler language generation for difficult images. Note that in the e-commerce scenario, meta-data are often available for free without expensive manual annotation.

Table 2. Evaluation results.

Method	Bleu1	Bleu2	Bleu3	Bleu4	Rouge	Meteor	CIDEr
Category+Fine-tune	**15.1**	**6.55**	2.58×10^{-5}	5.56×10^{-8}	**12.9**	**4.69**	**11.2**
Category+Pre-train	9.43	3.72	1.65×10^{-5}	3.74×10^{-8}	9.74	3.70	8.01
Fine-tune	8.95	3.94	$\mathbf{2.03 \times 10^{0}}$	$\mathbf{3.06 \times 10^{-4}}$	4.98	2.24	1.88
Pre-train	8.77	3.26	1.50×10^{-5}	3.50×10^{-8}	9.32	3.36	6.87
MS-COCO	1.01	2.13	8.30×10^{-6}	1.70×10^{-8}	8.31	2.40	2.79

The difference between Pre-train and Fine-tune models explains how domain-specific image feature contributes to better learning and helps the model to generate high-quality descriptions. The result indicates that a pre-trained CNN is not sufficient to capture the visual patterns in Etsy dataset. MS-COCO baseline is performing significantly worse than other models, indicating that the general description generated by MS-COCO is far from attribute-centric description in product images. There is a significant difference between descriptions in the MS-COCO dataset and our Etsy corpus. The former tends to be complete, grammatically correct descriptions focusing on the relationship between entities in the scene, whereas Etsy product titles tend to omit a verb and often do not require recognizing spatial arrangement of multiple entities. A product description can be a fragment of phrases as seen in the actual data, and a long description can look rather unnatural.

5.5   Qualitative Results

Table 3 shows examples of generated descriptions by different models as well as the original product titles. Fine-tune+category model seems to have generated better expressions while other methods sometimes fail to generate meaningful description (e.g., *custom made to order*). MS-COCO model is generating significantly different descriptions, and always tries to generate a description explaining types of objects and the relationship among them.

Our model generates somewhat attribute-centric expressions such as *needle felted* or *primitive*. Especially the latter expression is relatively abstract. These examples confirms that at least we are able to automatically learn attribute expressions from noisy online data. The descriptions tend to be noun phrases. This tendency is probably due to the characteristics of e-commerce data containing phrases rather than a long, grammatically complete sentences. Our genera-

Table 3. Comparison of generated descriptions.

 CUSTOM iPad SLEEVE 1, 2, New, 3 Black Chevron Lime Green Personalized Monogram

 CUSTOM iPad SLEEVE 1, 2, New, 3 Light Pink Chevron Gray Fancy Script PERSONALIZED Monogram

 CUSTOM iPad SLEEVE 1, 2, New, 3 Black Damask Hot Pink Personalized Monogram

 CUSTOM iPad SLEEVE 1, 2, New, 3 Dark Blue Lattice Lime Green PERSONALIZED Monogram

 CUSTOM iPad SLEEVE 1, 2, New, 3 Blue Diagonal Green PERSONALIZED Monogram

 CUSTOM iPad SLEEVE 1, 2, New, 3 Light Blue Orange Floral Pattern Teal PERSONALIZED Monogram

| reclaimed wood coffee table | handmade journal notebook | watercolor painting of moai statues at sunset |

| crochet barefoot sandals with flower | hand painted ceramic mug | I'm going to be a big brother t-shirt |

Fig. 6. Examples of generated descriptions. Our model correctly identifies attribute expressions (left 2 columns). The rightmost column shows failure cases due to corpus issues.

tion results correctly reflect this characteristics. Figure 6 shows some examples of generated descriptions by our model (category+fine-tune). Our model predicts attribute expressions such as *reclaimed wood* or *hand-painted ceramic*. We observe failure due to corpus quality in some categories. For example, the painting or the printed t-shirts in Fig. 6 suffer from bias towards specific types of products in the category. Sometimes our model gives a better description than the original title. For example, The middle in Table 3 shows a product entitled *Rooted*, but it is almost impossible to guess the kind of product from the name, or maybe even from the appearance. Our model produces *art print* for this example, which seems to be much easier to understand the product kind and closer to our intuition, even if the result is not accurate.

6 Discussion

In this paper, we used a product image, a title and category information to generate a description. However, there is other useful information in the e-commerce data, such as tags, materials, or popularity metrics [31]. Especially, a product description is likely to have more detailed information about the product, with many attribute-like expressions having plenty of abstract or subjective words. E-commerce dataset looks promising in this respect since sellers are trying to attract more customers by "good" phrases which have a ring to it.

If we are able to find attribute expressions in the longer product description, we can expand our image-text corpus to a considerably larger scale. The challenge here is that we then need to identify which description is relevant to the given image, because product descriptions contain irrelevant information also. For example, in Etsy, a product often contains textual description on shipping information. For a preliminary study, we applied a syntactic parser on Etsy product descriptions, but often observed an error in a parse tree, due to grammatically broken descriptions in item listings. Identifying which description is relevant or irrelevant seems like an interesting vision-language problem in the e-commerce scenario.

Finally, in this paper we left tags and material information in the item listings in Etsy dataset. These meta-data could be useful to learn a conventional attribute or material classifier given an image, or to identify attribute-specific expressions in the long product description.

7   Conclusion and Future Work

We studied the natural language generation from product images. In order to learn a generation model, we collected product images from the online market Etsy, and built a corpus to learn a generation model by applying dataset cleansing procedure based on syntactic analysis and near-duplicate removal. We also collected human descriptions for evaluation of the generated descriptions. The empirical results suggest that our generation model fine-tuned on Etsy data with categorical input successfully learns from noisy online data, and produces the best language expression for the given product image. The result also indicates a huge gap between the task nature of attribute-centric language generation and a general scene description.

In the future, we wish to improve our automatic corpus construction from noisy online data. We have left potentially-useful product meta-data in this study. We hope to incorporate additional information such as product description or tags to improve language learning process, as well as to realize a new application such as automatic title and keywords suggestion to shop owners. Also, we are interested in improving the deep learning architecture to the generate attribute expressions.

Acknowledgement. This work was supported by JSPS KAKENHI Grant Numbers JP15H05919 and JP15H05318.

References

1. Antol, S., Agrawal, A., Lu, J., Mitchell, M., Batra, D., Lawrence Zitnick, C., Parikh, D.: VQA: visual question answering. In: International Conference on Computer Vision (ICCV) (2015)

2. Banerjee, S., Lavie, A.: Meteor: an automatic metric for MT evaluation with improved correlation with human judgments. In: Proceedings of the ACL Workshop on Intrinsic and Extrinsic Evaluation Measures for Machine Translation and/or Summarization, vol. 29, pp. 65–72 (2005)

3. Berg, T.L., Berg, A.C., Shih, J.: Automatic attribute discovery and characterization from noisy web data. In: Daniilidis, K., Maragos, P., Paragios, N. (eds.) ECCV 2010. LNCS, vol. 6311, pp. 663–676. Springer, Heidelberg (2010). doi:10. 1007/978-3-642-15549-9_48

4. Chen, D., Manning, C.D.: A fast and accurate dependency parser using neural networks. In: EMNLP, pp. 740–750 (2014)

5. Chen, X., Shrivastava, A., Gupta, A.: Neil: extracting visual knowledge from web data. In: ICCV, pp. 1409–1416, December 2013

6. Devlin, J., Cheng, H., Fang, H., Gupta, S., Deng, L., He, X., Zweig, G., Mitchell, M.: Language models for image captioning: the quirks and what works. In: Association for Computational Linguistics (ACL), pp. 100–105 (2015)

7. Di, W., Bhardwaj, A., Jagadeesh, V., Piramuthu, R., Churchill, E.: When relevance is not enough: promoting visual attractiveness for fashion e-commerce. arXiv preprint arXiv:1406.3561 (2014)

8. Di, W., Sundaresan, N., Piramuthu, R., Bhardwaj, R.: Is a picture really worth a thousand words?:-on the role of images in e-commerce. In: Proceedings of the 7th ACM International Conference on Web Search and Data Mining, pp. 633–642. ACM (2014)

9. Divvala, S., Farhadi, A., Guestrin, C.: Learning everything about anything: webly-supervised visual concept learning. In: CVPR (2014)

10. Ester, M., Kriegel, H.-P., Sander, J., Xu, X.: A density-based algorithm for discovering clusters in large spatial databases with noise. In: KDD (1996)

11. Hodosh, M., Young, P., Hockenmaier, J.: Framing image description as a ranking task: data, models and evaluation metrics. J. Artif. Intell. Res. **47**, 853–899 (2013)

12. Karpathy, A., Fei-Fei, L.: Deep visual-semantic alignments for generating image descriptions. In: Proceedings of the IEEE Conference on Computer Vision and Pattern Recognition, pp. 3128–3137 (2015)

13. Hadi Kiapour, M., Han, X., Lazebnik, S., Berg, A.C., Berg, T.L.: Where to buy it: matching street clothing photos in online shops. In: ICCV (2015)

14. Kovashka, A., Parikh, D., Grauman, K.: Whittlesearch: image search with relative attribute feedback. In: 2012 IEEE Conference on Computer Vision and Pattern Recognition (CVPR), pp. 2973–2980. IEEE (2012)

15. Krizhevsky, A., Sutskever, I., Hinton, G.E.: Imagenet classification with deep convolutional neural networks. In: Advances in Neural Information Processing Systems, pp. 1097–1105 (2012)

16. Lampert, C.H., Nickisch, H., Harmeling, S.: Attribute-based classification for zero-shot visual object categorization. IEEE Trans. Pattern Anal. Mach. Intell. **36**(3), 453–465 (2014)

17. Lin, C.-Y.: Rouge: a package for automatic evaluation of summaries. In: Text Summarization Branches Out: Proceedings of the ACL-04 Workshop, vol. 8 (2004)

18. Lin, T.-Y., Maire, M., Belongie, S., Hays, J., Perona, P., Ramanan, D., Dollár, P., Zitnick, C.L.: Microsoft COCO: common objects in context. In: Fleet, D., Pajdla, T., Schiele, B., Tuytelaars, T. (eds.) ECCV 2014. LNCS, vol. 8693, pp. 740–755. Springer, Heidelberg (2014). doi:10.1007/978-3-319-10602-1_48

19. Liu, S., Song, Z., Liu, G., Xu, C., Lu, H., Yan, S.: Street-to-shop: cross-scenario clothing retrieval via parts alignment and auxiliary set. In: 2012 IEEE Conference on Computer Vision and Pattern Recognition (CVPR), pp. 3330–3337. IEEE (2012)
20. Mathews, A.P., Xie, L., He, X.: Senticap: generating image descriptions with sentiments. CoRR, abs/1510.01431 (2015)
21. Ordonez, V., Kulkarni, G., Berg, T.L.: Im2text: describing images using 1 million captioned photographs. In: Advances in Neural Information Processing Systems, pp. 1143–1151 (2011)
22. Papineni, K., Roukos, S., Ward, T., Zhu, W.-J.: BLEU: a method for automatic evaluation of machine translation. In: Proceedings of the 40th Annual Meeting on Association for Computational Linguistics, pp. 311–318. Association for Computational Linguistics (2002)
23. Parikh, D., Grauman, K.: Relative attributes. In: Metaxas, D.N., Quan, L., Sanfeliu, A., Van Gool, L.J. (eds.) ICCV, pp. 503–510. IEEE Computer Society, Washington, D.C (2011)
24. Sun, C., Gan, C., Nevatia, R.: Automatic concept discovery from parallel text and visual corpora. In: ICCV, pp. 2596–2604 (2015)
25. Thomee, B., Shamma, D.A., Friedland, G., Elizalde, B., Ni, K., Poland, D., Borth, D., Li, L.-J.: The new data and new challenges in multimedia research. arXiv preprint arXiv:1503.01817 (2015)
26. Vedantam, R., Lawrence Zitnick, C., Parikh, D.: Cider: consensus-based image description evaluation. In: Proceedings of the IEEE Conference on Computer Vision and Pattern Recognition, pp. 4566–4575 (2015)
27. Vinyals, O., Toshev, A., Bengio, S., Erhan, D.: Show and tell: a neural image caption generator. In: Proceedings of the IEEE Conference on Computer Vision and Pattern Recognition, pp. 3156–3164 (2015)
28. Xu, K., Ba, J., Kiros, R., Courville, A., Salakhutdinov, R., Zemel, R., Bengio, Y.: Show, attend and tell: neural image caption generation with visual attention. arXiv preprint arXiv:1502.03044 (2015)
29. You, Q., Luo, J., Jin, H., Yang, J.: Robust image sentiment analysis using progressively trained and domain transferred deep networks. arXiv preprint arXiv:1509.06041 (2015)
30. Zakrewsky, S., Aryafar, K., Shokoufandeh, A.: Item popularity prediction in e-commerce using image quality feature vectors. arXiv e-prints, May 2016
31. Zakrewsky, S., Aryafar, K., Shokoufandeh, A.: Item popularity prediction in e-commerce using image quality feature vectors. arXiv preprint arXiv:1605.03663 (2016)
32. Zaremba, W., Sutskever, I., Vinyals, O.: Recurrent neural network regularization. CoRR, abs/1409.2329 (2014)

phi-LSTM: A Phrase-Based Hierarchical LSTM Model for Image Captioning

Ying Hua Tan and Chee Seng Chan(✉)

Center of Image and Signal Processing, Faculty of Computer Science and Information Technology, University of Malaya, Kuala Lumpur, Malaysia
tanyinghua@siswa.um.edu.my, cs.chan@um.edu.my

Abstract. *A picture is worth a thousand words.* Not until recently, however, we noticed some success stories in understanding of visual scenes: a model that is able to detect/name objects, describe their attributes, and recognize their relationships/interactions. In this paper, we propose a phrase-based hierarchical Long Short-Term Memory (phi-LSTM) model to generate image description. The proposed model encodes sentence as a sequence of combination of phrases and words, instead of a sequence of words alone as in those conventional solutions. The two levels of this model are dedicated to (i) learn to generate image relevant noun phrases, and (ii) produce appropriate image description from the phrases and other words in the corpus. Adopting a convolutional neural network to learn image features and the LSTM to learn the word sequence in a sentence, the proposed model has shown better or competitive results in comparison to the state-of-the-art models on Flickr8k and Flickr30k datasets.

1 Introduction

Automatic caption/description generation from images is a challenging problem that requires a combination of visual information and linguistic as illustrated in Fig. 1. In other words, it requires not only complete image understanding, but also sophisticated natural language generation [1–4]. This is what makes it such an interesting task that has been embraced by both the computer vision and natural language processing communities.

Image:

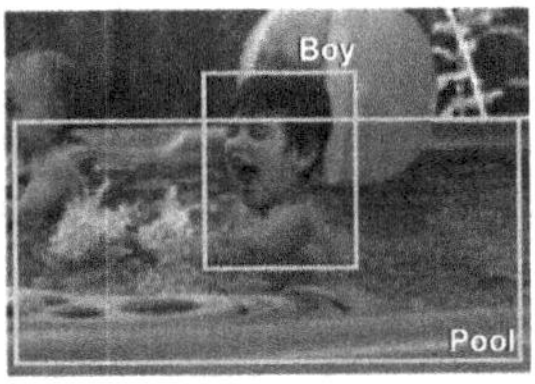

Our proposed: *A little boy is playing in the pool.*

Groundtruth: A boy with a beach ball behind him playing in a pool.

Fig. 1. Complete visual scene understanding is a holy grail in computer vision.

One of the most common models applied for automatic caption generation is a neural network model that composes of two sub-networks [5–10], where a convolutional neural network (CNN) [11] is used to obtain feature representation

S.-H. Lai et al. (Eds.): ACCV 2016, Part V, LNCS 10115, pp. 101–117, 2017.
DOI: 10.1007/978-3-319-54193-8_7

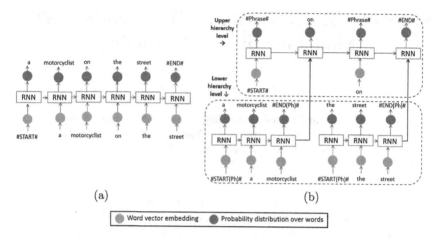

(a)   (b)

Fig. 2. Model comparison: (a) Conventional RNN language model, and (b) Our proposed phrase-based model.

of an image; while a recurrent neural network (RNN)[1] is applied to encode and generate its caption description. In particular, Long Short-Term Memory (LSTM) model [12] has emerged as the most popular architecture among RNN, as it has the ability to capture long-term dependency and preserve sequence. Although sequential model is appropriate for processing sentential data, it does not capture any other syntactic structure of language at all. Nevertheless, it is undeniable that sentence structure is one of the prominent characteristics of language, and Victor Yngve - an influential contributor in linguistic theory stated in 1960 that *"language structure involving, in some form or other, a phrase-structure hierarchy, or immediate constituent organization"* [13]. Moreover, Tai et al. [14] proved that a tree-structured LSTM model that incorporates syntactic interpretation of sentence structure, can learn the semantic relatedness between sentences better than a pure sequential LSTM alone. This gives rise to question of whether is it a good idea to disregard other syntax of language in the task of generating image description.

In this paper, we would like to investigate the capability of a phrase-based language model in generating image caption as compared to the sequential language model such as [6]. To this end, we design a novel phrase-based hierarchical LSTM model, namely **phi-LSTM** to encode image description in three stages - chunking of training caption, image-relevant phrases composition as a vector representation and finally, sentence encoding with image, words and phrases. As opposed to those conventional RNN language models which process sentence as a sequence of words, our proposed method takes noun phrase as a unit in the sentence, and thus processes the sentential data as a sequence of combination of both words and phrases together. Figure 2 illustrates the difference between

---

[1] RNN is a popular choice due to its capability to process arbitrary length sequences like language where words sequence governing its semantic is order-sensitive.

the conventional RNN language model and our proposal with an example. Both phrases and sentences in our proposed model are learned with two different sets of LSTM parameters, each models the probability distribution of word conditions on previous context and image. Such design is motivated by the observation that some words are more prone to appear in phrase, while other words are more likely to be used to link phrases. In order to train the proposed model, a new perplexity based cost function is defined. Experimental results using two publicly available datasets (Flickr8k [15] and Flickr30k [16]), and a comparison to the state-of-the-art results [5–7,9,17] have shown the efficacy of our proposed method.

2   Related Works

The image description generation task is generally inspired by two lines of research, which are (i) the learning of cross-modality transition or representation between image and language, and (ii) the description generation approaches.

2.1   Multimodal Representation and Transition

To model the relationship between image and language, some works associate both modalities by embedding their representations into a common space [18–21]. First, they obtain the image features using a visual model like CNN [19,20], as well as the representation of sentence with a language model such as recursive neural network [20]. Then, both of them are embedded into a common multimodal space and the whole model is learned with ranking objective for image and sentence retrieval task. This framework was also tested at object level by Karpathy et al. [21] and proved to yield better results for the image and sentence bi-directional retrieval task. Besides that, there are works that learn the probability density over multimodal inputs using various statistical approaches. These include Deep Boltzmann Machines [22], topic models [23], log-bilinear neural language model [8,24] and recurrent neural networks [5–7] etc. Such approaches fuse different input modalities together to obtain a unified representation of the inputs. It is notable to mention that there are also some works which do not explicitly learn the multimodal representation between image and language, but transit between modalities with retrieval approach. For example, Kuznetsova et al. [25] retrieve images similar to the query image from their database, and extract useful language segments (such as phrases) from the descriptions of the retrieved images.

2.2   Description Generation

On the other hand, caption generation approaches can be grouped into three categories in general as below:

Template-Based. These approaches generate sentence from a fixed template [26–30]. For example, Farhadi et al. [26] infer a single triplet of object, action and scene from an image and convert it into a sentence with fixed template. Kulkarni et al. [27] use complex graph of detections to infer elements in sentence with conditional random field (CRF), but the generation of sentences is still based on the template. Mitchell et al. [29] and Gupta et al. [30] use a more powerful language parsing model to produce image description. In overall, all these approaches generate description which is syntactically correct, but rigid and not flexible.

Composition Method. These approaches extract components related to the images and stitch them up to form a sentence [25,31,32]. Description generated in such manner is broader and more expressive compared to the template-based approach, but is more computationally expensive at test time due to its non-parametric nature.

Neural Network. These approaches produce description by modeling the conditional probability of a word given multimodal inputs. For instance, Kiros et al. [8,24] developed multimodal log-bilinear neural language model for sentence generation based on context and image feature. However, it has a fixed window context. The other popular model is recurrent neural network [5–7,9,33], due to its ability to process arbitrary length of sequential inputs such as sequence of words. This model is usually connected with a deep CNN that generates image features. The variants on how this sub-network is connected to the RNN have been investigated by different researchers. For instance, the multimodal recurrent neural network proposed by Mao et al. [5] introduces a multimodal layer at each time step of the RNN, before the softmax prediction of words. Vinyals et al. [6] treat the sentence generation task as a machine translation problem from image to English, and thus image feature is employed in the first step of the sequence trained with their LSTM RNN model.

2.3   Relation to Our Work

Automatic image caption generated via template-based [26–30] and composition methods [25,31,32] are typically two-stage approaches, where relevant elements such as objects (noun phrases) and relations (verb and prepositional phrases) are generated first before a full descriptive sentence is formed with the phrases. With the capability of LSTM model in processing long sequence of words, neural network based method that uses a two-stage approach deem unnecessary. However, we are still interested to find out how sequential model with phrase as a unit of sequence performs. The closest work related to ours is the one proposed by Lebret et al. [17]. They obtain phrase representation with simple word vector addition and learn its relevancy with image by training with negative samples. Sentence is then generated as a sequence of phrases, predicted using a statistical framework conditioned on previous phrases and its chunking tags. While their aim was to design a phrase-based model that is simpler than RNN, we intend to compare RNN phrase-based model with its sequential counterpart. Hence, our proposed model generates phrases and recomposes them into sentence with two sub-networks of LSTM, which are linked to form a hierarchical structure as shown in Fig. 2(b).

3   Our Proposed phi-LSTM Model

This section details how the proposed method encodes image description in three stages - (i) chunking of image description, (ii) encode words and phrases into distributed representations, and finally (iii) encodes sentence with the phi-LSTM model.

3.1   Phrase Chunking

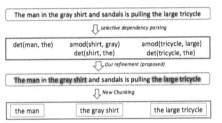

Fig. 3. Phrase chunking from dependency parse.

A quick overview on the structure of image descriptions reveals that, key elements which made up the majority of captions are usually noun phrases that describe the content of the image, which can be either objects or scene. These elements are linked with verb and prepositional phrases. Thus, noun phrase essentially covers over half of the corpus in a language model trained to generate image description. And so, in this paper, our idea is to partition the learning of noun phrase and sentence structure so that they can be processed more evenly, compared to extracting all phrases without considering their part of speech tag.

To identify noun phrases from a training sentence, we adopt the dependency parse with refinement using Stanford CoreNLP tool [34], which provides good semantic representation over a sentence by providing structural relationships between words. Though it does not chunk sentence directly as in constituency parse and other chunking tools, the pattern of noun phrase extracted is more flexible as we can select desirable structural relations. The relations we selected are:

- determiner relation (*det*),
- numeric modifier (*nummod*),
- adjectival modifier (*amod*),
- adverbial modifier (*advmod*), but is selected only when the meaning of adjective term is modified, e.g. *"dimly lit room"*,
- compound (*compound*),
- nominal modifier for possessive alteration (*nmod:of* & *nmod:poss*).

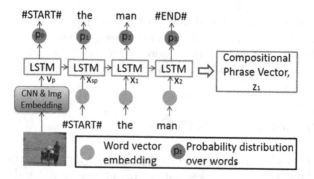

Fig. 4. Composition of phrase vector representation in the phi-LSTM model.

Note that the dependency parse only extracts triplet made up of a governor word and a dependent word linked with a relation. So, in order to form phrase chunk with the dependency parse, we made some refinements as illustrated in Fig. 3. The triplets of selected relations in a sentence are first located, and those consecutive words (as highlighted in the figure, e.g. "the", "man") are grouped as a single phrase, while the standalone word (e.g. "in") will remain as a unit in the sentence.

3.2   Compositional Vector Representation of Phrase

This section describes how compositional vector representation of a phrase is computed, given an image.

Image Representation. A 16-layer VggNet [35] pre-trained on ImageNet [36] classification task is applied to learn image feature in this work. Let $\mathbf{I} \in \mathbb{R}^D$ be an image feature, it is embedded into a K-dimensional vector, $\mathbf{v_p}$ with image embedding matrix, $\mathbf{W_{ip}} \in \mathbb{R}^{K \times D}$ and bias $\mathbf{b_{ip}} \in \mathbb{R}^K$.

$$\mathbf{v_p} = \mathbf{W_{ip}} \mathbf{I} + \mathbf{b_{ip}}. \tag{1}$$

Word Embedding. Given a dictionary $\mathcal{W}$ with a total of V vocabulary, where word $w \in \mathcal{W}$ denotes word in the dictionary, a word embedding matrix $\mathbf{W_e} \in \mathbb{R}^{K \times V}$ is defined to encode each word into a K-dimensional vector representation, $\mathbf{x}$. Hence, an image description with words $w_1 \cdots w_M$ will correspond to vectors $\mathbf{x}_1 \cdots \mathbf{x}_M$ accordingly.

Composition of Phrase Vector Representation. For each phrase extracted from the sentence, a LSTM-based RNN model similar to [6] is used to encode its sequence as shown in Fig. 4. Similar to [6], we treat the sequential modeling from image to phrasal description as a machine translation task, where the embedded image vector is inputted to the RNN on the first time step, followed by a start token $\mathbf{x_{sp}} \in \mathbb{R}^K$ indicating the translation process. It is trained to predict the next word at each time step by outputting $\mathbf{p_{t_p+1}} \in \mathbb{R}^{K \times V}$, which is modeled as the probability distribution over all words in the corpus. The last word of the phrase will predict an end token. So, given a phrase P which is made up by L words, the input $\mathbf{x_{t_p}}$ at each time step are:

$$\mathbf{x_{t_p}} = \begin{cases} \mathbf{v_p}, & \text{if } t_p = -1 \\ \mathbf{x_{sp}}, & \text{if } t_p = 0 \\ \mathbf{W}_e w_{t_p}, & \text{for } t_p = 1...L. \end{cases} \tag{2}$$

For a LSTM unit at time step t_p, let $\mathbf{i}_{t_p}, \mathbf{f}_{t_p}, \mathbf{o}_{t_p}, \mathbf{c}_{t_p}$ and $\mathbf{h}_{t_p}$ denote the input gate, forget gate, output gate, memory cell and hidden state at the time step respectively. Thus, the LSTM transition equations are:

$$\mathbf{i}_{t_p} = \sigma(\mathbf{W_i x}_{t_p} + \mathbf{U_i h}_{t_p-1}), \tag{3}$$

$$\mathbf{f}_{t_p} = \sigma(\mathbf{W_f x}_{t_p} + \mathbf{U_f h}_{t_p-1}), \tag{4}$$

$$\mathbf{o}_{t_p} = \sigma(\mathbf{W_o x}_{t_p} + \mathbf{U_o h}_{t_p-1}), \tag{5}$$

$$\mathbf{u}_{t_p} = tanh(\mathbf{W_u x}_{t_p} + \mathbf{U_u h}_{t_p-1}), \tag{6}$$

$$\mathbf{c}_{t_p} = \mathbf{i}_{t_p} \odot \mathbf{u}_{t_p} + \mathbf{f}_{t_p} \odot \mathbf{c}_{t_p-1}, \tag{7}$$

$$\mathbf{h}_{t_p} = \mathbf{o}_{t_p} \odot tanh(\mathbf{c}_{t_p}), \tag{8}$$

$$\mathbf{p}_{t_p+1} = \text{softmax}(\mathbf{h}_{t_p}). \tag{9}$$

Here, σ denotes a logistic sigmoid function while $\odot$ denotes elementwise multiplication. The LSTM parameters $\{\mathbf{W_i, W_f, W_o, W_u, U_i, U_f, U_o, U_u}\}$ are all matrices with dimension of $\mathbb{R}^{K \times K}$. Intuitively, each gating unit controls the extent of information updated, forgotten and forward-propagated while the memory cell holds the unit internal memory regarding the information processed up to current time step. The hidden state is therefore a gated, partial view of the memory cell of the unit. At each time step, the probability distribution of words outputted is equivalent to the conditional probability of word given the previous words and image, $P(w_t|w_{1:t-1}, I)$. On the other hand, the hidden state at the last time step L is used as the compositional vector representation of the phrase, $\mathbf{z} \in \mathbb{R}^K$, where $\mathbf{z} = \mathbf{h}_L$.

3.3 Encoding of Image Description

Once the compositional vector of phrases are obtained, they are linked with the remaining words in the sentence using another LSTM-based RNN model as shown in Fig. 5. Another start token $\mathbf{x_{ss}} \in \mathbb{R}^K$ and image representation $\mathbf{v_s} \in \mathbb{R}^K$ are introduced, where

$$\mathbf{v_s} = \mathbf{W_{is}I} + \mathbf{b_{is}}, \tag{10}$$

with $\mathbf{W_{is}} \in \mathbb{R}^{K \times D}$ and bias $\mathbf{b_{is}} \in \mathbb{R}^K$ as embedding parameters. Hence, the input units of the LSTM in this level will be the image representation $\mathbf{v_s}$, start

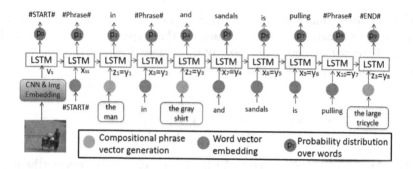

Fig. 5. Sentence encoding using the phi-LSTM model.

token $\mathbf{x_{ss}}$, followed by either compositional vector of phrase $\mathbf{z}$ or word vector $\mathbf{x}$ in accordance to the sequence of its description.

For simplicity purpose, the arranged input sequence will be referred as $\mathbf{y}$. Therefore, given the example in Figs. 4 and 5, the LSTM input sequence of the sentence will be $\{\mathbf{v_s}, \mathbf{x_{ss}}, y_1 \dots y_N\}$ where $N = 8$, and it is equivalent to sequence $\{\mathbf{v_s}, \mathbf{x_{ss}}, \mathbf{z}_1, \mathbf{x}_3, \mathbf{z}_2, \mathbf{x}_7, \mathbf{x}_8, \mathbf{x}_9, \mathbf{x}_{10}, \mathbf{z}_3\}$, as in Fig. 5. Note that a phrase token is added to the vocabulary, so that the model can predict it as an output when the next input is a noun phrase.

The encoding of the sentence is similar to the phrase vector composition. Equations 3–9 are applied here using $\mathbf{y_{t_s}}$ as input instead of $\mathbf{x_{t_p}}$, where t_p and t_s represent time step in phrase and sentence respectively. A new set of model parameters with same dimensional size is used in this hierarchical level.

4   Training the phi-LSTM Model

The proposed phi-LSTM model is trained with log-likelihood objective function computed from the perplexity[2] of sentence conditioned on its corresponding image in the training set. Given an image $\mathbf{I}$ and its description $\mathbf{S}$, let R be the number of phrases of the sentence, P_i correspond to the number of LSTM blocks processed to get the compositional vector of phrase i, Q is the length of composite sequence of sentence $\mathbf{S}$, while $\mathbf{p_{t_p}}$ and $\mathbf{p_{t_s}}$ are the probability output of LSTM block at time step $t_p - 1$ and $t_s - 1$ for phrase and sentence level respectively. The perplexity of sentence $\mathbf{S}$ given its image $\mathbf{I}$ is

$$\log_2 \mathcal{PPL}(\mathbf{S}|\mathbf{I}) = -\frac{1}{N}\left[\sum_{t_s=-1}^{Q} \log_2 \mathbf{p_{t_s}} + \sum_{i=1}^{R}\left[\sum_{t_p=-1}^{P_i} \log_2 \mathbf{p_{t_p}}\right]\right], \quad (11)$$

where

$$N = Q + \sum_{i=1}^{R} P_i. \quad (12)$$

---

[2] Perplexity is a standard approach to evaluate language model.

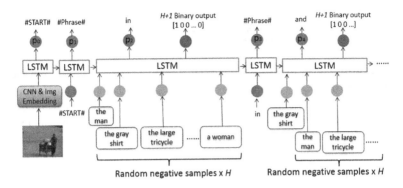

Fig. 6. Upper hierarchy of the phi-LSTM model with phrase selection objective.

Hence, with M number of training samples, the cost function of our model is:

$$\mathcal{C}(\theta) = -\frac{1}{L} \sum_{j=1}^{M} [N_j \log_2 \mathcal{PPL}(\mathbf{S_j}|\mathbf{I_j})] + \lambda_\theta \cdot \parallel \theta \parallel_2^2, \qquad (13)$$

where

$$L = M \times \sum_{j=1}^{M} N_j. \qquad (14)$$

It is the average log-likelihood of word given their previous context and the image described, summed with a regularization term, $\lambda_\theta \cdot \parallel \theta \parallel_2^2$, average over the number of training samples. Here, θ is the parameters of the model.

This objective however, does not discern on the appropriateness of different inputs at each time step. So, given multiple possible inputs, it is unable to distinguish which phrase is the most probable input at that particular time step during the decoding stage. That is, when a phrase token is inferred as the next input, all possible phrases will be inputted in the next time step. The candidate sequences are then ranked according to their perplexity up to this time step, where only those with high probability are kept. Unfortunately, this is problematic because subject in an image usually has much lower perplexity as compared to object and scene. Thus, such algorithm will end up generating description made up of only variants of subject noun phrases.

To overcome this limitation, we introduce a phrase selection objective during the training stage. At all time steps when an input is a phrase, H number of randomly selected phrases that are different from the ground truth input is feed into the phi-LSTM model as shown in Fig. 6. The model will then produce two outputs, which are the next word prediction solely based on the actual input, and a classifier output that distinguishes the actual one from the rest. Though the number of inputs at these time steps increases, the memory cell and hidden state that is carried to the next time step keep only information of the actual input. The cost function for phrase selection objective of a sentence is

$$\mathcal{C}_{PS} = \sum_{t_s \in \mathcal{P}} \sum_{k=1}^{H+1} \kappa_{t_s k} \sigma(1 - y_{t_s k} h_{t_s k} \mathbf{W_{ps}}). \tag{15}$$

where $\mathcal{P}$ is the set of all time steps where the input is phrase, $h_{t_s k}$ is the hidden state output at time step t_s from input k, and $y_{t_s k}$ is its label which is +1 for the actual input and -1 for the false inputs. $\mathbf{W_{ps}} \in \mathbb{R}^{K \times 1}$ is trainable parameters for the classifier while $\kappa_{t_s k}$ scales and normalizes the objective based on the number of actual and false inputs at each time step. The overall objective function is then

$$\mathcal{C}_F(\theta) = -\frac{1}{L} \sum_{j=1}^{M} [N_j \log_2 \mathcal{PPL}(\mathbf{S_j}|\mathbf{I_j}) + \mathcal{C}_{PSj}] + \lambda_\theta \cdot \| \theta \|_2^2 . \tag{16}$$

This cost function is minimized and backpropagated with RMSprop optimizer [37] and trained in a minibatch of 100 image-sentence pair per iteration. We cross-validate the learning rate and weight decay depending on dataset, and dropout regularization [38] is employed over the LSTM parameters during training to avoid overfitting.

5   Image Caption Generation

Generation of textual description using the phi-LSTM model given an image is similar to other statistical language models, except that the image relevant phrases are generated first in the lower hierarchical level of the proposed model. Here, embedded image feature of the given image followed by the start token of phrase are inputted into the model, acting as the initial context required for phrase generation. Then, the probability distribution of the next word over the vocabulary is obtained at each time step given the previous contexts, and the word with the maximum probability is picked and fed into the model again to predict the subsequent word. This process is repeated until the end token for phrase is inferred. As we usually need multiple phrases to generate a sentence, beam search scheme is applied and the top K phrases generated are kept as the candidates to form the sentence. To generate a description from the phrases, the upper hierarchical level of the phi-LSTM model is applied in a similar fashion. When a phrase token is inferred, K phrases generated earlier are used as the inputs for the next time step. Keeping only those phrases which generate positive result with the phrase selection objective, inference on the next word given the previous context and the selected phrases is performed again. This process iterates until the end token is inferred by the model.

Some constraints are added here, which are (i) each predicted phrase may only appears once in a sentence, (ii) maximum number of unit (word or phrase) that made up a sentence is limited to 20, (iii) maximum number of words forming a phrase is limited to 10, and (iv) generated phrases with perplexity higher than threshold T are discarded.

6   Experiment

6.1   Datasets

The proposed phi-LSTM model is tested on two benchmark datasets - Flickr8k [15] and Flickr30k [16], and compared to the state-of-the-art methods [5–7,9,17]. These datasets consist of 8000 and 31000 images respectively, each annotated with five ground truth descriptions from crowd sourcing. For both datasets, 1000 images are selected for validation and another 1000 images are selected for testing; while the rest are used for training. All sentences are converted to lower case, with frequently occurring punctuations removed and word that occurs less than 5 times (Flickr8k) or 8 times (Flickr30k) in the training data discarded. The punctuations are removed so that the image descriptions are consistent with the data shared by Karpathy and Fei-Fei [7].

6.2   Results Evaluated with Automatic Metric

Sentence generated using the phi-LSTM model is evaluated with automatic metric known as the bilingual evaluation understudy (BLEU) [39]. It computes the n-gram co-occurrence statistic between the generated description and multiple reference sentences by measuring the n-gram precision quality. It is the most commonly used metric in this literature.

Table 1 shows the performance of our proposed model in comparison to the current state-of-the-art methods. NIC [6] which is used as our baseline is a reimplementation, and thus its BLEU score reported here is slightly different from the original work. Our proposed model performs better or comparable to the state-of-the-art methods on both Flickr8k and Flickr30k datasets. In particular, we outperform our baseline on both datasets, as well as PbIC [17] - a work that is very similar to us on Flickr30k dataset by at least 5–10%.

Table 1. BLEU score of generated sentence on Flickr8k and Flickr30k dataset.

(a)

Flickr8k				
Models	B-1	B-2	B-3	B-4
NIC [6][3]	60.2(63)	40.4	25.9	16.5
DeepVs [7]	57.9	38.3	24.5	16.0
phi-LSTM	**63.6**	**43.6**	**27.6**	**16.6**

(b)

Flickr30k				
Models	B-1	B-2	B-3	B-4
mRNN [5]	60	41	28	**19**
NIC [6][4]	66.3(66)	42.3	27.7	18.3
DeepVS [7]	57.3	36.9	24.0	15.7
LRCNN [9]	58.7	39.1	25.1	16.5
PbIC [17]	59	35	20	12
phi-LSTM	**66.6**	**45.8**	**28.2**	17.0

[3]The BLEU score reported here is computed from our implementation of NIC [6], and the bracketed value is the reported score by the author.

[4]The BLEU score reported here is cited from [7], and the bracketed value is the reported score by the author.

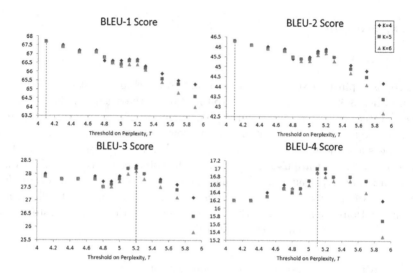

Fig. 7. Effect of the perplexity threshold, T and maximum number of phrases used for generating sentence, K on the BLEU score (best viewed in colour).

Table 2. Vocab size, word occurrence and average caption length in training data, test data, and generated description in Flickr8k dataset.

	Train data		Test data				Gen. caption	
Number of sentence	30000		5000		1000		1000	
	Actual	Trained	Actual	Trained	Actual	Trained	NIC [6]	phi-LSTM
Size of vocab	7371	2538	3147	1919	1507	1187	128	154
Number of words	324481	316423	54335	52683	11139	10806	8275	6750
Avg. caption length	10.8	10.5	10.9	10.5	11.1	10.8	8.3	6.8

As mentioned in Sect. 5, we generate K phrases from each image and discard those with perplexity higher than a threshold value T, when generating the image caption. In order to understand how these two parameters affect our generated sentence, we use different K and T to generate the image caption with our proposed model trained on the Flickr30k dataset. Changes of the BLEU score against T and K are plotted in Fig. 7. It is shown that K does not have a significant effect on the BLEU score, when T is set to below 5.5. On the other hand, unigram and bi-gram BLEU scores improve with lower perplexity threshold, in contrast to tri-gram and 4-gram BLEU scores that reach an optimum

Table 3. Top 5 (a) least trained word found, and (b) most trained word missing, from the generated captions in the Flickr8k dataset.

(a)

NIC [6]		phi-LSTM	
Word	Occurrence	Word	Occurrence
obstacle	93	*overlooking*	81
surfer	127	*obstacle*	93
bird	148	*climber*	96
woods	155	*course*	106
snowboarder	166	*surfer*	127

(b)

NIC [6]		phi-LSTM	
Word	Occurrence	Word	Occurrence
to	2306	*while*	1443
his	1711	*green*	931
while	1443	*by*	904
three	1052	*one*	876
small	940	*another*	713

value when T=5.2. This is because the initial (few) generated phrases with the lowest perplexity are usually different variations of phrase describing the same entity, such as '*a man*' and '*a person*'. Sentence made with only such phrases has higher chance to match with the reference descriptions, but it would hardly get a match on tri-gram and 4-gram. In order to avoid generating caption made from only repetition of similar phrases, we select T and K which yield the highest 4-gram BLEU score, which are T=6.5 and K=6 on Flickr8k dataset, and T=5.2 and K=5 on Flickr30k dataset. A few examples are shown in Fig. 8.

6.3 Comparison of the phi-LSTM Model with Its Sequence Model Counterpart

To compare the differences between a phrase-based hierarchical model and a pure sequence model in generating image caption, the phi-LSTM model and NIC [6] are both implemented using the same training strategy and parameter tuning. We are interested to know how well the corpus is trained by both models. Using the Flickr8k dataset, we computed the corpus information of (i) the training data, (ii) the reference sentences in the test data and (iii) the generated captions as tabulated in Table 2. We remove words that occur less than 5 times in the training data, and it results in 4833 words being removed. However, this reduction in term of word count is only 2.48%. Furthermore, even though the model is evaluated in comparison to all reference sentences in the test data, there are actually 1228 words within the references that are not in our training corpus. Thus, it is impossible for the model to predict those words, and this is a limitation on scoring with references in all language models. For a better comparison with the 1000 generated captions, we also compute another reference corpus based on the first sentence of each test image. From Table 2, it can be seen that even though there are at least 1187 possible words to be inferred with images in the test set, the generated descriptions are made up from only 128 and 154 words in NIC [6] and phi-LSTM model, respectively. These numbers show that the actual number of words learned by these two models are barely 10%, suggesting more research is necessary to improve the learning efficiency in this field. Nevertheless, it shows that introducing the phrase-based structure in sequential model still improves the diversity of caption generated.

Fig. 8. Example of phrases generated from images using the lower hierarchical level of the phi-LSTM model. Red fonts indicate that the perplexity of that phrase is below threshold T.

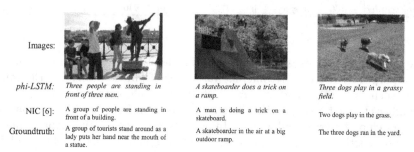

Images:			
phi-LSTM:	*Three people are standing in front of three men.*	*A skateboarder does a trick on a ramp.*	*Three dogs play in a grassy field.*
NIC [6]:	A group of people are standing in front of a building.	A man is doing a trick on a skateboard.	Two dogs play in the grass.
Groundtruth:	A group of tourists stand around as a lady puts her hand near the mouth of a statue.	A skateboarder in the air at a big outdoor ramp.	The three dogs ran in the yard.

Fig. 9. Examples of caption generated with the phi-LSTM model, in comparison to NIC [6].

To get further insight on how the word occurrence in the training corpus affects the word prediction when generating caption, we record the top five, most trained words that are missing from the corpus of generated captions, and the top five, least trained words that are predicted by both models when generating description, as shown in Table 3. We consider only those words that appear in the reference sentences to ensure that these words are related to the images in the test data. It appears that the phrase-based model is able to infer more words which are less trained, compared to the sequence model. Among the top five words that are not predicted, even though they have high occurrence in the training corpus, it can be seen that those words are either not very observable in the images, or are more probable to be described with other alternative. For example, *the* is a more probable alternative of *another*.

A few examples of the image description generated with our proposed model and NIC model [6] are shown in Fig. 9. It can be seen that both models are comparable qualitatively. An interesting example is shown in the first image where our model mis-recognizes the statue as a person, but is able to infer the total number of "persons" within the image. The incorrect recognition stems from insufficient training data on the word *statue* in the Flickr8k dataset, as it only occurs for 48 times, which is about 0.015% in the training corpus.

7   Conclusion

In this paper, we present the phi-LSTM model, which is a neural network model trained to generate reasonable description on image. The model consists of a

12

CNN sub-network connected to a two-hierarchical level RNN, in which the lower level encodes noun phrases relevant to the image; while the upper level learns the sequence of words describing the image, with phrases encoded in the lower level as a unit. A phrase selection objective is coupled when encoding the sentence. It is designed to aid the generation of caption from relevant phrases. This design preserves syntax of sentence better, by treating it as a sequence of phrases and words instead of a sequence of words alone. Such adaptation also splits the content to be learned by the model into two, which are stored in two sets of parameters. Thus, it can generate sentence which is more accurate and with more diverse corpus, as compared to a pure sequence model.

Acknowledgement. This research is supported by the FRGS MoHE Grant FP070-2015A. Also, we would like to thank NVIDIA for the GPU donation.

References

1. Sadeghi, M.A., Farhadi, A.: Recognition using visual phrases. In: CVPR, pp. 1745–1752 (2011)
2. Gupta, A., Verma, Y., Jawahar, C.: Choosing linguistics over vision to describe images. In: AAAI, pp. 606–612 (2012)
3. Bernardi, R., Cakici, R., Elliott, D., Erdem, A., Erdem, E., Ikizler-Cinbis, N., Keller, F., Muscat, A., Plank, B.: Automatic description generation from images: a survey of models, datasets, and evaluation measures. J. Artif. Intell. Res. **55**, 409–442 (2016)
4. Rasiwasia, N., Costa Pereira, J., Coviello, E., Doyle, G., Lanckriet, G.R., Levy, R., Vasconcelos, N.: A new approach to cross-modal multimedia retrieval. In: ACM-MM, pp. 251–260 (2010)
5. Mao, J., Xu, W., Yang, Y., Wang, J., Huang, Z., Yuille, A.: Deep captioning with multimodal recurrent neural networks (M-RNN). arXiv preprint arXiv:1412.6632 (2014)
6. Vinyals, O., Toshev, A., Bengio, S., Erhan, D.: Show and tell: a neural image caption generator. In: CVPR, pp. 3156–3164 (2015)
7. Karpathy, A., Fei-Fei, L.: Deep visual-semantic alignments for generating image descriptions. In: CVPR, pp. 3128–3137 (2015)
8. Kiros, R., Salakhutdinov, R., Zemel, R.: Unifying visual-semantic embeddings with multimodal neural language models. arXiv preprint arXiv:1411.2539 (2014)
9. Donahue, J., Anne Hendricks, L., Guadarrama, S., Rohrbach, M., Venugopalan, S., Saenko, K., Darrell, T.: Long-term recurrent convolutional networks for visual recognition and description. In: CVPR, pp. 2625–2634 (2015)
10. Xu, K., Ba, J., Kiros, R., Cho, K., Courville, A., Salakhudinov, R., Zemel, R., Bengio, Y.: Show, attend and tell: neural image caption generation with visual attention. In: Proceedings of the 32nd International Conference on Machine Learning (ICML 2015), pp. 2048–2057(2015)
11. Krizhevsky, A., Sutskever, I., Hinton, G.: Imagenet classification with deep convolutional neural networks. In: NIPS, pp. 1097–1105 (2012)
12. Hochreiter, S., Schmidhuber, J.: Long short-term memory. Neural Comput. **9**, 1735–1780 (1997)

13. Yngve, V.: A model and an hypothesis for language structure. Proc. Am. Philos. Soc. **104**, 444–466 (1960)
14. Tai, K.S., Socher, R., Manning, C.: Improved semantic representations from tree-structured long short-term memory networks. arXiv preprint arXiv:1503.00075 (2015)
15. Rashtchian, C., Young, P., Hodosh, M., Hockenmaier, J.: Collecting image annotations using Amazon's mechanical turk. In: NAACL: Workshop on Creating Speech and Language Data with Amazon's Mechanical Turk, pp. 139–147 (2010)
16. Young, P., Lai, A., Hodosh, M., Hockenmaier, J.: From image descriptions to visual denotations: new similarity metrics for semantic inference over event descriptions. Trans. Assoc. Comput. Linguist. **2**, 67–78 (2014)
17. Lebret, R., Pinheiro, P.H., Collobert, R.: Phrase-based image captioning. In: International Conference on Machine Learning (ICML). Number EPFL-CONF-210021 (2015)
18. Hodosh, M., Young, P., Hockenmaier, J.: Framing image description as a ranking task: data, models and evaluation metrics. J. Artif. Intell. Res. **47**, 853–899 (2013)
19. Frome, A., Corrado, G.S., Shlens, J., Bengio, S., Dean, J., Mikolov, T., et al.: Devise: a deep visual-semantic embedding model. In: NIPS, pp. 2121–2129 (2013)
20. Socher, R., Karpathy, A., Le, Q.V., Manning, C.D., Ng, A.: Grounded compositional semantics for finding and describing images with sentences. Trans. Assoc. Comput. Linguist. **2**, 207–218 (2014)
21. Karpathy, A., Joulin, A., Fei-Fei, L.: Deep fragment embeddings for bidirectional image sentence mapping. In: NIPS, pp. 1889–1897 (2014)
22. Srivastava, N., Salakhutdinov, R.: Multimodal learning with deep Boltzmann machines. In: NIPS, pp. 2222–2230 (2012)
23. Jia, Y., Salzmann, M., Darrell, T.: Learning cross-modality similarity for multinomial data. In: ICCV, pp. 2407–2414 (2011)
24. Kiros, R., Salakhutdinov, R., Zemel, R.: Multimodal neural language models. In: ICML, pp. 595–603 (2014)
25. Kuznetsova, P., Ordonez, V., Berg, T., Choi, Y.: Treetalk: composition and compression of trees for image descriptions. Trans. Assoc. Computat. Linguist. **2**, 351–362 (2014)
26. Farhadi, A., Hejrati, M., Sadeghi, M.A., Young, P., Rashtchian, C., Hockenmaier, J., Forsyth, D.: Every picture tells a story: generating sentences from images. In: Daniilidis, K., Maragos, P., Paragios, N. (eds.) ECCV 2010. LNCS, vol. 6314, pp. 15–29. Springer, Heidelberg (2010). doi:10.1007/978-3-642-15561-1_2
27. Kulkarni, G., Premraj, V., Ordonez, V., Dhar, S., Li, S., Choi, Y., Berg, A., Berg, T.: Babytalk: understanding and generating simple image descriptions. IEEE Trans. Pattern Anal. Mach. Intell. **35**, 2891–2903 (2013)
28. Yang, Y., Teo, C.L., Daumé III, H., Aloimonos, Y.: Corpus-guided sentence generation of natural images. In: EMNLP, pp. 444–454 (2011)
29. Mitchell, M., Han, X., Dodge, J., Mensch, A., Goyal, A., Berg, A., Yamaguchi, K., Berg, T., Stratos, K., Daumé III, H.: Midge: generating image descriptions from computer vision detections. In: EACL, pp. 747–756 (2012)
30. Gupta, A., Mannem, P.: From image annotation to image description. In: Huang, T., Zeng, Z., Li, C., Leung, C.S. (eds.) ICONIP 2012. LNCS, vol. 7667, pp. 196–204. Springer, Heidelberg (2012). doi:10.1007/978-3-642-34500-5_24
31. Li, S., Kulkarni, G., Berg, T., Berg, A., Choi, Y.: Composing simple image descriptions using web-scale n-grams. In: CoNLL, pp. 220–228 (2011)

32. Kuznetsova, P., Ordonez, V., Berg, A., Berg, T., Choi, Y.: Collective generation of natural image descriptions. In: ACL, pp. 359–368 (2012)

33. Chen, X., Zitnick, L.: Learning a recurrent visual representation for image caption generation. arXiv preprint arXiv:1411.5654 (2014)

34. Manning, C., Surdeanu, M., Bauer, J., Finkel, J., Bethard, S., McClosky, D.: The Stanford CoreNLP natural language processing toolkit. In: ACL, pp. 55–60 (2014)

35. Simonyan, K., Zisserman, A.: Very deep convolutional networks for large-scale image recognition. arXiv preprint arXiv:1409.1556 (2014)

36. Deng, J., Dong, W., Socher, R., Li, L.J., Li, K., Fei-Fei, L.: Imagenet: a large-scale hierarchical image database. In: CVPR, pp. 248–255 (2009)

37. Hinton, G., Srivastava, N., Swersky, K.: Lecture 6a overview of mini-batch gradient descent (2012). Coursera Lecture slides https://class.coursera.org/neuralnets-2012-001/lecture

38. Srivastava, N., Hinton, G., Krizhevsky, A., Sutskever, I., Salakhutdinov, R.: Dropout: a simple way to prevent neural networks from overfitting. J. Mach. Learn. Res. **15**, 1929–1958 (2014)

39. Papineni, K., Roukos, S., Ward, T., Zhu, W.J.: BLEU: a method for automatic evaluation of machine translation. In: ACL, pp. 311–318 (2002)

Deep Relative Attributes

Yaser Souri[1]([⊠]), Erfan Noury[2], and Ehsan Adeli[3]

[1] Sobhe, Tehran, Iran
souri@sobhe.ir
[2] Sharif University of Technology, Tehran, Iran
[3] University of North Carolina at Chapel Hill, Chapel Hill, USA

Abstract. Visual attributes are great means of describing images or scenes, in a way both humans and computers understand. In order to establish a correspondence between images and to be able to compare the strength of each property between images, relative attributes were introduced. However, since their introduction, hand-crafted and engineered features were used to learn increasingly complex models for the problem of relative attributes. This limits the applicability of those methods for more realistic cases. We introduce a deep neural network architecture for the task of relative attribute prediction. A convolutional neural network (ConvNet) is adopted to learn the features by including an additional layer (ranking layer) that learns to rank the images based on these features. We adopt an appropriate ranking loss to train the whole network in an end-to-end fashion. Our proposed method outperforms the baseline and state-of-the-art methods in relative attribute prediction on various coarse and fine-grained datasets. Our qualitative results along with the visualization of the saliency maps show that the network is able to learn effective features for each specific attribute. Source code of the proposed method is available at https://github.com/yassersouri/ghiaseddin.

1 Introduction

Visual attributes are linguistic terms that bear semantic properties of (visual) entities, often shared among categories. They are both human understandable and machine detectable, which makes them appropriate for better human machine communications. Visual attributes have been successfully used for many applications, such as image search [1], interactive fine-grained recognition, [2,3] and zero-shot learning [4,5].

Traditionally, visual attributes were treated as binary concepts [6,7], as if they are present or not, in an image. Parikh and Grauman [5] introduced a more natural view on visual attributes, in which pairs of visual entities can be compared, with respect to their relative strength of any specific attribute. With a set of human assessed relative orderings of image pairs, they learn a global ranking function for each attribute that can be used to compare a pair of two novel images respective to the same attribute (Fig. 1). While binary visual attributes relate properties to entities (*e.g.*, a dog being furry), relative attributes

© Springer International Publishing AG 2017
S.-H. Lai et al. (Eds.): ACCV 2016, Part V, LNCS 10115, pp. 118–133, 2017.
DOI: 10.1007/978-3-319-54193-8_8

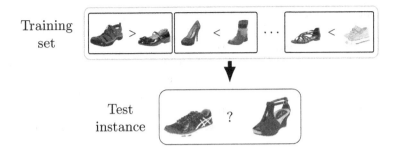

Fig. 1. Visual Relative Attributes. This figure shows samples of training pairs of images from the UT-Zap50K dataset, comparing shoes in terms of the *comfort* attribute (top). The goal is to compare a pair of two novel images of shoes, respective to the same attribute (bottom).

make it possible to relate entities to each other in terms of their properties (*e.g.*, a bunny being furrier than a dog).

Many have tried to build on the seminal work of Parikh and Grauman [5] with more complex and task-specific models for ranking, while still using hand-crafted visual features, such as GIST [8] and HOG [9]. Recently, Convolutional Neural Networks (ConvNets) have proved to be successful in various visual recognition tasks, such as image classification [10], object detection [11] and image segmentation [12]. Many ascribe the success of ConvNets to their ability to learn multiple layers of visual features from the data.

In this work, we propose to use a ConvNet-based architecture comprising of a feature learning and extraction and ranking portions. This network is used to learn the ranking of images, using relatively annotated pairs of images with similar and/or different strengths of some particular attribute. The network learns a series of visual features, which are known to perform better than the engineered visual features for various tasks [13]. These layers could simply be learned through gradient descent. As a result, it would be possible to learn (or fine-tune) the features through back-propagation, while learning the ranking layer. Interweaving the two processes leads to a set of learned features that appropriately characterizes each single attribute. Our qualitative investigation of the learned feature space further confirms this assumption. This escalates the overall performance and is the main advantage of our proposed method over previous methods. Furthermore, our proposed model can effectively utilize pairs of images with equal annotated attribute strength. The equality relation can happen quite frequently when humans are qualitatively deciding about the relations of attributes in images. In previous works, this is often overlooked and mainly inequality relations are exploited. Our proposed method incorporates an easy and elegant way to deal with equality relations (*i.e.*, an attribute is similarly strong in two images). In addition, it is noteworthy to pinpoint that by exploiting the saliency maps of the learned features for each attribute, similar to [14], we can discover the pixels which contribute the most towards an attribute in the image. This can be used to coarsely localize the specific attribute.

Our approach achieves very competitive results and improves the state-of-the-art (with a large margin in some datasets) on major publicly available datasets for relative attribute prediction, both coarse and fine-grained, while many of the previous works targeted only one of the two sets of problems (coarse or fine-grained), and designed a method accordingly.

The rest of the paper is organized as follows: Sect. 2 discusses the related works. Section 3 illustrates our proposed method. Then, Sect. 4 exhibits the experimental setup and results, and finally, Sect. 5 concludes the paper.

2   Related Works

We usually describe visual concepts with their attributes. Attributes are, there-fore, mid-level representations for describing objects and scenes. In an early work on attributes, Farhadi et al. [7] proposed to describe objects using mid-level attributes. In another work [15], the authors described images based on a semantic triple "object, action, scene". In the recent years, attributes have shown great performance in object recognition [7,16], action recognition [17,18] and event detection [19]. Lampert et al. [4] predicted unseen objects using a zero-shot learning framework, incorporating the binary attribute representation of the objects.

Although detection and recognition based on the presence of attributes appeared to be quite interesting, comparing attributes enables us to easily and reliably search through high-level data derived from e.g., documents or images. For instance, Kovashka et al. [20] proposed a relevance feedback strategy for image search using attributes and their comparisons. In order to establish the capacity for comparing attributes, we need to move from binary attributes towards describing attributes relatively. In the recent years, relative attributes have attracted the attention of many researchers. For instance, a linear relative comparison function is learned in [5], based on RankSVM [21] and a non-linear strategy in [22]. In another work, Datta et al. [23] used trained rankers for each facial image feature and formed a global ranking function for attributes.

For the process of learning the attributes, different types of low-level image features are often incorporated. For instance, Parikh and Grauman [5] used 512-dimensional GIST [8] descriptors as image features, while Jayaraman et al. [24] used histograms of image features, and reduced their dimensionality using PCA. Other works tried learning attributes through e.g., local learning [25] or fine-grained comparisons [26]. Yu and Grauman [26] proposed a local learning-to-rank framework for fine-grained visual comparisons, in which the ranking model is learned using only analogous training comparisons. In another work [27], they proposed a local Bayesian model to rank images, which are hardly distinguishable for a given attribute. However, none of these methods leverage the effectiveness of feature learning methods and only use engineered and hand-crafted features for predicting relative attributes.

As could be inferred from the literature, it is very hard to decide what low-level image features to use for identifying and comparing visual attributes.

Recent studies show that features learned through the convolutional neural networks (CNNs) [28] (also known as deep features) could achieve great performance for image classification [10] and object detection [29]. Zhang *et al.* [30] utilized CNNs for classifying binary attributes. In other works, Escorcia *et al.* [31] proposed CCNs with attribute centric nodes within the network for establishing the relationships between visual attributes. Shankar *et al.* [32] proposed a weakly supervised setting on convolutional neural networks, applied for attribute detection. Khan *et al.* [33] used deep features for describing human attributes and thereafter for action recognition, and Huang *et al.* [34] used deep features for cross-domain image retrieval based on binary attributes.

Neural networks have also been extended for learning-to-rank applications. One of the earliest networks for ranking was proposed by Burges *et al.* [35], known as RankNet. The underlying model in RankNet maps an input feature vector to a Real number. The model is trained by presenting the network pairs of input training feature vectors with differing labels. Then, based on how they should be ranked, the underlying model parameters are updated. This model is used in different fields for ranking and retrieval applications, *e.g.*, for personalized search [36] or content-based image retrieval [37]. In another work, Yao *et al.* [38] proposed a ranking framework for videos for first-person video summarization, through recognizing video highlights. They incorporated both spatial and temporal streams through 2D and 3D CNNs and detect the video highlights.

3 Proposed Method

We propose to use a ConvNet-based deep neural network that is trained to optimize an appropriate ranking loss for the task of predicting relative attribute strength. The network architecture consists of two parts, the *feature learning and extraction* part and the *ranking* part.

The feature learning and extraction part takes a fixed size image, I_i, as input and outputs the learned feature representation for that image $\psi_i \in \mathbb{R}^d$. Over the past few years, different network architectures for computer vision problems have been developed. These deep architectures can be used for extracting and learning features for different applications. For the current work, outputs of an intermediate layer, like the last layer before the probability layer, from a ConvNet architecture (*e.g.*, AlexNet [10], VGGNet [39] or GoogLeNet [40]) can be incorporated. In our experiments we use the VGG-16 architecture [39] with the last fully connected layer (the class probabilities) removed. This architecture takes as input a 224×224 RGB image and consists of 13, 3×3 convolutional layers with max pooling layers in between. In addition, it has 2 fully connected layers on top of the convolutional layers. For details on the architecture see [39].

One of the most widely used models for relative attributes in the literature is RankSVM [21]. However, in our case, we seek a neural network-based ranking procedure, to which relatively ordered pairs of feature vectors are provided during training. This procedure should learn to map each feature vector to an absolute ranking, for testing purpose. Burges *et al.* [35] introduced such a neural network based ranking procedure that exquisitely fits our needs. We adopt a similar strategy and thus, the ranking part of our proposed network architecture is analogous to [35] (referred to as RankNet).

During training for a minibatch of image pairs and their target orderings, the output of the feature learning and extraction part of the network is fed into the ranking part and a ranking loss is computed. The loss is then back-propagated through the network, which enables us to simultaneously learn the weights of both feature learning and extraction (ConvNet) and ranking (RankNet) parts of the network. Further with back-propagation we can calculate the derivative of the estimated ordering with respect to the pixel values. In this way, we can generate saliency maps for each attribute (see Sect. 4.6). These saliency maps exhibit interesting properties, as they can be used to localize the regions in the image that are informative about the attribute.

3.1   RankNet: Learning to Rank Using Gradient Descent

This section briefly overviews the RankNet procedure in our context. Given a set (of size n) of pairs of sample feature vectors $\{(\psi_1^{(k)}, \psi_2^{(k)})|k \in \{1, \ldots, n\}\} \in \mathbb{R}^{d \times d}$, and target probabilities $\{t_{12}^{(k)}|k \in \{1, \ldots, n\}\}$, which indicate the probability of sample $\psi_1^{(k)}$ being ranked higher than sample $\psi_2^{(k)}$. We would like to learn a ranking function $f : \mathbb{R}^d \mapsto \mathbb{R}$, such that f specifies the ranking order of a set of features. Here, $f(\psi_i) > f(\psi_j)$ indicates that the feature vector ψ_i is ranked higher than ψ_j, denoted by $\psi_i \triangleright \psi_j$. The RankNet model [35] provides an elegant procedure based on neural networks to learn the function f from a set of pairs of samples and target probabilities.

Denoting $r_i \equiv f(\psi_i)$, RankNet models the mapping from rank estimates to posterior probabilities $p_{ij} = P(\psi_i \triangleright \psi_j)$ using a logistic function

$$p_{ij} := \frac{1}{1 + e^{-(r_i - r_j)}}. \tag{1}$$

The loss for the sample pair of feature vectors (ψ_i, ψ_j) along with target probability t_{ij} is defined as

$$C_{ij} := -t_{ij} \log(p_{ij}) - (1 - t_{ij}) \log(1 - p_{ij}), \tag{2}$$

which is the binary cross entropy loss. Figure 2 (left) plots the loss value C_{ij} as a function of $r_i - r_j$ for three values of target probability $t_{ij} \in \{0, 0.5, 1\}$. This function is quite suitable for ranking purposes, as it acts differently compared to regression functions. Specifically, we are not interested in regression instead of ranking for two reasons: First, we cannot regress the absolute rank of images, since the annotations are only available in pairwise ordering for each attribute, in relative attribute datasets (see Sect. 4.1). Second, regressing the difference $r_i - r_j$ to t_{ij} is inappropriate. To understand this, let's consider the squared loss

$$R_{ij} = \left[(r_i - r_j) - t_{ij}\right]^2, \tag{3}$$

which is typically used for regression, illustrated in Fig. 2 (right). We observe that the regression loss forces the difference of rank estimates to be a specific value and disallows over-estimation. Furthermore, its quadratic natures makes it sensitive to noise. This sheds light into why regression objective is the wrong objective to optimize when the goal is ranking.

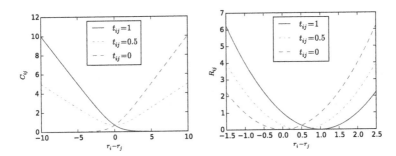

Fig. 2. The ranking loss value for three values of the target probability (left). The squared loss value for three values of the target probability, typically used for regression (right).

Note that when $t_{ij} = 0.5$, and no information is available about the relative rank of the two samples, the ranking cost becomes symmetric. This can be used as a way to train on patterns that are desired to have similar ranks. This is somewhat not much studied in the previous works on relative attributes. Furthermore, this model asymptotically converges to a linear function which makes it more appropriate for problems with noisy labels.

Training this model is possible using stochastic gradient descent or its variants like RMSProp. While testing, we only need to estimate the value of $f(\psi_i)$, which resembles the absolute rank of the testing sample. Using $f(\psi_i)$s, we can easily infer both absolute or relative ordering of the testing pairs.

3.2 Deep Relative Attributes

Our proposed model is depicted in Fig. 3. The model is trained separately, for each attribute. During training, pairs of images (I_i, I_j) are presented to the network, together with the target probability t_{ij}. If for the attribute of interest $I_i \triangleright I_j$ (image i exhibits more of the attribute than image j), then t_{ij} is expected to be larger than 0.5 depending on our confidence on the relative ordering of I_i and I_j. Similarly, if $I_i \triangleleft I_j$, then t_{ij} is expected to be smaller than 0.5, and if it is desired that the two images have the same rank, t_{ij} is expected to be 0.5. Because of the nature of the datasets, we chose t_{ij} from the set $\{0, 0.5, 1\}$, according to the available annotations in the dataset.

The pair of images then go though the feature learning and extraction part of the network (ConvNet). This procedure maps the images onto feature vectors ψ_i and ψ_j, respectively. Afterwards, these feature vectors go through the

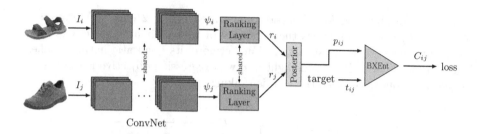

Fig. 3. The overall schematic view of the proposed method during training. The network consists of two parts, the *feature learning and extraction* part (labeled ConvNet in the figure), and the *ranking* part (the Ranking Layer). Pairs of images are presented to the network with their corresponding target probabilities. This is used to calculate the loss, which is then back-propagated through the network to update the weights.

ranking layer, as described in Sect. 3.1. We choose the ranking layer to be a fully connected neural network layer with linear activation function, a single output neuron and weights w and b. It maps the feature vector ψ_i to the estimated absolute rank of that feature vector, $r_i \in \mathbb{R}$, where

$$r_i := w^T \psi_i + b. \tag{4}$$

The two estimated ranks r_i and r_j, for the two images I_i and I_j in comparison, are then combined (using Eq. (1)) to output the estimated posterior probability $p_{ij} = P(I_i \triangleright I_j)$. This estimated posterior probability is used along with the target probability t_{ij} to calculate the loss, as in Eq. (2). This loss is then back-propagated through the network and is used to update the weights of the whole network, including both the weights of the feature learning and extraction sub-network and the ranking layer.

During testing (Fig. 4), we need to calculate the estimated absolute rank r_k for each testing image I_k. Using these estimated absolute ranks, we can then easily infer both the relative or absolute attribute ordering, for all testing pairs.

Fig. 4. During testing, we only need to evaluate r_k for each testing image. Using this value, we can infer the relative or absolute ordering of testing images, for the attribute of interest.

4   Experiments

To evaluate our proposed method, we quantitatively compare it with the state-of-the-art methods, as well as an informative baseline on all publicly available benchmarks for relative attributes to our knowledge. Furthermore, we perform multiple qualitative experiments to demonstrate the capability and superiority of our method.

4.1 Datasets

To assess the performance of the proposed method, we have evaluated it on all publicly available datasets to our knowledge: **Zappos50K** [26] (both coarse and fine-grained versions), **LFW-10** [41] and for the sake of completeness and comparison with previous works, on **PubFig** and **OSR** datasets of [5].

UT-Zap50K [26] dataset is a collection of images with annotations for relative comparison of 4 attributes. This dataset contains two collections: Zappos50K-1, in which relative attributes are annotated for coarse pairs, where the comparisons are relatively easy to interpret, and Zappos50K-2, where relative attributes are annotated for fine-grained pairs, for which making the distinction between them is hard according to human annotators. Training set for Zappos50K-1 contains approximately 1500 to 1800 annotated pairs of images for each attribute. These are divided into 10 train/test splits which are provided alongside the dataset and used in this work. Meanwhile, Zappos50K-2 only contains a test set of approximately 4300 pairs, while its training set is the combination of training and testing sets of Zappos50K-1.

We have also conducted experiments on the **LFW-10** [41] dataset. This dataset has 2000 images of faces of people and annotations for 10 attributes. For each attribute, a random subset of 500 pairs of images have been annotated for each training and testing set.

PubFig [5] dataset (a set of public figure faces), consists of 800 facial images of 8 random subjects, with 11 attributes. **OSR** [5] dataset contains 2688 images of outdoor scenes in 8 categories, for which 6 relative attributes are defined. The ordering of samples in both PubFig and OSR datasets are annotated in a category level, *i.e.*, all images in a specific category may be ranked higher, equal, or lower than all images in another category, with respect to an attribute. This sometimes causes annotation inconsistencies [41]. In our experiments, we have used the provided training/testing split of PubFig and OSR datasets.

4.2 Experimental Setup

We train our proposed model (described in Sect. 3) for each attribute, separately. In our proposed model, it is possible to train multiple attributes at the same time, however, this is not done due to the structure of the datasets, in which for each training pair of images only a certain attribute is annotated.

We have used the Lasagne [42] deep learning framework to implement our model. In all our experiments, for the feature learning and extraction part of the network, we use the VGG-16 model of [39] and trim out the probability layer (all layers up to fc7 are used, only the probability layer is not included). We initialize the weights of the model using a pretrained model on ILSVRC 2014 dataset [43] for the task of image classification. These weights are fine-tuned as the network learns to predict the relative attributes (see Sect. 4.5). The weights w of the ranking layer are initialized using the Xavier method [44], and the bias is initialized to 0.

For training, we use stochastic gradient descent with RMSProp [45] updates and minibatches of size 32 (16 pair of images). We set the learning rate of the feature learning and extraction layers of the network to 10^{-5} and the ranking layer to 10^{-4} for all experiments initially, then RMSProp changes the learning rates dynamically during training. We have also used weight decay (ℓ_2 norm regularization), with a fixed 10^{-5} multiplier. Furthermore, when calculating the binary cross entropy loss, we clip the estimated posterior p_{ij} to be in the range $[10^{-7}, 1 - 10^{-7}]$. This is used to prevent the loss from diverging.

In each epoch, we randomly shuffle the training pairs. The number of epochs of training were chosen to reflect the training size. For Zappos50K and LFW-10 datasets, we train for 25 and 40 epochs, respectively. For PubFig and OSR datasets, we train for 2 epochs due to the large number of training sample pairs. When performing evaluation on OSR the total number of pairs is too large (around 3 million pairs) we only evaluate on a 5% random subset of them.

4.3   Baseline

As a baseline, we have also included results for the RankSVM method (as in [5]), when the features given to the method were computed from the output of the VGG-16 pretrained network on ILSVRC 2014.

Using this baseline we can evaluate the extent of effectiveness of off-the-shelf ConvNet features [13] for the task of ranking. In a sense, comparing this baseline with our proposed method reveals the effect of features fine-tuning, for the task.

4.4   Quantitative Results

Following [5, 26, 41], we report the accuracy in terms of the percentage of correctly ordered pairs. For our proposed method, we report the mean accuracy and standard deviation over 3 separate runs.

Tables 1 and 2 shows our results on the OSR and PubFig dataset respectively. Our method outperforms the baseline and the state-of-the-art on this dataset by a considerable margin, on most attributes. These are relatively easy datasets but have their own challenges. Specifically the OSR dataset contains attributes

Table 1. Results for the OSR dataset

Method	Natural	Open	Perspective	Large size	Diag	ClsDepth	Mean
Relative attributes [5]	95.03	90.77	86.73	86.23	86.50	87.53	88.80
Relative forest [22]	95.24	92.39	87.58	88.34	89.34	89.54	90.41
Fine-grained comparison [26]	95.70	94.10	90.43	91.10	92.43	90.47	92.37
VGG16-fc7 (baseline)	98.00	94.46	92.92	94.08	94.91	95.02	94.90
RankNet (ours)	99.40 ($\pm$0.10)	97.44 ($\pm$0.16)	96.88 ($\pm$0.13)	96.79 ($\pm$0.32)	98.43 ($\pm$0.23)	97.65 ($\pm$0.16)	**97.77** ($\pm$0.10)

Table 2. Results for the PubFig dataset

Method	Male	White	Young	Smiling	Chubby	Forehead	Eyebrow	Eye	Nose	Lip	Face	Mean
Relative Attributes [5]	81.80	76.97	83.20	79.90	76.27	87.60	79.87	81.67	77.40	79.17	82.33	80.56
Relative Forest [22]	85.33	82.59	84.41	83.36	78.97	88.83	81.84	83.15	80.43	81.87	86.31	83.37
Fine-grained Comparison [26]	91.77	87.43	91.87	87.00	87.37	94.00	89.83	91.40	89.07	90.43	86.70	89.72
VGG16-fc7 (baseline)	85.56	80.59	85.20	84.81	82.56	88.50	83.50	83.11	81.52	85.67	86.23	84.30
RankNet (ours)	95.50	94.60	94.33	95.36	92.32	97.28	94.53	93.19	94.24	93.62	94.76	**94.52**
	(± 0.36)	(± 0.55)	(± 0.36)	(± 0.56)	(± 0.36)	(± 0.49)	(± 0.64)	(± 0.51)	(± 0.24)	(± 0.20)	(± 0.24)	(± 0.08)

Table 3. Results for the LFW-10 dataset

Method	Bald	DkHair	Eyes	GdLook	Mascu.	Mouth	Smile	Teeth	FrHead	Young	Mean
Fine-grained Comparison [22]	67.9	73.6	49.6	64.7	70.1	53.4	59.7	53.5	65.6	66.2	62.4
Relative Attributes [5]	70.4	75.7	52.6	68.4	71.3	55.0	54.6	56.0	64.5	65.8	63.4
Global + HOG [46]	78.8	72.4	70.7	67.6	84.5	67.8	67.4	71.7	79.3	68.4	72.9
Relative Parts [41]	71.8	80.5	90.5	77.6	67.0	77.6	81.3	76.2	80.2	82.4	78.5
Spatial Extent [47]	83.21	88.13	82.71	72.76	93.68	88.26	88.16	88.46	90.23	75.05	**84.66**
VGG16-fc7 (baseline)	72.26	79.23	55.64	62.85	90.80	62.42	66.38	59.38	64.45	66.31	67.97
RankNet (ours)	81.14	88.92	74.44	70.28	98.08	85.46	82.49	82.77	81.90	76.33	82.18
	(± 3.39)	(± 0.75)	(± 5.97)	(± 0.54)	(± 0.33)	(± 0.70)	(± 1.41)	(± 2.15)	(± 2.00)	(± 0.43)	(± 1.08)

like "Perspective" which are very generic, high level and global in the image, which might not correspond easily to local low level image features. We think that our proposed method is specially well suited for such cases.

Table 3 shows our results on the LFW-10 dataset. On this dataset, our method performs competitive with respect to the state-of-the-art, but cannot outperform it. We think this might be due to label noise in this dataset and due to the fact that most of the attributes in this dataset are highly local and methods that outperform us on this dataset look locally on regions of the image instead of the whole image.

Tables 4 and 5 show the results on Zappos50K-1 and Zappos50K-2 datasets, respectively. Our method, again, achieves the state-of-the-art accuracy on both coarse-grained and fine-grained datasets. Our proposed method learns appropriate features for the task, given the large amount of training data available in this dataset.

4.5   Qualitative Results

Our proposed method uses a deep network with two parts, the feature learning and extraction part and the ranking part. During training, not only the weights

Table 4. Results for the UT-Zap50K-1 (coarse) dataset

Method	Open	Pointy	Sporty	Comfort	Mean
Relative attributes [5]	87.77	89.37	91.20	89.93	89.57
Fine-grained comparison [26]	90.67	90.83	92.67	92.37	91.64
Spatial extent [47]	95.03	94.80	96.47	95.60	95.47
VGG16-fc7 (baseline)	89.67	90.67	91.67	91.00	90.75
RankNet (ours)	95.37 (±0.82)	94.43 (±0.75)	97.30 (±0.81)	95.57 (±0.97)	**95.67** (±0.49)

Table 5. Results for the UT-Zap50K-2 (fine-grained) dataset

Method	Open	Pointy	Sporty	Comfort	Mean
Relative attributes [5]	60.18	59.56	62.70	64.04	61.62
Fine-grained comparison [26]	74.91	63.74	64.54	62.51	66.43
LocalPair + ML + HOG [46]	76.2	65.3	64.8	63.6	67.5
VGG16-fc7 (baseline)	64.82	64.51	67.31	67.01	65.91
RankNet (ours)	73.45 ($\pm$1.23)	68.20 ($\pm$0.18)	73.07 ($\pm$0.75)	70.31 ($\pm$1.50)	**71.26** ($\pm$0.50)

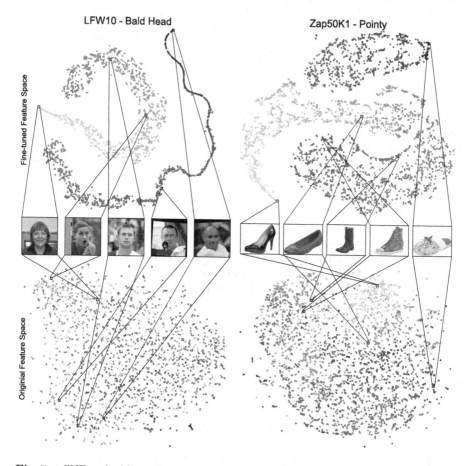

Fig. 5. t-SNE embedding of images in fine-tuned feature space (top) and original feature space (bottom). The set of visualizations on the left are for the *Bald Head* attribute of the LFW-10 dataset, while the visualizations on the right are for the *Pointy* attribute of the Zappos50K-1 dataset. Images in the middle row show a number of samples from the feature space. In the fine-tuned feature space, it is clear that images are ordered according to their value of the attribute. Each point is colored according to its value of the respective attribute, to discriminate images according to their value of the attribute.

for the ranking part are learned, but also the weights for the feature learning and extraction part of the network, which were initialized using a pretrained network, are fine-tuned. By fine-tuning the features, our network learns a set of features that are more appropriate for the images of that particular dataset, along with the attribute of interest. To show the effectiveness of fine-tuning the features of the feature learning and extraction part of the network, we have projected them (features before and after fine-tuning) into 2-D space using the t-SNE [48], as can be seen in Fig. 5. The visualizations on the top of each figure show the images projected into 2-D space from the fine-tuned feature space, while the visualizations on the bottom show the images from the original feature space. Each image is displayed as a point and colored according to its attribute strength. It is clear from these visualizations that fine-tuned feature space is better in capturing the ordering of images with respect to the respective attribute. Since t-SNE embedding is a non-linear embedding, relative distances between points in the high-dimensional space and the low-dimensional embedding space are preserved, thus close points in the low-dimensional embedding space are also close to each other in the high-dimensional space. It can, therefore, be seen that fine-tuning indeed changes the feature space such that images with similar values of the respective attribute get projected into a close vicinity of the feature space. However, in the original feature space, images are projected according to their visual content, regardless of their value of the attribute.

Another property of our network is that it can achieve a total ordering of images, given a set of pairwise orderings. In spite of the fact that training samples are pairs of images annotated according to their relative value of the attribute, the network can generalize the relativity of attribute values to a global ranking of images. Figure 6 shows some images ordered according to their value of the respective attribute.

Fig. 6. Sample images from different datasets, ordered according to the predicted value of their respective attribute.

LFW10 - Bald Head

LFW10 - Good Looking

OSR - Natural

Zappos50k1 - Pointy

Fig. 7. Saliency maps obtained from the network. First we feed two test images into the network and compute the derivative of the estimated posterior with respect to the pair of input images and use the method of [14] to visualize salient pixels with Gaussian smoothing. In each row, the two input images from the a dataset's test set with their corresponding overlaid saliency maps are shown (the warmer the color of the overlay image, the more salient that pixel is).

4.6   Saliency Maps and Localizing the Attributes

We have also used the method of [14] to visualize the saliency of each attribute. Giving two image as inputs to the network, we take the derivative of the estimated posterior with respect to the input images and visualize them. Figure 7 shows some sample visualization for some test pairs. To generate this figure we have applied Gaussian smoothing to the saliency map.

These saliency maps visualize the pixels in the images which contributed most to the ranking predicted by the network. Sometimes these saliency maps are easily interpretable by humans and they can be used to localize attributes using the same network that was trained to rank the attributes in an unsupervised manner, *i.e.*, although we haven't explicitly trained our network to localize the salient and informative regions of the image, it has implicitly learned to find these regions. We see that this technique is able to localize both easy to localize attributes such as "Bald Head" in the LFW10 dataset and abstract attributes such as "Natural" in the OSR dataset.

5   Conclusion

In this paper, we introduced an approach for relative attribute prediction on images, based on convolutional neural networks. Unlike previous methods that use engineered or hand-crafted features, our proposed method learns attribute-specific features, on-the-fly, during the learning procedure of the ranking function. Our results achieve state-of-the-art performance in relative attribute prediction on various datasets both coarse- and fine-grained. We qualitatively show that the feature learning and extraction part, effectively learns appropriate features for each attribute and dataset. Furthermore, we show that one can use a trained model for relative attribute prediction to obtain saliency maps for each attribute in the image.

Acknowledgments. We would like to thank Computer Engineering Department of Sharif University of Technology and HPC center of IPM for their support with computational resources.

References

1. Kovashka, A., Parikh, D., Grauman, K.: Whittlesearch: image search with relative attribute feedback. In: CVPR (2012)
2. Branson, S., Beijbom, O., Belongie, S.: Efficient large-scale structured learning. In: CVPR (2013)
3. Branson, S., Wah, C., Schroff, F., Babenko, B., Welinder, P., Perona, P., Belongie, S.: Visual recognition with humans in the loop. In: Daniilidis, K., Maragos, P., Paragios, N. (eds.) ECCV 2010. LNCS, vol. 6314, pp. 438–451. Springer, Heidelberg (2010). doi:10.1007/978-3-642-15561-1_32
4. Lampert, C., Nickisch, H., Harmeling, S.: Attribute-based classification for zero-shot visual object categorization. IEEE TPAMI **36**, 453–465 (2014)

5. Parikh, D., Grauman, K.: Relative attributes. In: CVPR, pp. 503–510 (2011)
6. Ferrari, V., Zisserman, A.: Learning visual attributes. In: NIPS, pp. 433–440 (2007)
7. Farhadi, A., Endres, I., Hoiem, D., Forsyth, D.: Describing objects by their attributes. In: CVPR (2009)
8. Oliva, A., Torralba, A.: Modeling the shape of the scene: A holistic representation of the spatial envelope. IJCV **42**, 145–175 (2001)
9. Dalal, N., Triggs, B.: Histograms of oriented gradients for human detection. In: CVPR, pp. 886–893 (2005)
10. Krizhevsky, A., Sutskever, I., Hinton, G.E.: Imagenet classification with deep convolutional neural networks. In: NIPS (2012)
11. Girshick, R., Donahue, J., Darrell, T., Malik, J.: Rich feature hierarchies for accurate object detection and semantic segmentation. In: CVPR (2014)
12. Long, J., Shelhamer, E., Darrell, T.: Fully convolutional networks for semantic segmentation. In: CVPR, pp. 3431–3440 (2015)
13. Razavian, A.S., Azizpour, H., Sullivan, J., Carlsson, S.: CNN features off-the-shelf: an astounding baseline for recognition. In: CVPRW, pp. 512–519 (2014)
14. Simonyan, K., Vedaldi, A., Zisserman, A.: Deep inside convolutional networks: visualising image classification models and saliency maps. arXiv preprint arXiv:1312.6034 (2013)
15. Farhadi, A., Hejrati, M., Sadeghi, M.A., Young, P., Rashtchian, C., Hockenmaier, J., Forsyth, D.: Every picture tells a story: generating sentences from images. In: Daniilidis, K., Maragos, P., Paragios, N. (eds.) ECCV 2010. LNCS, vol. 6314, pp. 15–29. Springer, Heidelberg (2010). doi:10.1007/978-3-642-15561-1_2
16. Tao, R., Smeulders, A.W., Chang, S.F.: Attributes and categories for generic instance search from one example. In: CVPR, pp. 177–186 (2015)
17. Khan, F., van de Weijer, J., Anwer, R., Felsberg, M., Gatta, C.: Semantic pyramids for gender and action recognition. IEEE TIP **23**, 3633–3645 (2014)
18. Liu, J., Kuipers, B., Savarese, S.: Recognizing human actions by attributes. In: CVPR, pp. 3337–3344 (2011)
19. Liu, J., Yu, Q., Javed, O., Ali, S., Tamrakar, A., Divakaran, A., Cheng, H., Sawhney, H.: Video event recognition using concept attributes. In: WACV, pp. 339–346 (2013)
20. Kovashka, A., Grauman, K.: Attribute pivots for guiding relevance feedback in image search. In: ICCV, pp. 297–304 (2013)
21. Joachims, T.: Optimizing search engines using clickthrough data. In: ACM KDD, pp. 133–142 (2002)
22. Li, S., Shan, S., Chen, X.: Relative forest for attribute prediction. In: Lee, K.M., Matsushita, Y., Rehg, J.M., Hu, Z. (eds.) ACCV 2012. LNCS, vol. 7724, pp. 316–327. Springer, Heidelberg (2013). doi:10.1007/978-3-642-37331-2_24
23. Datta, A., Feris, R., Vaquero, D.: Hierarchical ranking of facial attributes. In: FG, pp. 36–42 (2011)
24. Jayaraman, D., Sha, F., Grauman, K.: Decorrelating semantic visual attributes by resisting the urge to share. In: CVPR, pp. 1629–1636 (2014)
25. Zhang, H., Berg, A., Maire, M., Malik, J.: SVM-KNN: discriminative nearest neighbor classification for visual category recognition. In: CVPR, vol. 2, pp. 2126–2136 (2006)
26. Yu, A., Grauman, K.: Fine-grained visual comparisons with local learning. In: CVPR (2014)
27. Yu, A., Grauman, K.: Just noticeable differences in visual attributes. In: ICCV (2015)

28. LeCun, Y., Boser, B.E., Denker, J.S., Henderson, D., Howard, R.E., Hubbard, W.E., Jackel, L.D.: Handwritten digit recognition with a back-propagation network. In: NIPS (1989)

29. Girshick, R., Donahue, J., Darrell, T., Malik, J.: Rich feature hierarchies for accurate object detection and semantic segmentation. In: CVPR, pp. 580–587 (2014)

30. Zhang, N., Paluri, M., Ranzato, M., Darrell, T., Bourdev, L.: PANDA: pose aligned networks for deep attribute modeling. In: CVPR, pp. 1637–1644 (2014)

31. Escorcia, V., Carlos Niebles, J., Ghanem, B.: On the relationship between visual attributes and convolutional networks. In: CVPR (2015)

32. Shankar, S., Garg, V.K., Cipolla, R.: Deep-carving: discovering visual attributes by carving deep neural nets. In: CVPR (2015)

33. Khan, F.S., Anwer, R.M., Weijer, J., Felsberg, M., Laaksonen, J.: Deep semantic pyramids for human attributes and action recognition. In: Paulsen, R.R., Pedersen, K.S. (eds.) SCIA 2015. LNCS, vol. 9127, pp. 341–353. Springer, Heidelberg (2015). doi:10.1007/978-3-319-19665-7_28

34. Huang, J., Feris, R.S., Chen, Q., Yan, S.: Cross-domain image retrieval with a dual attribute-aware ranking network. In: ICCV (2015)

35. Burges, C., Shaked, T., Renshaw, E., Lazier, A., Deeds, M., Hamilton, N., Hullender, G.: Learning to rank using gradient descent. In: ICML, pp. 89–96 (2005)

36. Song, Y., Wang, H., He, X.: Adapting deep ranknet for personalized search. In: WSDM (2014)

37. Wan, J., Wang, D., Hoi, S.C.H., Wu, P., Zhu, J., Zhang, Y., Li, J.: Deep learning for content-based image retrieval: a comprehensive study. In: ACM MM, pp. 157–166 (2014)

38. Yao, T., Mei, T., Rui, Y.: Highlight detection with pairwise deep ranking for first-person video summarization. In: CVPR (2016)

39. Simonyan, K., Zisserman, A.: Very deep convolutional networks for large-scale image recognition. arXiv preprint arXiv:1409.1556 (2014)

40. Szegedy, C., Liu, W., Jia, Y., Sermanet, P., Reed, S., Anguelov, D., Erhan, D., Vanhoucke, V., Rabinovich, A.: Going deeper with convolutions. In: CVPR (2015)

41. Sandeep, R.N., Verma, Y., Jawahar, C.V.: Relative parts: distinctive parts for learning relative attributes. In: CVPR (2014)

42. Dieleman, S., Schlter, J., Raffel, C., Olson, E., Snderby, S.K., Nouri, D., Maturana, D., Thoma, M., Battenberg, E., Kelly, J., Fauw, J.D., Heilman, M., diogo149, McFee, B., Weideman, H., takacsg84, peterderivaz, Jon, instagibbs, Rasul, D.K., CongLiu, Britefury, Degrave, J.: Lasagne: first release (2015)

43. Russakovsky, O., Deng, J., Su, H., Krause, J., Satheesh, S., Ma, S., Huang, Z., Karpathy, A., Khosla, A., Bernstein, M., et al.: Imagenet large scale visual recognition challenge. Int. J. Comput. Vision **115**, 211–252 (2015)

44. Glorot, X., Bengio, Y.: Understanding the difficulty of training deep feedforward neural networks. In: AISTATS, pp. 249–256 (2010)

45. Tieleman, T., Hinton, G.: Lecture 6.5–RmsProp: divide the gradient by a running average of its recent magnitude. COURSERA: Neural Netw. Mach. Learn. (2012)

46. Verma, Y., Jawahar, C.V.: Exploring locally rigid discriminative patches for learning relative attributes. In: BMVC (2015)

47. Xiao, F., Jae Lee, Y.: Discovering the spatial extent of relative attributes. In: CVPR (2015)

48. Van der Maaten, L., Hinton, G.: Visualizing data using t-SNE. JMLR **9**, 85 (2008)

Fast Fashion Guided Clothing Image Retrieval: Delving Deeper into What Feature Makes Fashion

Yuhang He and Long Chen[✉]

School of Data and Computer Science, Sun Yat-sen University,
Guangzhou, People's Republic of China
chenl46@mail.sysu.edu.cn

Abstract. Clothing fashion represents human's aesthetic appreciation towards their outfits and reflects the development status of society, humanitarian and economics. Modelling fashion via machine is extremely difficult due to the fact that fashion is too abstract to be efficiently described by machine. In this paper, we delve into two fashion related problems: what type of image feature best describes fashion and how can we fast retrieve the fashionably similar images with any given query fashion image. To address these two problems, we first conduct extensive experiments on various image features, ranging from traditional low-level hand-crafted features, mid-level style aware features to current high-level powerful deep learning based features, to find the feature best describes clothing fashion. To test each candidate feature's performance, we further design a fast fashion guided clothing image retrieval framework by efficiently converting float formatted features into binary codes, with which we can achieve much faster image retrieval without much accuracy reduction. Finally, we validate our proposed framework on two publicly available datasets. Experimental results on both intra-domain and cross-domain fashion clothing image retrieval show that deep learning based image features with explicit fashion prior knowledge guidance best describe fashion, and feature binarization scheme also achieves comparable results in terms of various fashion clothing image retrieval tasks.

1 Introduction

Fashion, primarily a visual art form, integrates aesthetics, art, science and design to create the work that reflects human's understanding and preference to the current world's forefront development trend. As fashion direct carrier, clothing fashion pushes the whole world forward in a way in which it affects our everyday lives and both fashion designers and laymen can join in. As the fashion designer, Marc Jacobs said, "clothing is a form of self-expression - there are hints about who you are in what you wear". In the meantime, clothing fashion trends are erratic and fluctuating. For example, the warm red color and chiffon were very popular in 2011, but the military green and taffeta came into burst in 2012 and 2013 respectively. Spring 2012 saw the instant emergence of neon color

© Springer International Publishing AG 2017
S.-H. Lai et al. (Eds.): ACCV 2016, Part V, LNCS 10115, pp. 134–149, 2017.
DOI: 10.1007/978-3-319-54193-8_9

that reminiscently belongs to 90s. Clothing fashion is also visually apparent, which allows us to analysis it through computer vision related methods. Yet modelling abstract fashion is a challenging task due to large gap between machine percepting an image as pixel-wise real values and we human's percepting fashion via extremely abstract manner.

Recent years witnessed significant clothing online shopping explosion. According to a study from the technology and market research firm Forrester[1], the number of online shoppers is expected to grow to 192 million, or 56% of U.S. population, by 2016, comparing by 53% in 2015. Large amount of clothing purchasing behaviours are driven by clothing fashion attribute. The huge potential market has catalyzed numerous research topics in fashion in last several years in both industry, like eBay[2] and Taobao, and academia, ranging from fashion trend prediction [1], fashion image ranking [2], detection [3] to fashion visual analysis [4–7], cross-domain visual matching [8,9] and recommendation [10]. As image feature representation is prerequisite, they turn to either traditional hand-crafted features (*i.e.* color, texture) or current powerful convolutional neural network (CNNs) feature driven by specific tasks [2,3]. However, these feature representations they depend on have been merely demonstrated to be helpful in non-fashion related tasks, such as object detection and classification. There is a lack of comprehensive study of what type of feature best describes fashion. Kiapour *et al.* [6] describe fashion clothing with 5 styles that easily understandable by humans but difficultly recognizable and processable by machines: hipster, bohemian, pinup, pretty and goth. Vittayakorn *et al.* [4] tried to figure out whether low-level image feature or mid-level attribute (they call *style* and *shape* feature) contribute more to fashion. They show mid-level features perform better than low-level features on their collected runaway dataset and paper doll dataset [11] in terms of fashion description. Still, they did not take high-level image features that are specially designed for fashion description into consideration, which we will show perform much better than both low-level and mid-level feature in this paper. Actually, fashion modelling is an extremely difficult problem. For example, all the 8 fashion images that share the same fashion property in Fig. 1 come from both the same brand name, year and fashion show season. However, they mutually keep large visual discrepancy in terms of color, texture and other common image features.

In this paper, we commit to answer two questions: what type of extracted image features best models the clothing fashion? and given the fashion feature representation extracted from a query image, how can we fast retrieve images with similar fashions attribute? In general, we assume clothing images coming from the same fashion show, same season and holding the same brand name share similar fashion attributes. Because, fashion designers would usually express only one unique fashion theme during their launched fashion show for a particular season. We argue that fashion, especially the clothing fashion, is a distinctive and

---

[1] see report in http://mashable.com/2012/02/27/ecommerce-327-billion-2016-study/ #kc.44t96Zqq3.
[2] see http://labs.ebay.com/tags/fashion.

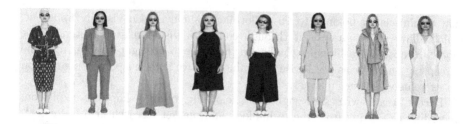

Fig. 1. Large visual discrepancy exists in fashion images. All the 8 images share the same fashion property as they derive from fashion brand Apiece Apart in 2013 Spring ready-to-wear season. However, they also hold large visual discrepancy between images, including color information, texture, even the extraneous shape feature.

habitual trend in the style that people practice in their everyday dresses. It is a kind of aesthetics that directly relates to people's outfits, including footwear, accessories, makeup, clothing. It integrates so many factors and a costume's parts to express a fashion theme that hardly can machines analyze it accurately and comprehensively, even though it is comparably much easier for human to percept and analyze. In this paper, we devote to bridge this gap to make the machine to be able to "understand" fashion. Thus, to tackle the two problems, we first devote to model fashion-aware image feature, ranging from low-level, mid-level features mentioned above to high-level features that are abstract and semantically expressive. We conduct comprehensive experiments to test various image features, including color, texture (low-level), shape, style feature (mid-level), and convolutional neural networks (CNNs), CNNs guided distance metric learning ranking (CNNs&DML) and AutoEncoder features (high-level). CNNs directly learn feature representation from a stack of non-linear neural networks. CNNs&DML works in the same way but the learning process is deliberately supervised by fashion similarity metrics via a triplet ranking loss. While CNNs and CNNs&DML are supervised learning, AutoEncoder is completely unsupervised learning. It learns feature representation by encoding and decoding a fashion image, guaranteeing the input image and decoded image are maximally the same.

Then, as a mean of testing each feature's performance, we design another fast image retrieval framework by converting the long float formatted feature vectors into binary codes, which allows us to fast calculate two image's fashion similarity. We follow the binarization approach proposed by Xia *et al.* [12] to transform all fashion features to binary codes, in which we introduce a sparsity encouraging regularizer and additive noise to reduce the accuracy loss caused by this binarization process. We will show in our experiment that this fast image retrieval framework dramatically improves retrieval speed without obvious retrieval accuracy loss, which enables real-time application.

To validate our proposed fashion features' performance, as well as to test the fast fashion image retrieval framework, we conduct experiments on two publicly available datasets: Runway dataset [4] and Paper Doll dataset [11]. Runway dataset [4] contains fashion images from various fashion shows launched by famous brands (*i.e.* Christian Dior), ranging from 2010 to 2014. Paper Doll dataset [11] contains clothing images people dress in their daily lives and shown in various fashion shows. These two datasets enable us to test our framework from different viewpoints. For example, retrieving images w.r.t. brand name, year, season, or cross-domain retrieval between people wearing clothing fashion show clothing. Overall, the main contributions of this paper lie in: 1. we conduct extensive experiments on various existing image features to find the feature that best describes fashion. The features we here use span from traditional low-level color, texture features, mid-level shape, style feature to high-level deep learning based features. To the best of our knowledge, we have covered most existing image features that have shown superiority in various tasks. 2. We design a novel fashion guided fast image retrieval framework which enables fast image retrieval according to different requests. Besides, our feature vector binarization scheme achieves real-time application without obvious accuracy loss.

2   Fashion Image Feature Pool

Efficient image feature representation is of vital importance for various vision tasks. Up to now, traditional hand-crafted features, semantically engineered feature as well as supervised learning based features have been proposed to address various vision problems. Yet, none of these features was initially designed for fashion description. To delve into what image features makes fashion, we take 7 kinds of features into consideration, namely color, texture, shape, style, Convolutional Neural Networks (CNNs) feature, CNNs supervised by distance metric learning feature (CNNs&DML) and AutoEncoder feature.

Color Feature. Color information is the most direct and intuitive visual information we receive from an image. Given an image, we extract two 512 dimensional histograms in both RGB space and Lab color space and further concatenate them together to form a 1024 dimensional feature vector. To avoid irrelevant background interference, we merely extract color feature in the regions parsed as foreground by [11]. (see Fig. 2 for parsing result).

Texture Feature. Texture captures an item's surface physical appearance and characteristics, such as roughness, topological structure and subtle color orientation. Clothing texture conveys fashion theme from a bottom-to-top scope. In this paper, texture feature consists of two bag-of-words (BoW) histograms from regions parsed as foreground (also by [11]). The first one derives from the histogram from MR8 response [13] quantized into 256 visual words. The second histogram derives from HOG descriptor [14] (8×8 blocks, 4 pixel step size, 9 orientations) quantized into 1000 words. These two histograms are also concatenated together to form one final feature vector.

Shape Feature. We follow the method proposed in [4] for shape feature extraction. Particularly, given a fashion image, we first apply the pose estimation algorithm [15] to find the body part, then we divide the body part into 9 subregions for head, chest, torso, left/right arm, between/left/right legs. For each subregion,

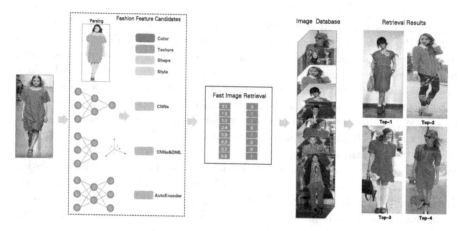

Fig. 2. Framework overview: given a fashion image, we first extract 7 fashion feature candidates, then we convert these float formatted feature vectors into binary codes, with which we can fast retrieve fashionably similar images.

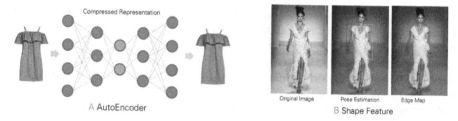

Fig. 3. AutoEncoder feature and Shape feature. A. AutoEncoder receives an input fashion image and feeds it to the encoding-decoding neural network to learn compressed representation (aka AutoEncoder feature). B. Shape feature calculation process: given an original image, we first estimate the nini-tina's pose with a bunch of bounding boxes. Then we gradually binarize all bounding boxes to calculate the edge map, which serves as shape feature.

we extract an edge map by an edge detection algorithm (see Fig. 3B). Finally, we binarize the edge map by minimizing the following loss function

$$\mathcal{L}_{shape} = \sum_{i \in x} d(x_i, \overline{x_j}) + \sum_{j \in \overline{x}} d(\overline{x_j}, x_i) \tag{1}$$

where $d(x_i, \overline{x_j})$ indicates the Euclidean distance of pixel i of the binary map x binarized at the threshold t to the nearest pixel j of the clothing contour $\overline{x}$.

Style Feature. Style feature is particularly introduced by Yamaguchi et al. [11]. For an image, we first extract 24 key points and further use these key points to create part-specific descriptors. Each descriptor builds on low-level features, such as RGB, Lab, MR8, HOG and boundary distance, skin hair distance. Skin-hair distance is calculated by using logistic regression for skin, hair, background and clothing at pixel-level. Finally, all spatial descriptors are concatenated together to form the style feature vector. PCA is also applied for reducing the dimensionality from 39,168 to 441.

CNNs Feature. Convolutional neural networks have shown state of the art performance on various vision tasks due to its super power to learn discriminative feature on large datasets. Tasks such as object detection [16], image semantic segmentation [17] and image similarity measurement [8,18] benefited much from the CNNs to learn discriminative feature through a layer-wise and highly non-linear neural network. While shallow layer in CNNs architecture learns the localized feature (*i.e.* edges, boundaries and textures), and intermediate layer learns mid-level feature (*i.e.* motif, object, attribute), deep layers learns global and abstract image feature. We exploit this advantage and extract CNNs final full connection (*fc*) layer and treat its activation value vector as CNNs fashion feature. Note that the CNNs feature discussed here is trained as a classification problem. For example, given a set of fashion images, we can train a deep convolution neural network to classify these images according their brands. Then we treat full connection feature before the softmax layer as CNNs fashion feature. By utilizing this feature, we test CNNs feature's generalization ability to interpret fashion.

CNNs&DML Feature. Instead of simply replying on CNNs feature alone, we want to go further to supervise CNNs feature learning process via distance metric learning (DML), anticipating the learned CNNs feature better fits for fashion description. Distance metric learning has already been extensively applied to image retrieval [19–23]. The key idea of distance metric learning is to find an optimal metric that minimizes the predefined distance of similar images but maximizes the distance of dissimilar images. In general, distance metric learning either learns a global metric by satisfying all constraints simultaneously or a local metric by merely satisfying partial constraints. In this paper, we adopt triple ranking loss to rank the fashion similarity of an image pair. Triplet ranking loss has been successfully applied to cross-domain clothing image retrieval [18] and content-based image retrieval [9]. Triplet ranking loss requires triple image pair input. Denoting I, I^+ and I^- the *anchor, positive* and *negative* input image, respectively, in which I and I^+ are fashionably similar but I^- is fashionably dissimilar to any image of the two. Our goal is to train the triplet ranking loss by forcing it minimizes distance $d_{(I,I^+)}$ between *anchor* and *positve* and, at the same time, maximizes the distance $d_{(I,I^-)}$ between *anchor* and *negative*. This constraint is achieved by letting $d_{(I,I^+)}$ to be larger than $d_{(I,I^-)}$ by a small pre-defined margin δ.

$$\mathcal{L}_{triplet} = \sum_{(I,I^+,I^-)} \max(0, \delta + d_{(I,I^+)} - d_{(I,I^-)}) \qquad (2)$$

AutoEncoder Feature. Note that all the aforementioned fashion features generation approaches require explicit heavy human engineering work, either in human deliberately involved feature quantization (low-level and mid-level feature) or human guided feature learning strategy (high-level feature). Actually, quantizing fashion image feature via human supervision cannot withstand scrutiny because there is current no general agreement upon what categories or labels are meaningful for clothing fashion modelling. One more appropriate way is to generate clothing fashion representation without explicit human intervention. Inspired by this motivation, we leverage the AutoEncoder scheme introduced by Torres [24] to create clothing fashion feature. AutoEncoder is initially designed for dimension reduction or feature compression via a sequence of symmetrical neural networks by maximally keeping all meaningful information in a much smaller domain. We take this advantage to train a neural network to automatically learn fashion representation which automatically strips away all irrelevant information in the original clothing image and stores the fashion features in a condensed vector.

AutoEncoder learns the fashion presentation of clothing images by first encoding them via a stack of neural networks with descending neutron number order, and then decoding the compressed representation through another stack of neural networks with ascending neuron number order (see Fig. 3A). The whole encoding-decoding neural network is trained through a sequence of forward and backward propagation by forcing the input image to be the same as the neural network's output image. The biggest advantage of AutoEncoder feature is that the neural networks automatically learn fashion representation without being explicitly told what these representations should be look like. Conventional AutoEncoder is notorious for being easily prone to be overfitting. To avoid this dilemma, we follow the variational AutoEncoder introduced by Torres [24] to introduce a regularization term and uncertain noise into the neural network. Specifically, rather than treating the encoded fashion feature vector as static numeric values, we interpret it under Bayesian framework and treat it as statistical distribution with multivariate normal and identity covariance so that we can draw samples from this distribution.

To be specific, given an input image I, we first forward propagate it through the encoding neural network, then compute compression layer's mean value μ and covariance σ^2. With the two values, we can resample the encoding vector through variational posterior $q(z) = \mathcal{N}(z; \mu, \sigma^2 I)$. After forward propagating the resampled feature to the decoding neural network to get the reconstructed image, we can calculate the prior distribution $p(z) = \mathcal{N}(z, 0, I)$. The final loss function consists of two parts: the mean square error and KL divergence of our trained posterior $q(z)$ and pre-constructed prior $p(z)$.

$$\mathcal{L} = MSE(I_i, I_o) + D_{KL}(q(z)||p(z)) \tag{3}$$

where MSE indicates the mean square error calculator. Involving regularization term and extra noise during the whole training process keeps the AutoEncoder from overfitting and thus guarantees fine fashion feature extraction in any test fashion clothing image.

3   Fast Fashion Image Retrieval with Binary Codes

After calculating feature vectors for all fashion image datasets, we can calculate any query image's similarity with each image stored in the database by similarity metrics such as Euclidean distance, Hamming distance and Cosine distance. However, since all the feature vectors are float formatted, distance calculation is too computationally heavy, which is intolerable for many real applications, especially when the database involves millions of images. An empirical way to reduce this computation burden is to convert the float formatted feature vector to binary code vectors without much information loss. With the binary code, the similarity distance can be fast calculated by XOR operation. To this end, we follow the method introduced by Xia *et al.* [12] to convert float fashion representation to binary fashion representation.

The basic theory of float vector binarization is simple: given an original float feature f_f, our goal is to train a matrix W to map f to new feature space $f_b = W \cdot f_f$. f_b is then binarized by thresholding. The whole process is supervised by minimizing the distortion and variants between f_f and f_b. More formally, we use $F \in \mathbb{R}^{d \times n}$ to denote the input float feature matrix, each column of which is a datum, our goal to train a projection matrix $W^{b \times d}$ which directly maps F into the target binary codes $B \in \mathbb{R}^{b \times n}$ by $B = \text{sign}(WF) \in \{-1, 1\}^{b \times n}$. The key challenges arising from this process include a lack of an effective regularizer for accurate mapping and high computation cost. We here introduce a sparsity encouraging regularizer to mitigate these challenge by reducing the number of parameters involved in projection operation. In sum, the objective functions goes as

$$\min_{W,B} = \|WF - B\|_F^2$$

$$s.t. \quad W^T W = I, \quad |W|_0 \leqslant m \tag{4}$$

where $|\cdot|_0$ indicates the number of non-zero elements in W. m is the sparsity controller. By optimizing Eq. 4, the float formatted fashion feature set can be mapped to an binary domain, with which we can fast compute similarity with Euclidean distance.

4   Experiment

We test our frameworks on two publicly available datasets: Runway dataset [4] and Paper Doll dataset [11]. Runway dataset consists of runway images from a wide variety of fashion shows. There are a total of 348,598 images which are collected from *style.com*, including 9,328 fashion shows from 2000 to 2014. Each image is tagged with a meta data describing the image's brand name (*i.e.* Christian Dior), show date, season (*i.e.* Resort 2007), city, the author name, as well as the short text description. The number of images of each individual brand ranges from 10 to 100. There are 8 seasons in total: spring ready to wear (S-RTW), spring menswear (S-MENS), Spring Couture (S-COUT), resort, pre-fall, fall read-to-wear (F-RTW), fall menswear (F-MENS), fall couture (F-COUT). In our experiment, we assume only fashion images sharing the same brand name, year and seasons share the same fashion property because each fashion designer always expresses one particular fashion theme for a fashion show and all clothing items serve to express this fashion theme. For example, in the 2008 pre-fall fashion show, Burberry Prorsum has expressed brands signature outerwear centered fashion through coutures "Coats were ruched, beaded, piped with patent leather. Underneath, there was a loosened-up silhouette: still fitted at the top, but fuller at the bottom". Therefore, an image is only considered as correct retrieval only if it comes from the same fashion show with the query image. With the Runway dataset, we can conduct experiments on intra-domain fashion image retrieval where "intra-domain" means all images coming from various fashion shows.

Paper Doll dataset [11] contains clothing images people wearing in their daily lives. There is a total of 339,797 images collected from the social network named Fashionista which focuses on Chictopia fashion. Each image in Paper Doll dataset [11] does not have a fashion tag, so we can not directly test each feature candidate's performance within our proposed fast fashion image retrieval framework regarding fashion metrics. Here we utilize Paper Doll dataset [11] to test each feature candidate's generalization capability in retrieving really clothing images (clothing people wearing everyday) for runway fashion images. We call it cross-domain fashion retrieval. Since there is no ground truth dataset for quantitative evaluation, we involve human subjective evaluation: for each runway query image's retrieval results, we ask 5 volunteers to label the fashion similarity between the query image and each retrieved realway image.

As for evaluation metrics, we adopt mean average precision (mAP), precision and recall rate at particular ranks ("P@K", "R@K") metrics that are often employed for many standard image retrieval applications. Note that mAP strikes a balance between precision and recall rate. Besides, it takes the retrieved image's location into consideration. The more forward an accurately retrieved fashion image ranks, the higher mAP values it achieves. Specifically, mAP is computed via the following equation,

$$mAP = \frac{1}{N} \sum_{i=2}^{N} (r_i - r_{i-1}) \cdot \frac{(p_{i-1} + p_i)}{2.0} \tag{5}$$

where r_i and p_i indicate the recall and precision rate of top-i retrieval results, respectively. A well-designed image retrieval framework often generates high mAP value.

We implement CNNs and CNNs&DML feature learning on the open source deep learning framework Caffe [25] with 4 Tesla K40 GPUs. The CNNs architecture we adopt here is the 18-layer residual network with identity mapping proposed by He et al. [16], which has shown promising performance on various vision tasks. In CNNs feature training, we classify the Runway dataset [4] according to their brand names (thus, the output layer is a softmax layer with 851 outputs). Note that other classification criterias truly exist, such as year-based and fashion season-based classification. We do not involve them here because we just want to test non-fashion guided CNNs feature's performance on fashion image retrieval. We divide Runway dataset [4] into 300,000 and 48,598 for train and testing, respectively. Data augmentation methods like scaling, vignetting, fish eye distortion are involved here and it takes 6 days to train the whole dataset.

We observe that the loss fluctuates slightly at the very beginning several epochs iteration and then gradually diminishes to a small value. For CNNs&DML feature training, we first collect fashion images of the same brand, year and season to form the *anchor* and *positive* pair, any image violating this similar-fashion criterion is treated as *negative*. Finally, we randomly generate 500,000 triplet pair, covering all the 9260 different fashion shows. The testing dataset for CNNs fashion feature is also adopted here for testing.

Intra-domain Experiment. We only consider top-10 retrieval results for Runway dataset [4] because, in some extreme situations, one query fashion image corresponds only up to 10 fashion images. Specifically, we calculate mAP, P(R)@3, P(R)@5, P(R)@7 and P(R)@10 metrics. The float formatted retrieval results are given in Table 1 and the binary codes retrieval results are given in Table 2. We can clearly observe that machine learning based features (CNNs, CNNs&DML and AutoEncoder) far outperform traditional hand-crafted features by a large margin on both float formatted retrieval and binary codes based retrieval. Traditional low-level features including color, texture, shape and style usually get fine retrieval result with a low ranking K, which means that fashion property shows correlation with low-level image features. However, while the retrieving number increases, traditional low-level features soon loss discrimination capability

Table 1. Fashion image retrieval results on runway dataset [4] (float formatted features).

Feature	mAP	P@3	P@5	P@7	P@10	R@3	R@5	R@7	R@10
Color	0.41	0.70	0.43	0.51	0.57	0.033	0.065	0.110	0.167
Texture	0.45	0.67	0.46	0.54	0.57	0.036	0.063	0.110	0.172
Shape	0.51	0.73	0.60	0.63	0.71	0.049	0.078	0.136	0.198
Style	0.53	0.76	0.70	0.63	0.69	0.048	0.080	0.148	0.213
CNNs	0.74	0.87	0.83	0.83	0.85	0.088	0.104	0.197	0.318
CNNs&DML	**0.87**	**0.89**	**0.93**	**0.87**	**0.92**	**0.094**	**0.138**	**0.210**	**0.321**
AutoEncoder	0.76	0.80	0.85	0.83	0.84	0.090	0.112	0.187	0.317

Table 2. Fashion image retrieval results on runway dataset [4] (binary codes features).

Feature	mAP	P@3	P@5	P@7	P@10	R@3	R@5	R@7	R@10
Color	0.37	0.68	0.40	0.43	0.50	0.030	0.055	0.082	0.154
Texture	0.45	0.67	0.40	0.47	0.50	0.029	0.047	0.090	0.164
Shape	0.47	0.68	0.52	0.60	0.68	0.042	0.071	0.126	0.189
Style	0.50	0.73	0.63	0.59	0.64	0.043	0.076	0.136	0.201
CNNs	0.70	0.86	0.81	0.82	0.80	0.087	0.104	0.193	0.310
CNNs&DML	**0.87**	**0.88**	**0.93**	**0.87**	**0.90**	**0.094**	**0.137**	**0.210**	**0.321**
AutoEncoder	0.74	0.74	0.83	0.82	0.84	0.088	0.109	0.183	0.300

regarding fashion description (which are testified by the fact that precision values with a higher K are much smaller than prevision values with a lower K). Therefore, hand-crafted image features only hold weak correlation with fashion description, much more robust and discriminative characteristics of fashion are still are incorporated by these hand-crafted features.

On the contrary, our proposed three machine learning based image features (CNNs, CNNs&DML, AutoEncoder) have managed to grasp these hidden characteristics of fashion. Their multi-layers perceptron and high non-linearity perception manner assist them to mine deeper semantic and more abstract characteristics of fashion. It in turn attests fashion integrates various visual information, both intuitive and abstract, to be fashionable. Among the three machine learning based methods, CNNs&DML performs the best (an average of 10 percent increasing in mAP). We learn that explicitly telling the neural network some side information about fashion, like what two images are fashionably similar but the other two are not in our experiments, dramatically assist machines to understand fashion. Unsupervised and self-explanatory AutoEncoder and non-fashion task guided neural network CNNs, to some extent, often fail to fully capture fashion properties. In addition, we also note that the three machine learning based image features are barely affected by the variation of ranking number K (almost stayed the same regardless of K changes). This shows that these three feature managed to jump over the fashion interpretation barrier showing in Fig. 1. They are better capable of understanding "what makes fashion".

Visual results are shown in the left side of Fig. 4, from which we can see that hand-crafted image features often treat images from different brands, years and seasons as fashionably similar. However, deep learning based methods can avoid this problem and find the truly fashionably similar images, even though they have dramatic visual difference. We do not provide the average processing time difference here between float formatted features and binary codes. The reason is that, on the one hand, the retrieval time for each query image heavily depends on the size of database. Direct comparison without taking the database size into consideration is somewhat meaningless. On the other hand, we observe that no obvious time difference in the 30W+ Runway database [4] between float formatted features and binary codes. However, when we applied the same feature binarization scheme to other image retrieval problem on a much larger database (*i.e.*, 6 million), processing time difference emerges: the average time to retrieve an image with float formatted feature is about 0.5 s, but 0.2 s with its corresponding binary code.

Cross-domain Experiment. An intuitive idea is to figure out whether the learned or hand-crafted fashion descriptors on runway scenarios (Runway dataset [4]) can successfully be applied to real life clothing items' fashion analysis. This can help us to test these features' generalization and transformation ability. Thus, we further conduct experiment on Paper Poll dataset [11]. As we discussed above, there is not ground truth for Paper Poll dataset [11] for quantitative evaluation. What we do here is to ask 5 volunteers to label the retrieval results. This helps us to understand the fashion from human perspective, even

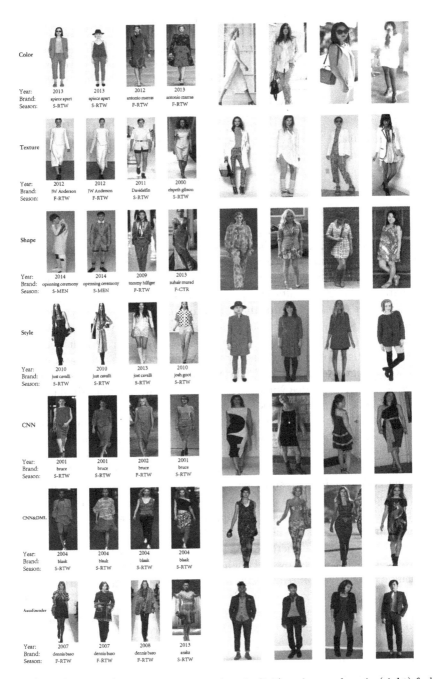

Fig. 4. Visual representation of both intra-domain (left) and cross-domain (right) fashion image retrieval results. The first image in each row for either domain is the runway query image. For the intra-domain fashion image retrieval, we further provide year, brand and season tags for better comparison.

Table 3. Human judgement results on cross-domain experiment. Given an query image from the Runway dataset [4], we retrieve fashionably similar images from Paper Doll dataset [11]. Then we ask 5 volunteers to label each retrieval result and assign the labelling result to one of the three classes according to the number of volunteers who agree on the retrieval result. Finally, we calculate the rate of the three classes on the testing dataset.

Classes	Color	Texture	Shape	Style	CNNs	CNNs&DML	AutoEncoder
Unanimity	0.43	0.28	0.30	0.38	0.40	0.48	0.43
Major	0.33	0.20	0.35	0.31	0.37	0.34	0.30
Some	0.24	0.52	0.35	0.31	0.23	0.18	0.27

though subjective personal preference and experience are heavily involved in this experiments. Specifically, we follow the scoring system provided by Vittayakorn *et al.* to ask the volunteers to label all query-retrieval image pair as fashionably similar or dissimilar. Then we calculate the number of volunteers who have given agreed labelling results and accordingly classify each retrieval results as one of the three classes: **Unanimity** which means all the agreed on the retrieving result. **Majority** which means more than or equal 3 volunteers agreed on the retrieving result, and **Some** which indicates less than 3 volunteers agreed on the retrieving result. To maximally reduce personal prejudice and unprofessional judgement, we first ask all the volunteers to carefully look at Runway dataset [4] to learn the idea what makes two images fashionably similar. Finally, we calculate the rate of the whole testing images being classified as the three classes regarding the 7 feature candidates. The result in shown in Table 3, from which we can get that pre-trained volunteers exhibit professional expertise to compare the fashion similarity for runway-realway image pair. They show compatible labelling result w.r.t intra-domain experiment. That is, machine learning based methods' retrieved results leads to larger unanimity rate. Hand-crafted features, especially the shape and style features, arise much controversy between volunteers. We think the reason behind it is that mid-level feature is neither fully human understandable nor machine discernible. So, none of them can lead to promising results on either the machine side or human judgement side. The low-level features, however, even though they are still machine discernible regarding fashion interpretation, can easily motivate human's perception and further inspire human to make judgement. This specially applies to color feature, from which we see a large unanimity rate. Color information reduces human's hesitation to make a judgement, but texture varies significantly according to different people.

The cross-domain visual result is given on the right side of Fig. 4. We can clearly see that texture feature can lead to large ambiguity between different clothing parts in an image (second row). The nini-tina's overcoat, trouser as well as bag hold large texture discrepancies, texture feature alone failed to interpret them hierarchically and efficiently. So the retrieved results exhibit large texture variations, even though all the three retrieved images have the white overcoat. However, large machine learning based methods, especially deep learning based methods, managed to strike a balance between them and accurately retrieve image according to their fashion property.

Overall, after comprehensive and extensive experiments on both intra-domain and cross-domain situation, we can see that deep learning based image features (CNNs, CNNs&DML and AntoEncoder) can be used to describe an image's fashion property in both intra-domain and cross-domain application. Explicit fashion guided training helps to learn better fashion features. Traditional features, including low-level features and mid-level features, cannot grasp discriminative and deep fashion properties. Besides, our proposed fast image retrieval framework helps to fast retrieve fashion images according to their fashion properties.

5 Conclusion

In this paper, we delve into what special image feature makes fashion. We conduct extensive experiments to test various existing image features' performances in terms of fashion aware image retrieval, assuming fashion images deriving from the same fashion show, same brand name as well as same season share similar fashion properties. The image features we exploit in this paper cover most famous low-level, mid-level and high-level image features that have been demonstrated to be useful in other vision tasks. Our comprehensive experimental results show that machine learning (especially deep learning) based image features better describe fashion than traditional hand-crafted image features. Among all machine learning generated image features, fashion-guided machine learning generated features (CNNs&DML) performs slightly better than both non-fashion task supervised machine learning generated image feature (i.e. classification task supervised CNNs image feature) and unsupervised machine learning generated image feature (i.e. AutoEncoder), which shows that fashion term is highly abstract and can be better described by telling the machine some side information about fashion, such as what two images share similar fashion property. Even though fashion is somewhat easily understandable by humans but difficultly processible by machines, we find that it can still be efficiently modelled by machines via current successful deep learning based methods.

In addition, to fast retrieve a fashion image, we propose to convert float formatted feature vectors into binary codes. The feature binarization process allows real-time fashion image retrieval application. Still, fashion is an open problem and many interesting problems such as fashion trend prediction and image fashion likelihood probability prediction, remain to be tackled.

Acknowledgement. This research is supported by the National Natural Science Foundation of China (NSFC) under grant No. 41401525, the Natural Science Foundation of Guangdong Province under grant No. 2014A030313209.

References

1. He, R., McAuley, J.: Ups and downs: modeling the visual evolution of fashion trends with one-class collaborative filtering. In: Proceedings of WWW (2016)
2. Wang, J., Nabi, A.A., Wang, G., Wan, C., Ng, T.: Towards predicting the likeability of fashion images. arXiv preprint arXiv:1511.05296 (2015)
3. Hara, K., Jagadeesh, V., Piramuthu, R.: Fashion apparel detection: the role of deep convolutional neural network and pose-dependent priors. In: Proceedings of WACV (2016)
4. Vittayakorn, S., Yamaguchi, K., Berg, A.C., Berg, T.L.: Runway to realway: visual analysis of fashion. In: Proceedings of WACV (2015)
5. Liu, S., Liu, L., Yan, S.: Fashion analysis: current techniques and future directions. In: IEEE MultiMedia (2014)
6. Kiapour, M.H., Yamaguchi, K., Berg, A.C., Berg, T.L.: Hipster wars: discovering elements of fashion styles. In: Fleet, D., Pajdla, T., Schiele, B., Tuytelaars, T. (eds.) ECCV 2014. LNCS, vol. 8689, pp. 472–488. Springer, Heidelberg (2014). doi:10.1007/978-3-319-10590-1_31
7. Simo-Serra, E., Fidler, S., Moreno-Noguer, F., Urtasun, R.: Neuroaesthetics in fashion: modeling the perception of fashionability. In: Proceedings of CVPR (2015)
8. Lin, L., Wang, G., Zuo, W., Feng, X., Zhang, L.: Cross-domain visual matching via generalized similarity measure and feature learning. In: Proceedings of T-PAMI (2016)
9. Wan, J., Wang, D., Hoi, S., Wu, P., Zhu, J., Zhang, Y., Li, J.: Deep learning for content-based image retrieval: a comprehensive study. In: Proceedings of ACM MM, pp. 157–166 (2014)
10. Jagadeesh, V., Piramuthu, R., Bhardwaj, A., Di, W., Sundaresan, D.: Large scale visual recommendations from street fashion images. In: Proceedings of KDD (2014)
11. Yamaguchi, K., Hadi Kiapour, M., Berg, T.L.: Paper doll parsing: retrieving similar styles to parse clothing items. In: Proceedings of ICCV (2013)
12. Xia, Y., He, K., Kohli, P., Sun, J.: Sparse projections for high-dimensional binary codes. In: Proceedings of CVPR (2015)
13. Varma, M., Zisserman, A.: A statistical approach to texture classification from single images. Proc. IJCV $62(1–2)$, 61–81 (2005)
14. Dalal, N., Triggs, B.: Histograms of oriented gradient for human detection. In: Proceedings of CVPR (2005)
15. Yang, Y., Ramanan, D.: Articulated pose estimation with flexible mixtures-of-parts. In: Proceedings of CVPR (2011)
16. He, K., Zhang, X., Ren, S.: Identity mappings in deep residual networks. arXiv preprint arXiv:1603.05027 (2016)
17. Noh, H., Hong, S., Han, B.: Learning deconvolution network for semantic segmentation. In: Proceedings of ICCV (2015)
18. Huang, J., Feris, R., Chen, Q., Yan, S.: Cross-domain image retrieval with a dual attribute-aware ranking network. In: Proceedings of ICCV (2015)
19. Guillaumin, M., Verbeek, J., Schmid, C.: Is that you? Metric learning approaches for face identification. In: Proceedings of ICCV (2009)
20. Weinberger, K.Q., Blitzer, J., Saul, L.K.: Distance metric learning for large margin nearest neighbor classification. In: Proceedings of NIPS (2005)
21. Wang, J., Song, Y., Leung, T., Rosenberg, C., Wang, J., Philbin, J., Chen, B., Wu, Y.: Learning fine-grained image similarity with deep ranking. In: Proceedings of CVPR (2014)

22. Schroff, F., Kalenichenko, D., Philbin, J.: Facenet: a unified embedding for face recognition and clustering. In: Proceedings of CVPR (2015)
23. Wang, X., Gupta, A.: Unsupervised learning of visual representations using videos. In: Proceedings of ICCV (2015)
24. Kingma, D.P., Welling, M.: Auto-encoding variational bayes. arXiv preprint arXiv:1312.6114
25. Jia, Y., Evan, D.J., Sergey, K., Jonathan, L., Ross, G., Sergio, G., Trevor, D.: Caffe: convolutional architecture for fast feature embedding. arXiv preprint arXiv:1408.5093 (2014)

Using Gaussian Processes to Improve Zero-Shot Learning with Relative Attributes

Yeshi Dolma$^{(\boxtimes)}$ and Vinay P. Namboodiri

Department of Computer Science and Engineering,
Indian Institute of Technology Kanpur, Kanpur, India
yeshidpn@gmail.com

Abstract. Relative attributes can serve as a very useful method for zero-shot learning of images. This was shown by the work of Parikh and Grauman [1] where an image is expressed in terms of attributes that are relatively specified between different class pairs. However, for zero-shot learning the authors had assumed a simple Gaussian Mixture Model (GMM) that used the GMM based clustering to obtain the label for an unknown target test example. In this paper, we contribute a principled approach that uses Gaussian Process based classification to obtain the posterior probability for each sample of an unknown target class, in terms of Gaussian process classification and regression for nearest sample images. We analyse different variants of this approach and show that such a principled approach yields improved performance and a better understanding in terms of probabilistic estimates. The method is evaluated on standard Pubfig and Shoes with Attributes benchmarks.

1 Introduction

Consider the task of recognizing a person at test time when we are not provided with any images of the person at training. This setting for classification is termed zero-shot learning, i.e. the classifier is provided with no training image for obtaining the classification. A technique used to recognize unseen classes is through the use of attributes [5]. These attributes describe a person in terms as the gender of a person, or type of hair that person has. However, as shown by Parikh and Grauman [1], a more natural description is obtained by describing the attributes of a person in relation to those that are known. For instance, we can say that 'Tracy Morgan's face is chubbier as compared to 'Anderson Cooper' but less as compared to 'Karl Rove'.

In this paper, we consider this problem of zero-shot recognition of different objects like faces or shoes using relative attributes. The initial work by Parikh and Grauman [1] used relative attributes in zero-shot recognition by using a Gaussian mixture model of the relative attributes. However, a simple Gaussian mixture model does not transfer the knowledge effectively in the model. In this

Electronic supplementary material The online version of this chapter (doi:10. 1007/978-3-319-54193-8_10) contains supplementary material, which is available to authorized users.

© Springer International Publishing AG 2017
S.-H. Lai et al. (Eds.): ACCV 2016, Part V, LNCS 10115, pp. 150–164, 2017.
DOI: 10.1007/978-3-319-54193-8_10

paper, we propose a more principled approach where we use a Gaussian Process prior over the relative attributes in order to obtain zero-shot recognition. This approach while being principled also enables us to model the variance in the samples. We further analyze different variants of using Gaussian process prior for obtaining zero-shot recognition of samples.

In our approach we use two stages of Gaussian processes. In the first stage, we use a Gaussian process based classifier to classify the set of classes that are known in training. In the second stage, we use Gaussian process based regression to obtain the zero shot recognition for samples in test that have no training examples. The two stages allow for effective knowledge transfer from known training samples of a fixed set of categories to unknown test samples of a set of categories for which no training samples are present.

The main contribution of this work is to demonstrate a two-stage framework using Gaussian process that allows us to obtain principled probabilistic estimates of the relative attributes for zero shot learning. We obtain in this framework not only the probablistic estimates of $p(y|x)$ where y is the class label and x is the feature set, but also the uncertainty in estimating $p(y|x)$ that is extremely relevant in the zero-shot setting. We demonstrate the efficacy of our method with detailed comparison to the previous work [1] on standard benchmark datasets.

The rest of the paper is organized as follows: In the next section we give a brief overview of the related work. In Sect. 3 we provide the background that briefly provides an overview of the relative attribute zero shot learning based setting. In Sect. 4 we provide detailed description of the proposed method and its variants. Section 5 discusses the experiments performed and the results obtained from the experiments and we finally conclude in Sect. 6 with directions for future work.

2   Related Works

The use of attributes for zero shot learning was initially proposed by Lampert et al. [5]. In their work they had shown that animals could be described in terms of binary attribute vectors that captured the properties of each class. This was then used to recognize an unseen class in terms of its attributes. Akata et al. [7] extend the work by considering the attribute representation problem as one of label embedding and learn the embedding instead of using a direct attribute presentation [6]. Further work has been undertaken where they consider that the attributes may be unreliable [8]. Another interesting line of work has been analysed by Elhoseiny et al. [10] where the authors analysed the use of pure textual descriptions instead of well defined attribute representations. A recent work explores the structure of the semantic manifold in terms of semantic class label graph for representing the distance [14]. Another explores the co-occurrence of visual concepts for zero shot classification [15].

These methods have addressed the attribute representation. However, in our work we address the method used for zero-shot recognition. The basic premise is that just using a clustering would not exploit the structure of the data for zero-shot recognition. Recently there has been interesting work by Yu and Grauman [9] where the authors show that using Bayesian local learning they are able to analyse when two images are indistinguishable for a specific attribute. In our work we jointly rely on multiple attributes and treat the problem of identifying the sample through Gaussian process regression.

The present work relies on relative attributes which were proposed by Parikh and Grauman [1]. In their work the authors introduced relative attributes and showed that they were applicable for a number of use-cases including zero-shot learning of unseen classes. Further, Berg [4] have shown that relative attributes could be coupled with relative feedback and this would be useful for image search cases such as searching for a shoe. These use-cases that extend relative attributes could also be applicable using the proposed method.

Gaussian process is extensively used in our work. This framework has been excellently presented by Rasmussen and Williams [2] in their book. This approach while primarily suited for regression has also been used for other related tasks such as multi-relational learning [11] and for one-shot recognition [12]. In our approach we use it in a two stage approach for classification and regression based on attribute data for zero-shot learning.

3   Background

Our method builds on the work of Devi Parikh and Grauman [1] where the classes are modelled as Gaussian Distributions using *relative attributes*, which depict the strength of an attribute as opposed to binary attributes which shows its presence or the absence in the image.

During training, given a set of training images X represented by N-dimensional feature vector, $x_i \in \mathbb{R}^N$, and a set of M attributes, A_m, the relation between the attribute strength of the seen classes are given as sets of ordered pair $O_m = \{(i,j)\}$ and similarity-pair $S_m = \{(i,j)\}$. These pairs are such that if $(i,j) \in O_m$, then image i has stronger attribute a_m than image j. Similarly, if the pair $(i,j) \in S_m$, image i and image j have similar strength of attribute a_m. Using these pairs as supervision, M ranking functions are learned for each attribute that maps an image to its attribute strength score. These functions transform the images $x_i \in \mathbb{R}^N \implies \mathbb{R}^M$. The images are now M-dimensional vector where mth dimension represents the attribute a_m's rank score. For the unseen classes, the supervision is given with respect to one or two seen classes. An unseen class c_k^u can be described relative to seen classes c_p^s and c_q^s, using all or a subset of M attributes, as $c_{pm}^s \prec c_{km}^u \prec c_{qm}^s$ or $c_{pm}^s \prec c_{km}^u$, or $c_{km}^u \prec c_{qm}^s$, where the unseen class c_k^u has mth attribute stronger than class c_p^s but weaker than class c_q^u.

Now given a novel image j to be classified into one of the seen or unseen classes, a generative model of all the seen classes in $\mathbb{R}^M$ is built. A seen class c_p^s is represented by a Gaussian distribution $c_p^s \sim \mathcal{N}(\mu_p^s, \Sigma_p^s)$ where mean is $\mu_p^s \in \mathbb{R}^M$ and Σ_p^s is $M \times M$ covariance matrix. The parameters of the generative model of the unseen classes U are described relative to the parameters of the seen classes, built according to the supervision given. For attribute a_m, if an unseen class c_k^u is described as $c_p^s \prec c_k^u \prec c_q^s$, the mth component of the mean of unseen class μ_{km}^u is set as $\frac{1}{2}\left(\mu_{pm}^s + \mu_{qm}^s\right)$. Similarly for the unseen classes described relative to just one seen class as $c_p^s \prec c_k^u$ or $c_k^u \prec c_q^s$, μ_{km}^u is described as $\mu_{pm}^s + d_m$ or $\mu_{qm}^s - d_m$ respectively, where d_m is the average of the distances between the sorted mean rank scores of seen classes for the mth attribute and the covariance Σ_k^u is $\frac{1}{S}\sum_{i=1}^{S}\sigma_i^s$.

Finally, maximum likelihood is computed and the test image j is assigned the label with the highest likelihood of a seen or an unseen class.

$$c^* = \underset{j\in\{1,..,N\}}{\arg\max}\ \mathcal{P}\left(\tilde{x}_i | \mu_j, \Sigma_j\right) \tag{1}$$

The description of the unseen classes as simply the mean of the related seen classes may not best represent the unseen class and hence a more accurate approach is proposed to represent the unseen class for recognition.

4 Approach

In this section, we first explain our approach to improve zero-shot recognition using Gaussian Processes by providing more accurate and systematic framework to describe the images of the unseen class. Second, we describe in Sect. 4.2, Gaussian-process based classifier for the seen classes and then, in Sect. 4.3, Gaussian Process (GP) based method that improves the accuracy of recognition for the unseen class using k-nearest training images. In Sect. 4.4, we explain a variant of our method that relies on multiple versions of distributions. This method is however subsumed in terms of performance by the GP-kNN algorithm.

4.1 Gaussian Processes for Zero-Shot Recognition

Gaussian Process is a distribution of random variables such that any finite number of distribution of these variables is jointly Gaussian. The observations in the process occur in a continuous domain. Any Gaussian process $f(x)$ can be specified as

$$f(x) \sim \mathcal{GP}\left(m(x), k(x^T, x)\right) \tag{2}$$

where the process's mean function and the covariance function are respectively:

$$m(x) = \mathbb{E}[f(x)], \quad k(x^T, x) = \mathbb{E}\left[\left(f(x) - m(x)\right)\left(f(x) - m(x)\right)\right]. \tag{3}$$

Let a regression model with Gaussian noise be given as

$$f(\mathbf{x}) = \mathbf{x}^T\mathbf{w}, \quad y = f(\mathbf{x}) + \mathcal{N}\left(0, \sigma_n^2\right) \tag{4}$$

where $\mathbf{x}$ is the input vector, $\mathbf{w}$ is the vector of weights (parameters) of the model and f is the function value. The outcome observed is represented by y, assuming that the additional noise term is an independent zero-mean Gaussian distribution. We assume a zero-mean Gaussian prior $w \sim \mathcal{N}\left(0, \Sigma_p\right)$. Given the model and the noise assumption, the *likelihood* and the *posterior*, given by combining the prior with the likelihood using the Bayes' rule, are respectively as follows.

$$\mathcal{P}(\mathbf{y}|X, \mathbf{w}) = \mathcal{N}(X^{\mathbf{w}}, \sigma_n^2 I) \tag{5}$$

$$\mathcal{P}(\mathbf{y}|X) = \int \mathcal{P}(\mathbf{y}|X, \mathbf{w})\mathcal{P}(\mathbf{w})d\mathbf{w} \tag{6}$$

Finally the predictive outcome f_* at x_* is given by

$$\mathcal{P}(f_*|\mathbf{x}_*, X, \mathbf{y}) = \mathcal{N}(\frac{1}{\sigma_n^2}\mathbf{x}_*^T A^{-1} X \mathbf{y}, \mathbf{x}_*^T A^{-1}\mathbf{x}_*) \tag{7}$$

Further details of the full Bayesian treatment for Gaussian process is presented by Rasmussen and Williams [2].

Our two-tier method uses Gaussian process (GP) based classifier in the first step and Gaussian process regression for a more accurate description of unseen class in the second step. In the first step, for each test image j, if the GP-based classifier outputs a prediction greater than a certain set threshold τ, the classifier corresponding to a seen-class c_p^s labels image j as 'class-p'. This takes care of those test images which are very similar to a seen class's training images, thus suggesting that the target unknown-class has higher probability to be one of the seen classes. The GP-based classifier for the seen classes is explained in Subsect. 4.2.

In the second step, for a test image j which is not labeled by any of the GP-classifiers of the first step, new Gaussian models representing the unseen classes are created by modeling more accurate description of the attribute value of the unseen class based on k sample images chosen from the training set which are nearest to the test image j. These new distributions are also taken into account, along with their initial Gaussian distribution, to represent the unseen classes. Based on the maximum likelihood computed for all the distributions the final label is assigned. The method is explained further in the following subsections.

4.2 Gaussian Process Based Classifier

During training, we are given a set of training images X belonging to S number of seen classes and a set of attributes, A_m. These training images are represented by $\mathbb{R}^N$ feature vector. Using the supervision given for the relative attributes between these seen classes, a ranking function is learnt which transforms the $\mathbb{R}^N$ image feature vector to $\mathbb{R}^M$ vector in attribute-space.

For all the training images j, Mahalanobis distance of the image from every seen class c_p^s is computed. This distance shows how many standard deviations away an image j is from a seen class. The distance comes out smaller for images similar to the seen class, according to the attributes, and larger for images that are dissimilar. By taking the average of these distances, Mahalanobis distance is calculated for each pair of seen classes.

For every seen class c_p^s, a Gaussian Process classifier is created, in the attribute space, with the training images from c_p^s and c_q^s as positive and negative samples, where class c_p^s and c_q^s are most distant from each other. The GPML tool box [13] is used for the computation.

These Gaussian process classifiers, each corresponding to a seen class, are used to find the posterior mean given the test image as the input. If the posterior mean of the prediction is greater than the set threshold τ, (experimentally set to 0.9), the test image j is labelled positive by the classifier. In case more than one classifier labels an image positive, the label by the classifier with a more positive mean is assigned.

4.3 Zero-Shot Recognition Using Gaussian Process - kNN Approach

In the previous approach, given a generative model for all the classes, each class is represented by a Gaussian distribution. The unseen classes are modeled using supervision given for all or a subset of M attributes (see Sect. 3). Every class-p, seen and unseen, has a set of parameters corresponding to the mean μ_p and the covariance Σ_p of the class. The label is assigned to the test image based on the highest likelihood value computed for each of the classes.

In our proposed approach to improve zero-shot recognition, for all the test images which are not labelled by any of the seen-classes' GP-based classifier, Gaussian process is used to improve the recognition in the following way as is shown in Fig. 1.

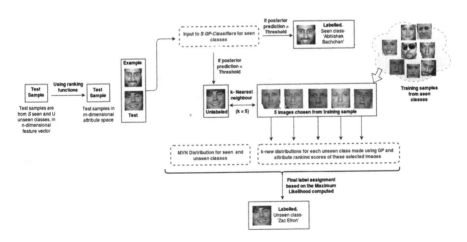

Fig. 1. *Basic outline of the proposed GP based method for Zero-shot recognition.* Test image in N-dimensional feature space is first transformed to M-dimensional attribute space using the ranking function learned for each attribute. These images are then given as an input to be labelled by GP-based classifiers for the seen classes, determined by a threshold for the predicted posterior. k-training samples from seen classes are then chosen according to their euclidean distance from the unlabeled test samples. Using Gaussian process, explained in Sect. 4.3, and the attribute rank scores of these chosen images to the GP, multivariate normal distributions (MVN) are modelled to represent the unseen class more accurately. The label corresponding to the distribution which gives the maximum likelihood, is assigned to the image.

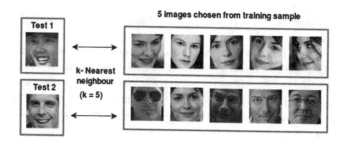

Fig. 2. k-nearest neighbours computed for two unlabelled test samples: Michelle Wie and Ben Stiller. From the training set of 5 seen classes, k-nearest neighbour (k = 5) based on the Euclidean distance from the test image is seen. The neighbors selected depends on attributes. Attributes like the shape of face and age is similar for the nearest neighbors in this example.

1. From the set of training images, k-nearest samples are chosen whose Euclidean distance is shortest from the test image j. These k images resemble the test image most closely, in the attribute space. (See example in the Fig. 2 for two test samples- Michelle Wie and Ben Stiller).
2. For every unseen class c_o^u, for an attribute a_m, if the supervision is given with respect to two seen classes c_p^s and c_q^s as $c_{pm}^s \prec c_{om}^u \prec c_{qm}^s$, then the mth component of the mean of the unseen class, μ_{om}^u is computed using Guassian process (GP) and the k nearest neighbours.

   The unseen class is represented by a set of k means and covariances, $(\mu_o^{iu}, \Sigma_o^{iu})$, $i \in \{1, ..k\}$. A GP is created with the rank scores of the mth attribute of the training images from seen classes c_p^s and c_q^s as positive and negative training samples respectively. Now, the mth component in each of the μ_o^{iu} is the posterior prediction mean output, with the mth attribute rank score of the ith-nearest training samples (chosen in Step 1) $i \in \{1, ..k\}$ as input to the above constructed GP.
3. For the attribute whose supervision is given with respect to just one seen class, as $c_{pm}^s \prec c_{om}^u$ or $c_{om}^u \prec c_{qm}^s$, the mth component of the mean of the unseen class is taken as $\mu_{pm}^s + d_m$ or $\mu_{qm}^s - d_m$ respectively. Here d_m is the average of the distances between the sorted mean rank scores of seen classes for the mth attribute.
4. To assign label to the test image, the likelihood score is computed by $\mathcal{P}(\tilde{x}_j | \mu_i, \Sigma_i)$, where μ_i and Σ_i is the mean and covariance of all the classes, including the k new sets of $(\mu_o^{iu}, \Sigma_o^{iu})$ constructed for the unseen classes in the previous step. The label is finally assigned based on the maximum likelihood value.

4.4 Tray of Multivariate Normal Distributions - A Variant of Our Proposed Method

We also experimented with a variant of our proposed GP-kNN method, and studied its performance in a subset of PubFig dataset. In this step, for all those test images which are not labelled by any of the GP-classifiers from the first step (Sect. 4.2), likelihood of the image belonging to each class is computed. If the likelihood of the test image to belong to a seen-class is highest, the label is assigned to it accordingly. However, if the likelihood of the test image to belong to one of the unseen classes is highest, instead of one set of mean μ_i and covariance Σ_i, multiple sets or a 'tray' of mean and covariances representing that class is dynamically created as we come across test samples. The image is labeled accordingly and a new distribution (μ_i', Σ_i), where the mth component of μ_i' is the posterior mean predicted with the test image's mth attribute score as input, is added to the tray. For subsequent test images, the likelihood for labeling, will be computed using all the earlier distributions representing the classes and those which are added to the tray.

In this approach rather than using GP regression, we had considered dynamic updation of the multi-variate normal distribution for the unseen classes. Keeping a dynamically increasing tray of multivariate normal distributions to compute the likelihood and assign label to the test image, accomodates the idea that a labeled test sample may improve the description of the unseen class, for the following test images, than the original Gaussian mixture model. However, improvement by this method is dependent on the order of test images which led to the development of more systematic algorithm (GP-kNN) for the unseen classes' description. Moreover, as shown in Sect. 5.4, this method does not perform as well as the GP kNN regression method.

5   Experiments

We evaluate our method for zero-shot recognition using GP-based classifier and k-nearest neighbors and compare our accuracy rate with the results obtained by GMM based clustering, as in the work of Parikh and Grauman [1]. We report results to demonstrate a more systematic and accurate description of the unseen class and validate the improvement achieved in recognition.

5.1   Setup

Our experiments used two datasets: a subset of **Public Figure Face Database** (PubFig) [3] and **Shoes with Attributes Dataset** [4]. The PubFig dataset consists of images of 60 different personalities, each image being represented by a 73-dimensional feature vector. Four sets of experiments were done on this dataset to validate our method where in each set, 8–10 classes of people are randomly chosen. The effect of changing the number of seen classes, the number of attributes to describe the classes and varying the supervision in terms of 10 different relative attributes is also demonstrated.

The experiment on Shoes with Attributes Dataset is done by taking 8 classes of shoes which are visibly distinct from each other, in terms of 10 relative attributes. The effect of varying supervision in terms of the number of classes seen is also presented. The images are represented as concatenation of the

960-dimensional gist descriptor with 30-dimensional color histogram image features. The feature vector was chosen to be same as the relative attributes work [1] to which it is being compared.

5.2 Zero-Shot Learning Results

Results of PubFig Dataset: Four sets of experiment are done on this dataset consisting of randomly chosen classes and 10 relative-attributes. Table 1 shows in detail the classes that were randomly chosen, the attributes taken into consideration and the partial ordering of the subset of relative attributes given as supervision for the unseen classes, in one of the experiments. (For example supervision '(8) : $J \prec S \prec H$' means that Scarlett Johansson has narrower eyes than Hugh Laurie and Jared Lato has narrower eyes than Scarlett Johansson). In Fig. 3 we show some examples where our proposed method does better than the GMM based. The green labels are correct labels assigned by our GP-based method and labels in red are the incorrect labels. In an example, for a test sample of class 'Miley Cyrus', both of the methods fail as the relative attribute supervision given is not sufficient to distinguish it from the class 'Alyssa Milano'.

By varying supervision in terms of attributes to relate classes, our method follows a general trend of increasing accuracy rate with increase in the number of seen classes. This is not only because with greater number of seen class the supervision is more elaborate but also because as the number of seen classes

Table 1. *Classes, relative attributes and supervision in one of the experiments with PubFig dataset.* Given four seen classes, and the unseen classes are described using relative attributes with respect to the seen classes. Note that supervision column marks the labels available for training.

Attributes	Classes	Supervision
Male (1)	Alex Rodriguez (A)	*seen-class*
White (2)	Clive Owen (C)	*seen-class*
Young (3)	Hugh Laurie (H)	*seen-class*
Smiling (4)	Jared Leto (J)	*seen-class*
Chubby (5)	Miley Cyrus (M)	(1): M <J (3): A <M (6): J <M <C (10): C <M <A
Visible Forehead (6)	Viggo Mortensen (V)	(3): V <A (4): V <C (5): V <J (10): H <V <C
Bushy Eyebrows (7)	Scarlett Johansson (S)	(1): S <J (3): S <M (8): J <S <H (9): C <S <H (10): A <S
Narrow Eyes (8)	Zac Efron (Z)	(1): Z <A (3): Z <J (5): H <Z <A (6): Z <C
Pointy Nose (9)		
Big Lips (10)		

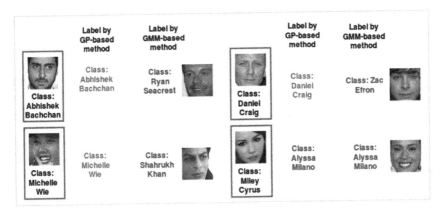

Fig. 3. The figure shows some examples of prediction using our GP-based method and GMM-based method. The color green shows the correct prediction and label in color red shows the incorrect prediction. (Color figure online)

increases, the number of test images that are labeled correctly in our first step by the GP-based classifiers also increases.

Secondly, our GP-based method, using the k-sample images nearest to the test image, provides a more accurate description of the unseen class as opposed to Gaussian mixture model of the classes where the unseen class is described as means of the seen classes. This can be clearly seen as our method outperforms the GMM based recognition. 120–150 test images uniformly belonging to each of the seen and unseen classes, are randomly taken for evaluation. The graphs below shows the accuracy curve obtained by GP-based method vs. GMM-based method.

Graph 1 (top-left) and Graph 2 (top-right) presents the performance curve of our proposed method vs the GMM based method. For 10 classes (seen and unseen), 10 attributes are used to relatively describe the classes for learning the ranking function and a subset of these attributes for unseen classes' supervision. The classes and the set of attributes vary for both the experiments. The classes are randomly selected and the attributes are such chosen that they are capable of representing these classes and vary well among the classes to make them distinct. To study the effect of supervision in terms of the proportion of seen classes, the number of seen classes are varied from 4 to 10, keeping the total number of classes same. It is seen that as we see more number of classes, the overall accuracy percentage increases for a test set of 150 images as the unseen classes can be related to more number of seen classes to make itself more distinguishable. The testset consists of randomly selected images, uniformly belonging to each of the classes.

Graph 3 (bottom-left) and Graph 4 (bottom-right) validates the performance of our method in the same way. Here, the 8 classes are randomly chosen and are represented by 10 relative attributes for both of the experiments. The proportion of seen classes are varied from 4 to 8 (all seen) and an increasing graph for accuracy in the recognition is observed. The test set consists of 120 randomly selected images, uniformly belonging to each of the seen and unseen classes.

Results of Shoes with Attributes Dataset: In the experiment to evaluate our method in shoes with attribute dataset, 8 distinct classes of the dataset with 6 attributes relating them were chosen. The relative attribute supervision is similar to that provided in the previous experiment. In Fig. 5 we show examples where our proposed method does better labeling than the GMM based method. The labels in green are correct labels for the test samples, assigned by our GP-based method and labels in red are the incorrect labels. For test sample of Rainboots, using the relative attributes chosen, it was difficult to distinguish 'rainboots' from 'boots'.

The performance result obtained in this dataset is very similar to the one obtained with the PubFig dataset. The classes in this dataset are chosen such that they can be humanly perceived as distinct from each other without confusion (*e.g.* keeping only 'Athletic shoes' and not -both Sneakers and Athletic shoes and keeping 'pumps' instead of both pumps and high-heels). The accuracy of our method increases as we increase the number of seen classes and outperforms the GMM-based method. In the graph of Fig. 6, the proportion of seen classes are varied keeping the total classes same.

5.3   Varying the Number of Attributes

Variation in the performance by varying the number of attributes to describe the seen and the unseen classes is seen. For a PubFig dataset consisting of 8 classes (5 seen and 3 unseen), the number of attributes used to describe these classes relatively, were varied. In the graph of Fig. 7, number of attributes to describe the classes are varied in the x-axis from 6 to 11. It is seen that greater the number of relative attributes learned to represent a class, the more descriptive it is of the class and hence the recognition rate increases. Our proposed GP-based method outperforms the GMM-based method for the recognition. The test set consisted of 120 images randomly chosen and uniformly belonging to all the classes (Fig. 4).

5.4   Comparing Performance of Various Methods for Zero-Shot Learning

Performance of proposed GP-based method is compared to GMM based method and MVN-tray method (See Sect. 4.4). The curve in Fig. 8, shows the accuracy achieved by different methods on 6 classes of PubFig dataset. The classes were chosen at random and 7 relative attributes were used to describe the classes. From left to right, while Gaussian Mixture Model (GMM) achieves an accuracy of 56.60%, a variant of our method of keeping a dynamically increasing tray of the mutivariate normal (MVN) distribution for each unseen class, as more test samples are seen, improves upon it. In this case, when more than one seen

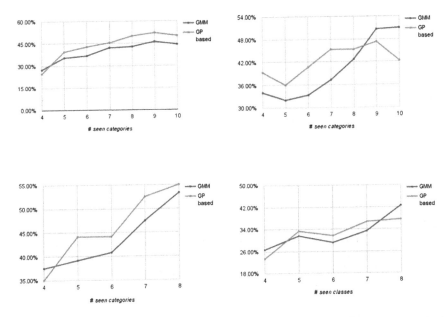

Fig. 4. *Performance curve for experiment with PubFig dataset.* The accuracy rate is presented for four different sets of experiments done on PubFig and changes in the accuracy for recognition as the proportion of seen classes is varied.

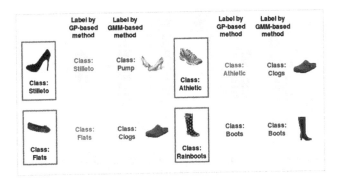

Fig. 5. The figure shows some examples of Prediction using our GP-based method and GMM-based method. The *green* color shows the correct prediction and label in color *red* shows the incorrect prediction. (Color figure online)

classes' classifier gives a positive output in the first step of our algorithm, the test image is not assigned any label.

Slight modification is done to this MVN-Tray method which improves the accuracy further. In case of a tie between two classifiers which outputs a positive prediction for the test image, label is assigned to the test image by the classifier with more positive prediction posterior as opposed to MVN-Tray where no label is assigned in such a case. This variant of MVN-Tray method is named as 'MVN-

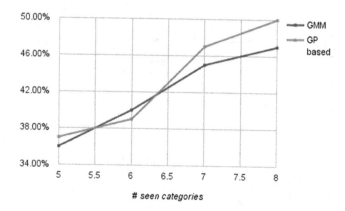

Fig. 6. Performance curve evaluated on Shoes with Attribute Dataset with 8 different categories of shoes represented by 6 relatively defined attributes. The accuracy of recognition increases as the number of seen classes increases from left to right. The accuracy is compared to GMM based method for recognition. The test set consisted of 100 images randomly chosen and belonging to all the classes.

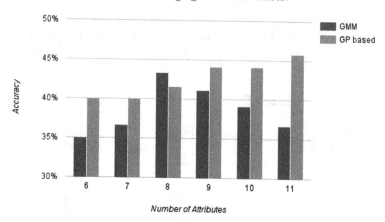

Fig. 7. The graph shows performance of our proposed method vs GMM-based method, as the number of attributes to describe the classes is varied. The setup is 8 randomly chosen classes from PubFig dataset with 5 seen and 3 unseen classes. The x-axis shows the number of attributes used to model a class.

Tray-Modified' in the figure. Finally, our proposed algorithm (GP-kNN) presents a more principled method using Gaussian process with k-nearest sample images, to improve the recognition of test images belonging to the seen classes, using GP-based classifiers, as well as the unseen classes by better description of the class using GP. The overall accuracy, using this method, increases to 63.33%. The test set for this experiment consisted of 90 randomly chosen images belonging to all the classes.

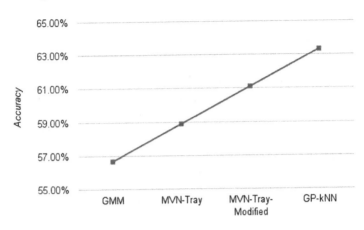

Fig. 8. *Accuracy curve for different approaches.* The curve depicts the accuracy of zero-shot recognition achieved by four different approaches. The accuracy of recognition increases as we go from left to right with GMM based method, MVN-Tray method, MVN-Tray-Modified for 'tie breaks' and our final proposed method using GP kNN.

6   Conclusion

In this paper we propose a two stage Gaussian process (GP) based zero-shot learning method using relative attributes. The method is extensively evaluated on two standard datasets. The results from the method show consistent improvement over the basic Gaussian mixture model based approach for zero-shot learning that was proposed earlier [1]. The method while being more accurate is also more descriptive. The GP based classifier allows us to estimate the uncertainty in a test sample to belong to one of the seen classes. The GP kNN based regression allows us to obtain reliable estimates of the attributes distribution for the unseen class in terms of the relative attribute representation. These allow us to obtain a better understanding of the mid-level representation obtained through relative attributes (Fig. 8).

In future we would like to undertake research to obtain structured attribute representations that are relative and are also structured with respect to the uncertainty or unreliability of the attribute. Further, it would be interesting to study the effect of the proposed method in the context of relative feedback.

References

1. Parikh, D., Grauman, K., Metaxas, D.N., Quan, L., Sanfeliu, A., Van Gool, L.J.: In: Proceedings of International Conference on Computer Vision (ICCV), pp. 503–510. IEEE Computer Society (2011)
2. Rasmussen, C.E., Williams, C.K.I.: Gaussian Processes for Machine Learning. The MIT Press, Cambridge (2006)
3. Biswas, A., Parikh, D.: Simultaneous active learning of classifiers and attributes via relative feedback. In: IEEE Conference on Computer Vision and Pattern Recognition (CVPR) (2013)

4. Berg, T.L., Berg, A.C., Shih, J.: Automatic attribute discovery and characterization from noisy web data. In: Daniilidis, K., Maragos, P., Paragios, N. (eds.) ECCV 2010. LNCS, vol. 6311, pp. 663–676. Springer, Heidelberg (2010). doi:10.1007/978-3-642-15549-9_48

5. Lampert, C.H., Nickisch, H., Harmeling, S.: Learning to detect unseen object classes by between class attribute transfer. In: IEEE International Conference on Computer Vision and Pattern Recognition (CVPR) (2009)

6. Lampert, C.H., Nickisch, H., Harmeling, S.: Attribute-based classification for zero-shot visual object categorization. IEEE Trans. Pattern Anal. Mach. Intell. **36**(3), 453–465 (2014)

7. Akata, Z., Perronnin, F., Harchaoui, Z., Schmid, C.: Label-embedding for attribute-based classification. In: Conference on Computer Vision and Pattern Recognition, Portland, OR, USA, 23–28 June 2013, pp. 819–826 (2013)

8. Jayaraman, D., Grauman, K.: Zero-shot learning with unreliable attributes. In: NIPS (2014)

9. Yu, A., Grauman, K.: Just noticeable differences in visual attributes. In: International Conference on Computer Vision (ICCV), December 2015

10. Elhoseiny, M., Saleh, B., Elgammal, A.: Write a classifier: zero-shot learning using purely textual descriptions. In: IEEE International Conference on Computer Vision (ICCV), December 2013

11. Zhao, X., Kersting, K., Tresp, V.: Multi-relational learning with Gaussian processes. In: Proceedings of the 21st International Joint Conference on Artificial Intelligence (IJCAI), Pasadena, California, USA, 11–17 July 2009, pp. 1309–1314 (2009)

12. Rodner, E., Denzler, J.: One-shot learning of object categories using dependent Gaussian processes. In: Goesele, M., Roth, S., Kuijper, A., Schiele, B., Schindler, K. (eds.) DAGM 2010. LNCS, vol. 6376, pp. 232–241. Springer, Heidelberg (2010). doi:10.1007/978-3-642-15986-2_24

13. Rasmussen, C.E., Nickisch, H.: Gaussian processes for machine learning (GPML) toolbox. J. Mach. Learn. Res. **11**, 3011–3015 (2010)

14. Zhen-Yong, F., Xiang, T.A., Kodirov, E., Gong, S.: Zero-shot object recognition by semantic manifold distance. In: IEEE Conference on Computer Vision and Pattern Recognition (CVPR 2015), Boston, MA, USA, 7–12 June 2015, pp. 2635–2644 (2015)

15. Mensink, T.E.J., Gavves, E., Snoek, C.G.M.: COSTA: co-occurrence statistics for zero-shot classification. In: IEEE Conference on Computer Vision and Pattern Recognition (2014)

MARVEL: A Large-Scale Image Dataset for Maritime Vessels

Erhan Gundogdu$^{(\boxtimes)}$, Berkan Solmaz, Veysel Yücesoy, and Aykut Koç

Intelligent Data Analytics Research Program Department,
Aselsan Research Center, Ankara, Turkey
{egundogdu,bsolmaz,vyucesoy,aykutkoc}@aselsan.com.tr

Abstract. Fine-grained visual categorization has recently received great attention as the volumes of the labelled datasets for classification of specific objects, such as cars, bird species, and aircrafts, have been increasing. The collection of large datasets has helped vision based classification approaches and led to significant improvements in performances of the state-of-the-art methods. Visual classification of maritime vessels is another important task assisting naval security and surveillance applications. In this work, we introduce a large-scale image dataset for maritime vessels, consisting of 2 million user uploaded images and their attributes including vessel identity, type, photograph category and year of built, collected from a community website. We categorize the images into 109 vessel type classes and construct 26 superclasses by combining heavily populated classes with a semi-automatic clustering scheme. For the analysis of our dataset, extensive experiments have been performed, involving four potentially useful applications; vessel classification, verification, retrieval, and recognition. We report encouraging results for each application. The introduced dataset is publicly available.

1 Introduction

The coastal and marine surveillance systems are mainly based on sensors such as radar and sonar, which allow detecting targets as well as taking counter measure actions. Vision based systems containing electro-optic imaging sensors can be exploited for the development of more effective systems. Categorization of maritime vessels is of utmost importance to improve the capabilities of maritime security systems. For a given image of a ship, the goal is to automatically identify its category using computer vision and machine learning techniques. Vessel images may include clues regarding different attributes such as vessel type, photo

---

E. Gundogdu—Please contact the corresponding author to download the dataset and its metadata or visit https://github.com/avaapm/marveldataset2016.

E. Gundogdu and B. Solmaz—These authors contributed equally to this work.

Electronic supplementary material The online version of this chapter (doi:10. 1007/978-3-319-54193-8_11) contains supplementary material, which is available to authorized users.

© Springer International Publishing AG 2017
S.-H. Lai et al. (Eds.): ACCV 2016, Part V, LNCS 10115, pp. 165–180, 2017.
DOI: 10.1007/978-3-319-54193-8_11

category, gross tonnage and draught. A large-scale dataset will be beneficial for extracting such clues and learning models from images containing several types of vessels.

Presence of benchmark datasets [1], with large quantities of images and labels with meaningful attributes, resulted in a significant increase in the performance of visual object classification by the use of appropriate machine learning methods such as deep architectures [2]. Moreover, powerful deep representations are employed in fine-grained visual categorization tasks by either training on the datasets from scratch [3], fine-tuning deep networks trained on large-scale datasets [4] or exploiting the previously trained architectures with specific modifications [5].

To classify images with a fine-grained resolution, a considerable amount of training data is required for a respectable model generalization. Thus, fine-grained datasets were published for specific object categories. Some examples are aircrafts datasets [6,7], bird species dataset UCSD [8] consisting of 12K images, car make and model datasets; Standford cars dataset [9] containing 16K car images and CompCars [10] dataset of 130K images. The only work related to marine vessel recognition is [11], where they utilized *Shipspotting* website[1] and trained a modified version of AlexNet [2] for the classification of vessel types with 130K random examples. In our dataset 140K images are utilized for vessel classification with 26 superclasses constructed using a semi-supervised clustering approach. Furthermore, our vessel superclasses are balanced; we force the training set to have equal number of examples in each superclass, i.e. we augment the data on the vessel classes with less number of examples than a predefined amount per class. However, there is a significant imbalance of examples between the classes in [11], which may result in a bias in classification towards the classes with dominant number of examples and makes it difficult to deduce a conclusion about the mean per class accuracy. Hence, in our work, we report the mean per class accuracy as the vessel type classification performance. In addition, we accomplish further important tasks with 400K vessel images and obtain pleasing results which will be described in details in the following sections.

In order to utilize the-state-of-the-art fine-grained visual classification methods for maritime vessel categorization, we collected a dataset consisting of 2M images downloaded from the *Shipspotting* website (See footnote 1), where hobby photographers upload images of maritime vessels with various annotations including vessel types, photo category, gross tonnage, draught, built year, International Maritime Organization (IMO) number, which uniquely identifies individual ships. To the best of our knowledge, our collected dataset, MARitime VEsseLs (MARVEL), is the largest-scale dataset for the fine-grained visual categorization, recognition, retrieval and verification tasks.

In addition to introducing a large-scale dataset of maritime vessel images and their corresponding annotations, our other major contributions are targeting visual vessel analysis from four different aspects: (1) *vessel classification,* (2) *vessel verification,* (3) *vessel retrieval,* and (4) *vessel recognition* which will

---

[1] www.shipspotting.com.

be discussed in Sect. 2.1. To verify the practicality of MARVEL and encourage researchers, we present baseline results for these tasks. We provide the relevant splits of the dataset for each application to form a comparison basis. Thus we hope that our structured dataset will be a benchmark for various visual processing tasks on maritime vessels. The researchers may also develop several other applications with the help of this dataset in addition to these four representative applications.

Our paper is organized as follows: Sect. 2 provides a description of the properties of our dataset. In Sect. 3, superclass generation from the *vessel types* is presented, and the superclass classification results of two state-of-the-art approaches are reported. Section 4 includes three maritime applications, *vessel verification, retrieval and recognition* in details, and experimental results are demonstrated. Finally, Sect. 5 concludes the paper with helpful remarks.

2   Dataset Properties

Our dataset consists of 2 million marine vessel images, collected from *Shipspotting* website (See footnote 1). For most of the images in our dataset, the following attributes are available: *Beam, build year, draught, flag, gross tonnage, IMO number, name, length, photo category, summer dwt, MMSI, vessel type. Beam* is the width of a ship at the widest cross section measured in the ship's waterline. *Draught* is the vertical distance between the bottom of the hull and the waterline. *Gross tonnage* is a unitless index calculated using the internal volume of the ship. *Summer dwt* is a measure of the carrying capacity of the ship. *MMSI* is an abbreviation of Maritime Mobile Service Identity, which is a series of nine digits to uniquely identify ship stations.

Besides the above attributes, we figure out that the most useful and visually meaningful categories are three fold: (1) *vessel type* (2) *photo category* and (3) *IMO number. Vessel type* is assigned based on the type of the cargo the vessel will be transporting. For instance, if the vessel carries passengers, its type is very likely to be a *Passengers Ship*. The dataset contains 1,607,190 images with annotated vessel types belonging to one of 197 categories. *Vessel type* histogram, highlighting the major categories, is depicted in Fig. 1(c). The second most important attribute is *photo category*, which is another vessel description. Examples of the *photo categories* with a significant amount are *chemical and products tankers, containerships built 2001–2010* and *Tugs* (please see Fig. 1(a)). All collected images have been assigned a *photo category* out of 185 categories in our dataset. The third category is *IMO number*, which is an abbreviation for International Maritime Organization number. Similar to the chassis numbers of cars, IMO numbers uniquely identify the registered ships to IMO regardless of any changes in the name, flag or owner of the ship. 1,628,056 of the collected images are annotated with IMO numbers (please refer to Fig. 1(b)). Moreover, there are 103,701 unique IMO numbers in our dataset.

2.1   Potential Computer Vision Tasks with MARVEL Dataset

Huge quantity of images existing in MARVEL makes it amenable to directly employ recent methods utilizing deep architectures such as AlexNet [2] for vessel categorization with the provided annotations in our dataset. One may choose a vessel attribute as *vessel type* or *photo category*, and apply classification methods to categorize the images according to the selected vessel attribute.

MARVEL has more than 8,000 unique vessels (i.e. a unique IMO number) with more than 50 examples as shown in Fig. 1(b). This makes it possible to use the dataset for maritime vessel verification and recognition, which could be an important part of a maritime security system, similar to scenarios for license plate recognition with a traffic security system.

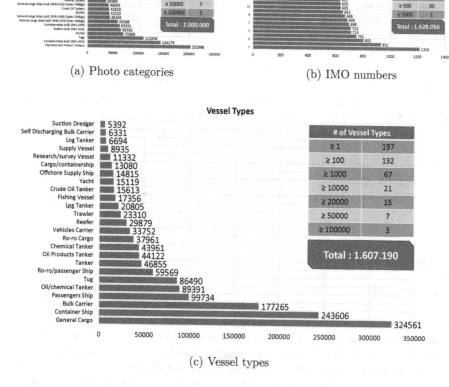

(a) Photo categories

(b) IMO numbers

(c) Vessel types

Fig. 1. Histograms of different categories.

The main foci of this study on MARVEL are four fold: (1) *vessel classification* since the content of the cargo that a ship carries, specifying its vessel type, is crucial for maritime surveillance, (2) *vessel verification* where the ultimate goal is whether the two vessel images belong to the same vessel with a unique IMO number, (3) *vessel retrieval* where a user might want to query a vessel image and retrieve an acceptable number of similar images, and finally (4) *vessel recognition* which is a challenging but interesting task which aims at recognizing a specific vessel within vessels of same type (This might be likened to a face recognition task.).

For **vessel classification**, we first generate a set of superclasses which may contain more than one *vessel type*, since some subsets of *vessel types* are not distinguishable even with a human supervision since the difference within the subsets arises from the invisible content of the cargo rather than the appearance of the ship. A concrete example of such a case is *vessel type* pair of *crude oil tanker* and *oil products tanker*, which is illustrated in Fig. 2. Although they have obvious functional differences, the visual discriminations are subtle especially when the images are far from the camera resulting in a small coverage of the image, and the deck of the ship is not visible from the camera view point. Therefore, we merge some of the vessel types to generate superclasses which are visually meaningful and discriminable. In the following section, we describe how specific *vessel types* are merged.

Vessel verification serves for deciding whether a pair of vessel images belong to the same vessel or not. This may be useful for a maritime surveillance application, where a specific vessel is required to be tracked using an electro-optic imaging system.

The task of **vessel retrieval** is similar to *vessel classification*, yet the user might want to retrieve more images than a single one to obtain a bunch of similar vessels from a database.

Fig. 2. Visual comparison of two very similar classes; *oil products tanker* and *crude oil tanker*.

Vessel recognition aims at finding the accurate identity of a vessel that a test image belongs to within a group of vessels of same category such as *vessel type* or *photo category.*

It is also notable that the attributes which exist for most of the images in MARVEL (e.g. gross tonnage, length, etc.) can be utilized to increase the recognition performance. Attribute-based computer vision tasks including object recognition [12], detection [13] and identification [14] have proven to increase the performance of the corresponding tasks. Moreover, we can learn hierarchical object categories within the fine-grained object recognition problem such as in [15] since MARVEL is constructed by merging relevant vessel types, and has a multi-level relevance information. Thus, we aim to exploit our fine-labeled dataset to further increase the performance of the particular tasks in the future.

3   Superclasses for Vessel Types

To generate superclasses from *vessel types*, first 50 *vessel types* containing largest amount of examples are selected and sorted according to their quantity. The *vessel type* with the largest amount of examples which is employed in our superclass generation, is *general cargo* class with 324,561 examples. The class with the smallest amount of examples is the *timber carrier* class with 1,837 examples. To analyze the visual similarities of the *vessel types*, a pretrained convolutional neural networks (CNN) architecture of VGG-F [16] is adopted to extract features using MatConvNet Toolbox [17] by resizing the vessel images to 224 × 224, the appropriate size of the network. The next to the last layer of VGG-F [16] activations are utilized as the visual representations of the images. Hence, each image is represented by a 4096-dimensional feature vector. By utilizing these feature vectors, we calculate a dissimilarity matrix for the selected 50 vessel classes. To generate superclasses, 1/10 of the collected 50 classes are used (approximately 130,000 images) and this data is used for estimating individual class statistics. Prior to calculating the dissimilarity matix, we first remove the outliers following the preprocessing step below.

Outlier Removal: Although annotations of the images in most of the categories are reliable and correctly labelled, indoor images of the vessels are also present in the dataset. Due to this reason and some other visual anomalies, we prune the outliers from individual *vessel types* to prevent their use while calculating the dissimilarity matrix. For this purpose, feature vector dimensions are reduced to 10 by principal component analysis (PCA) using all examples of the 50 classes, since Kullback-Leibler divergence is utilized in the dissimilarity calculation between two classes and the determinant of a very high dimensional matrix becomes unbounded. After the dimensionality reduction, each class is processed independently where a Gaussian distribution is assumed. Mean and covariance of each class are estimated. The feature vectors of the corresponding classes are whitened to obtain unit variance within each class. Since our aim is to prune the unlikely examples of the dataset to obtain a more clear dissimilarity matrix, the examples which are unlikely should be identified. Hence, we

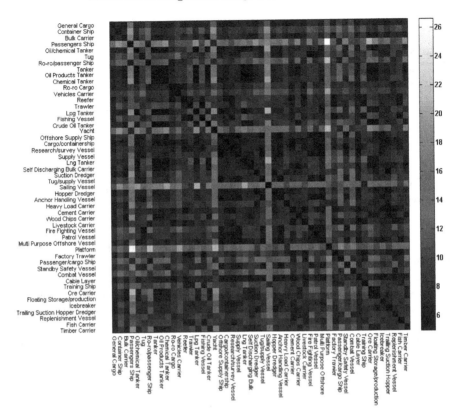

Fig. 3. Dissimilarity matrix for 50 classes. Lower values indicate more similarity.

utilize χ^2 distribution since the dataset is already whitened. For each example in individual classes, the sum of the square values of the 10-dimensional feature vectors are used as samples drawn from the χ^2 distribution with 10-degrees of freedom. Cumulative distribution function (cdf) value for each sample is calculated and removed from the class set if the cdf value is greater than 0.95, which corresponds to the samples drawn from the 5% tail of the χ^2 distribution.

Dissimilarity Matrix and Superclass Generation: Once the outliers are removed from each class by the above procedure, the remaining examples are used to compute the dissimilarity matrix. We use the symmetrised divergence as the dissimilarity index. Symmetrised divergence $D_S(P,Q)$ of two classes, namely P and Q, is defined as $D_S(P,Q) = \frac{1}{2}D_{KL}(P||Q) + \frac{1}{2}D_{KL}(Q||P)$, where $D_{KL}(.||.)$ stands for Kullback-Liebler divergence of two multivariate Gaussian distributions. The dissimilarity matrix is depicted in Fig. 3.

By exploiting the computed dissimilarity matrix, we merge the similar classes using a threshold. Prior to this thresholding, we apply spectral clustering methods with the help of the dissimilarity matrix. Nevertheless, the resulting groups were not semantically meaningful. Hence, we opt to continue by increasing the threshold for the similarities of the pairs of classes (i.e. this corresponds to each

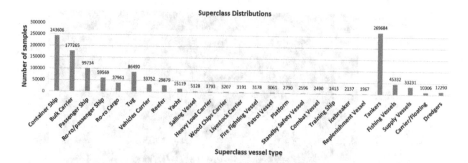

Fig. 4. Distribution of the *vessel types*. In total, 1,190,169 images are available for vessel type superclass classification.

entry of the dissimilarity matrix). If the dissimilarity index of a pair of classes is below a threshold, the pair is assigned to the same superclass. We increase the threshold until a point where semantically irrelevant classes (human supervision is adopted here) start to merge, and we define it as the final threshold for clustering. The majority of the resulting superclasses contain reasonable classes. The superclasses with more than one *vessel type* are: (1) *tankers* (which contains *oil products tanker, oil/chemical tanker, tanker, chemical tanker, crude oil tanker, lpg tanker, lng tanker, ore carrier*), (2) *carrier/floating* (which contains *timber carrier, floating storage production, self discharging bulk carrier*), (3) *supply vessels* (which contain *offshore supply ship, supply vessel, tug/supply vessel, anchor handling vessel, multi purpose offshore vessel*) (4) *fishing vessels* (which contains *trawler, fishing vessel, factory trawler, fish carrier*), (5) *dredgers* (which contains *suction dredger, hopper dredger*). Finally, hand-crafted marginal adjustments are done to make all superclasses as meaningful as possible. These adjustments include merging the superclass containing only *trailing suction hopper dredger* with the superclass consisting of *Suction Dredger* and *Hopper Dredger*. In addition, seven *vessel types* are removed entirely from the set of superclasses. The classes to be eliminated are decided according to the average dissimilarity of the classes to the rest. The salient overall dissimilarity scores are detected manually. The removed classes are namely; (1) *general cargo* (it is significantly confusing with the *container ship* and *ro-ro cargo*), (2) *cargo/containership*, (3) *research/survey vessel*, (4) *cement carrier*, (5) *multi purpose offshore vessel*, (6) *passenger/cargo ship* and (7) *cable layer*. The removed classes both visually and functionally contain more than at least two separate classes, i.e. *passenger/cargo ship* involve both passenger vessels and general cargo vessels. The merged classes with thresholding also contain visually very meaningful *vessel types*, i.e. all of the fish related vessels are clustered within the same superclass. The distribution of final 26 superclasses can be viewed in Fig. 4.

3.1   Superclass Classification

As demonstrated in Fig. 4, there exists an imbalance between superclasses. Nevertheless, even the superclass with the least amount of examples has a significant quantity of examples. Therefore; to classify superclasses of the *vessel types*, we train a deep CNN architecture AlexNet [2] implementation of the MatConvNet Toolbox [17] by using the default and recommended parameters without batch normalization. To avoid the imbalance between the superclasses, we select equal numbers of samples per class for both training and testing as 8192 and 1024, respectively. For the superclasses with examples less than the required quantity, we generate more examples by data augmentation (using different croppings of images). Hence, our training and test sets contain 212,992 and 26,624 examples, respectively, though we have 140K unique examples. We should also note that, no images of the same vessel are employed in both training and test sets. The classification performance is measured by the help of the normalized confusion matrix [7]. The practical performance metric for a fine-grained classification task can be the class-normalized average classification accuracy, which is obtained as the average of the diagonal elements of the normalized confusion matrix, C, and

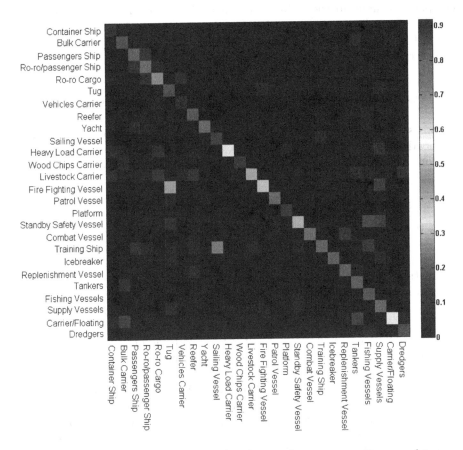

Fig. 5. Normalized confusion matrix of the 26 superclasses representing *vessel types*.

each of its entry is calculated as follows [6]:

$$C_{pq} = \frac{|\{i : \hat{y}_i = q \wedge y_i = p\}|}{|\{i : y_i = p\}|}, \tag{1}$$

where $|.|$ denotes the cardinality of the set and $\hat{y}_i$ indicates the estimated class label and y_i is the actual correct label for the ith training example. The final performance measure is the mean of the diagonal elements of the matrix C. This value for 26 superclasses is 73.14% for the normalized confusion matrix in Fig. 5. To emphasize the validity and efficacy of the learned network, we also compare it with another method utilizing multi-class Support Vector Machine (SVM) with the Crammer and Singer multi-class SVM [18] implementation of [19] in LIBLINEAR [20] library. The feature vectors for SVM are extracted from the VGG-F network of [16], their dimensionality is reduced to 256, and PCA whitening is applied. Since the memory requirements and computational complexity complicate the optimization, we use half of the training set. We report the class-normalized average classification accuracy as 53.89%. Compared to the use of the prelearned VGG-F weights with an SVM classifier, AlexNet trained from scratch has 35% improvement in accuracy.

4   Vessel Verification, Retrieval and Recognition

In this section, we make use of our dataset, MARVEL, for potential maritime applications; vessel verification, retrieval, and recognition. In the following subsections, these applications and necessary experimental settings are explained.

4.1   Vessel Verification

Akin to face verification [21], car model verification is applied in CompCars dataset [10] to serve for conceivable purposes in transportation systems. That kind of task is claimed to be more difficult compared to face verification, since the unconstrained viewpoints make car model verification more challenging. Accordingly, we perform maritime *vessel verification* where the attribute to be verified is the vessel identity. *Please note that our task is more challenging compared to identifying other attributes such as photo category or vessel type.* Furthermore, this problem is more challenging than both car model and face verification tasks, since it is desired to identify/verify pairs of individual vessels by looking only at their appearances.

We follow training and testing strategies as in [10]. First, 8000 vessels with unique IMO numbers are selected such that each vessel will have 50 examples, resulting in a total of 400K examples. This data is divided into two splits as training and testing. The training set consists of 4035 vessels (201,750 training examples in total), and the test set contains 3965 vessels (198,250 test examples in total). There exist 109 vessel types among 400K examples, and the training and test sets are split such that the number of vessel types are identical in each set. In the rest of the paper, we will call the training split of this subset as IMO training set, and the test split as IMO test set.

Prior to verification task, we learn a deep CNN representation from IMO training set by making use of *vessel type* labels. We train the same architecture of [2] as in *vessel classification* task except for the last layer since we have 109 classes rather than 26. Deep representations for each example are extracted as the penultimate layer activations of the trained network (as in the superclass generation part in Sect. 3) with 4096 dimensions. Since more discriminative features are preferable, we extract the penultimate layer activations prior to the rectified linear unit (ReLU) layer, which carry more information than the layer after ReLU since the negative values are cast to zero after ReLU. This choice makes our vessel verification performance better than the case with the deep representations after ReLU case.

After acquiring the deep representations, 50K positive pairs (belonging to same vessels) and 50K negative pairs (belonging to different vessels) are selected randomly from both the training and test splits out of the 201,750 training examples and 198,250 test examples, respectively[2]. For the total 400K training and testing examples, feature vector dimensionality is reduced to 100 by PCA exploited with only the training examples. Moreover, all 100-dimensional examples are PCA whitened since whitening increases the performance of SVM classifier. Concatenation of two 100-dimensional vectors are utilized for describing pairs during the verification experiments. Finally, we train SVM with RBF kernel on the training set by using LIBSVM library [22]. The precision recall curve varying the classification threshold is plotted in Fig. 6. We also compare the performance of SVM with nearest neighbour (NN) classification. The resulting precision and recall values of SVM and NN classifier are presented in Table 1. Accuracies of both classifiers are above 85%, which is very promising and quite satisfactory for a real world verification application.

Table 1. Vessel verification results on 50K positive pairs and 50K negative pairs of vessels for nearest neighbour and SVM classifiers by utilizing the feature vectors learned in IMO training set, which does not contain any images of the vessels in IMO test set.

	TP	TN	FP	FN	ACC	Precision	Recall
NN	44,978	40,198	9,802	5,022	85.18%	82.11%	89.96%
SVM	45,503	45,422	4,578	4,497	90.93%	90.86%	91.01%

4.2 Vessel Retrieval

Compelling amount of research efforts [23–25] have been spent for content based image retrieval (CBIR) as the image databases have been dramatically growing. Particularly, vessel retrieval is another promising application that may be

---

[2] A negative pair indicates a pair of different vessel images, whereas a positive pair corresponds to a pair of vessel images belonging to a unique vessel.

potentially required in a maritime security system, where the user would like to query the system with a test vessel and retrieve similar results. It could also be useful for annotation of vessels that are uploaded to a database with no meta-data. Hence, the system should be responsible for retrieving the similar vessels sharing the same content from a database. In our application, this content is not chosen as either the superclasses of *vessel types* that we constructed as the coarse attribute in Sect. 3.1, or the IMO number (aiming to identify the exact vessel), which is so fine for a retrieval task and appropriate for *vessel recognition* (This is studied as the recognition problem and is explained in the next subsection.). Instead, we use 109 *vessel types* of the 8000 unique vessels with 50 different examples, as the content of the retrieval task. Euclidean and χ^2 distance of two different representations are compared for the content based vessel retrieval.

The first representation is the 109-dimensional classifier output of the network which is trained in the verification task (Sect. 4.1) on IMO training set. On the other hand, we also would like to compare these learned deep representations (employing the content information) with another recent and effective representation. Hence, we use the prelearned VGG-F weights to extract the 4096-dimensional features (The dimension is also reduced to 20 similar to the *vessel verification* task). We train a multi-class SVM to obtain the classifier for the 109 vessel types by again using the IMO training set. For each example, classifier responses of dual combinations of 109 classes (generated during the multi-class SVM phase) are utilized as $\binom{109}{2}$ dimensional feature vectors. By utilizing these two representations, the results are retrieved with both Euclidean as well as χ^2 distance threshold. Mean average precision curves for both methods are shown in Fig. 7. Here, the deep representation learned specifically on the maritime vessels dataset significantly outperforms the generic deep representation learned for general object classification with 1000 classes [2,16] for both of the distance types. In addition, χ^2 distance has a significant superiority over the Euclidean distance for VGG-F features. For AlexNet features trained on our dataset, both of the distance types perform comparably well.

4.3   Vessel Recognition

The recognition problem is one of the crucial topics of computer vision. Especially, face recognition has been studied extensively, and state-of-the-art methods [26,27], which perform effectively on the benchmark datasets [28–30], have been proposed. Since encouraging performance results are obtained with the recent methods, the final application that we perform utilizing MARVEL is *vessel recognition* task, where ultimate goal is to find a vessel's identity from its visual appearance. It might be meaningless for object types other than vessels or faces such as cars since same car models of same color have no discriminative appearances and are not distinguishable. Nevertheless, individual vessels generally carry distinctive features, as the shapes of the vessels from the same *vessel types* significantly vary due to their customized construction processes.

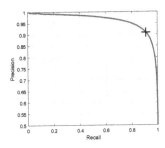

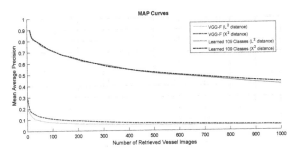

Fig. 6. Precision-recall curve of vessel verification for different classification thresholds of the trained SVM.

Fig. 7. Vessel type based retrieval curves for both the feature vectors of pre-trained VGG-F network and AlexNet network trained on our 109 vessel classes.

For representing vessels, 100-dimensional feature vectors are utilized as in *vessel verification* task. *Vessel identification* is performed among individual classes separately, e.g. vessels belonging to the *Passenger Ships* class are trained and tested within the *Passenger Ships* vessel type class, since there exist 3965 vessels with 109 different *vessel types* and it would be computationally infeasible to train all of these vessels with 3965 labels. Hence, we opt to perform vessel recognition among individual vessel types.

Among the 3965 vessels in IMO test set, there exist 29 *vessel types* that have at least 10 unique vessels, and each vessel has 50 examples. For recognition, we first divide the examples of each vessel into 5 folds where each fold has 10 examples per vessel. The training and testing set contains 4 folds (40 examples) and 1 fold (10 examples) per vessel, respectively. We make five-fold cross-validation for generating experimental results. For training, a multi-class SVM is employed where the number of classes is the number of unique vessels of the particular *vessel type*. In Table 2, the recognition performances are illustrated for each *vessel type*. Among the vessel types, *Supply Vessel* is the most distinguishable one

Table 2. Average recognition accuracies computed within each of the 29 vessel types.

Vessel Types	General Cargo	Containership	Oil/chemical Tanker	Bulk Carrier	Passengers Ship
Recognition Accuracy	34.2	27.88	47.8	39.34	42.5
# of unique vessels	965	851	295	196	196
Vessel Types	Ro-ro/passenger Ship	Tug	Ro-ro Cargo	Chemical Tanker	Vehicles Carrier
Recognition Accuracy	64.65	52.00	58.80	55.23	46.61
# of unique vessels	178	176	132	127	101
Vessel Types	Reefer	Oil Products Tanker	Tanker	Cargo/containership	Lpg Tanker
Recognition Accuracy	49.46	52.20	57.12	51.37	65.48
# of unique vessels	92	91	84	57	46
Vessel Types	Self Discharging Bulk Carrier	Crude Oil Tanker	Research/survey Vessel	Trawler	Offshore Supply Ship
Recognition Accuracy	49.13	45.24	85.47	73.68	80.11
# of unique vessels	23	21	19	19	19
Vessel Types	Yacht	Hopper Dredger	Suction Dredger	Sailing Vessel	Heavy Load Carrier
Recognition Accuracy	69.44	81.13	80.88	57.47	77.54
# of unique vessels	18	18	16	16	15
Vessel Types	Lng Tanker	Supply Vessel	Tug/supply Vessel	Fire Fighting Vessel	—
Recognition Accuracy	64.77	88.33	73.09	62.88	—
# of unique vessels	13	12	11	10	—

with a recognition accuracy of 88.33% for 12 different vessels whereas vessels of *Containership* have subtle differences and a recognition accuracy of 27.88% for 851 vessels. As the number of unique vessels increases within the dataset, the recognition performance decreases as expected. Yet, recognition accuracies over 50% can be obtained even though the number of unique vessels exceeds a hundred, such as in *Ro-ro Cargo* and *Chemical Tanker* vessel types.

5   Discussions

In this work, we introduce a large-scale dataset, MARVEL, for maritime vessels. With the help of this study, we aim to aid visual analysis tasks by adopting effective learning methods, providing a massive number of examples as well as the required labels to be used in corresponding tasks. Moreover, we merge the vessel types by making use of deep features and obtain semantically consistent superclasses. Upon this clustering, baseline classification results for the generated superclasses are reported to form a basis for further comparisons. We obtained promising results for vessel classification, 73% top-1 accuracy for 26 superclasses. We further utilize MARVEL for the vessel verification, retrieval, and recognition applications, and provide promising results. For vessel verification, we achieved an accuracy of 91%. Finally, we show that learning over 109 vessel types improves the performance over the representations learned for generic objects.

Acknowledgement. We would like to thank to Koray Akçay for his invaluable support and special consultancy for maritime vessels.

References

1. Russakovsky, O., Deng, J., Su, H., Krause, J., Satheesh, S., Ma, S., Huang, Z., Karpathy, A., Khosla, A., Bernstein, M., Berg, A.C., Fei-Fei, L.: ImageNet large scale visual recognition challenge. Int. J. Comput. Vis. (IJCV) **115**, 211–252 (2015)
2. Krizhevsky, A., Sutskever, I., Hinton, G.E.: Imagenet classification with deep convolutional neural networks. In: Advances in Neural Information Processing Systems (2012)
3. Lin, D., Shen, X., Lu, C., Jia, J.: Deep LAC: deep localization, alignment and classification for fine-grained recognition. In: 2015 IEEE Conference on Computer Vision and Pattern Recognition (CVPR), pp. 1666–1674 (2015)
4. Xie, S., Yang, T., Wang, X., Lin, Y.: Hyper-class augmented and regularized deep learning for fine-grained image classification. In: 2015 IEEE Conference on Computer Vision and Pattern Recognition (CVPR), pp. 2645–2654 (2015)
5. Liu, L., Shen, C., van den Hengel, A.: The treasure beneath convolutional layers: cross-convolutional-layer pooling for image classification. CoRR abs/1411.7466 (2014)
6. Maji, S., Rahtu, E., Kannala, J., Blaschko, M.B., Vedaldi, A.: Fine-grained visual classification of aircraft. CoRR abs/1306.5151 (2013)

7. Vedaldi, A., Mahendran, S., Tsogkas, S., Maji, S., Girshick, R., Kannala, J., Rahtu, E., Kokkinos, I., Blaschko, M.B., Weiss, D., Taskar, B., Simonyan, K., Saphra, N., Mohamed, S.: Understanding objects in detail with fine-grained attributes. In: 2014 IEEE Conference on Computer Vision and Pattern Recognition, pp. 3622–3629 (2014)

8. Wah, C., Branson, S., Welinder, P., Perona, P., Belongie, S.: The Caltech-UCSD Birds-200-2011 Dataset. Technical report CNS-TR-2011-001, California Institute of Technology (2011)

9. Krause, J., Stark, M., Deng, J., Fei-Fei, L.: 3d object representations for fine-grained categorization. In: 2013 IEEE International Conference on Computer Vision Workshops (ICCVW), pp. 554–561 (2013)

10. Yang, L., Luo, P., Loy, C.C., Tang, X.: A large-scale car dataset for fine-grained categorization and verification. In: 2015 IEEE Conference on Computer Vision and Pattern Recognition (CVPR), pp. 3973–3981 (2015)

11. Dao-Duc, C., Xiaohui, H., Morère, O.: Maritime vessel images classification using deep convolutional neural networks. In: Proceedings of the Sixth International Symposium on Information and Communication Technology, SoICT 2015, pp. 276–281. ACM, New York (2015)

12. Farhadi, A., Endres, I., Hoiem, D., Forsyth, D.: Describing objects by their attributes. In: Proceedings of the IEEE Computer Society Conference on Computer Vision and Pattern Recognition (CVPR) (2009)

13. Lampert, C.H., Nickisch, H., Harmeling, S.: Learning to detect unseen object classes by between-class attribute transfer. In: 2009 IEEE Conference on Computer Vision and Pattern Recognition, CVPR 2009, pp. 951–958 (2009)

14. Sun, Y., Bo, L., Fox, D.: Attribute based object identification. In: 2013 IEEE International Conference on Robotics and Automation, Karlsruhe, Germany, 6–10 May 2013, pp. 2096–2103 (2013)

15. Zhang, X., Zhou, F., Lin, Y., Zhang, S.: Embedding label structures for fine-grained feature representation. In: The IEEE Conference on Computer Vision and Pattern Recognition (CVPR) (2016)

16. Chatfield, K., Simonyan, K., Vedaldi, A., Zisserman, A.: Return of the devil in the details: delving deep into convolutional nets. In: British Machine Vision Conference (2014)

17. Vedaldi, A., Lenc, K.: Matconvnet - convolutional neural networks for MATLAB. CoRR abs/1412.4564 (2014)

18. Crammer, K., Singer, Y.: On the learnability and design of output codes for multiclass problems. Mach. Learn. **47**, 201–233 (2002)

19. Keerthi, S.S., Sundararajan, S., Chang, K.W., Hsieh, C.J., Lin, C.J.: A sequential dual method for large scale multi-class linear SVMs. In: Proceedings of the 14th ACM SIGKDD International Conference on Knowledge Discovery and Data Mining, KDD 2008, pp. 408–416. ACM, New York (2008)

20. Fan, R.E., Chang, K.W., Hsieh, C.J., Wang, X.R., Lin, C.J.: LIBLINEAR: a library for large linear classification. J. Mach. Learn. Res. **9**, 1871–1874 (2008)

21. Sun, Y., Wang, X., Tang, X.: Deep learning face representation from predicting 10,000 classes. In: Proceedings of the 2014 IEEE Conference on Computer Vision and Pattern Recognition (2014)

22. Chang, C.C., Lin, C.J.: LIBSVM: a library for support vector machines. ACM Trans. Intell. Syst. Technol. **27**(2), 1–27 (2011). Software available at http://www.csie.ntu.edu.tw/cjlin/libsvm

23. Guo, J.M., Prasetyo, H.: Content-based image retrieval using features extracted from halftoning-based block truncation coding. IEEE Trans. Image Process. **24**, 1010–1024 (2015)
24. Qiu, G.: Color image indexing using BTC. IEEE Trans. Image Process. **12**, 93–101 (2003)
25. Lai, C.C., Chen, Y.C.: A user-oriented image retrieval system based on interactive genetic algorithm. IEEE Trans. Instrum. Meas. **60**, 3318–3325 (2011)
26. Lai, J., Jiang, X.: Classwise sparse and collaborative patch representation for face recognition. IEEE Trans. Image Process. **25**, 3261–3272 (2016)
27. Gong, D., Li, Z., Tao, D., Liu, J., Li, X.: A maximum entropy feature descriptor for age invariant face recognition. In: 2015 IEEE Conference on Computer Vision and Pattern Recognition (CVPR), pp. 5289–5297 (2015)
28. Lee, K.C., Ho, J., Kriegman, D.J.: Acquiring linear subspaces for face recognition under variable lighting. IEEE Trans. Pattern Anal. Mach. Intell. **27**, 684–698 (2005)
29. Sim, T., Baker, S., Bsat, M.: The CMU pose, illumination, and expression database. IEEE Trans. Pattern Anal. Mach. Intell. **25**, 1615–1618 (2003)
30. Ricanek, K., Tesafaye, T.: Morph: a longitudinal image database of normal adult age-progression. In: 7th International Conference on Automatic Face and Gesture Recognition (FGR 2006), pp. 341–345 (2006)

'Part'ly First Among Equals: Semantic Part-Based Benchmarking for State-of-the-Art Object Recognition Systems

Ravi Kiran Sarvadevabhatla[1(✉)], Shanthakumar Venkatraman[2], and R. Venkatesh Babu[1]

[1] Video Analytics Lab, CDS, Indian Institute of Science, Bangalore 560012, India
ravikiran@grads.cds.iisc.ac.in, venky@cds.iisc.ac.in
[2] Indian Institute of Technology - Hyderabad, Hyderabad 502285, India
shanthakumar792@gmail.com

Abstract. An examination of object recognition challenge leaderboards (ILSVRC, PASCAL-VOC) reveals that the top-performing classifiers typically exhibit small differences amongst themselves in terms of error rate/mAP. To better differentiate the top performers, additional criteria are required. Moreover, the (test) images, on which the performance scores are based, predominantly contain fully visible objects. Therefore, 'harder' test images, mimicking the challenging conditions (e.g. occlusion) in which humans routinely recognize objects, need to be utilized for benchmarking. To address the concerns mentioned above, we make two contributions. *First*, we systematically vary the level of local object-part content, global detail and spatial context in images from PASCAL VOC 2010 to create a new benchmarking dataset dubbed PPSS-12. *Second*, we propose an object-part based benchmarking procedure which quantifies classifiers' robustness to a range of visibility and contextual settings. The benchmarking procedure relies on a semantic similarity measure that naturally addresses potential semantic granularity differences between the category labels in training and test datasets, thus eliminating manual mapping. We use our procedure on the PPSS-12 dataset to benchmark top-performing classifiers trained on the ILSVRC-2012 dataset. Our results show that the proposed benchmarking procedure enables additional differentiation among state-of-the-art object classifiers in terms of their ability to handle missing content and insufficient object detail. Given this capability for additional differentiation, our approach can potentially supplement existing benchmarking procedures used in object recognition challenge leaderboards.

1 Introduction

The performance of an object recognition system is typically measured in terms of error rate averaged over the object categories covered. In this respect, various deep-learning based classifiers have shown state-of-the-art performance on large-scale object recognition challenges in recent times. In fact, recognition challenge

© Springer International Publishing AG 2017
S.-H. Lai et al. (Eds.): ACCV 2016, Part V, LNCS 10115, pp. 181–197, 2017.
DOI: 10.1007/978-3-319-54193-8_12

leaderboards [1,2] typically list classifiers which show minuscule differences in the performance scores, particularly among the top-most performers. Moreover, the scores typically correspond to test images sourced from the same master image set used for training. Using such test images causes the well-documented phenomenon of dataset-bias [3–5] to creep into performance scores, thereby presenting a distorted picture of the classifiers' generalization ability. In the face of such observations, an important question arises: how else can these competing systems be differentiated?

The Holy Grail is, of course, human-like level of performance [6,7]. But, for a recognition system to claim it is within grasping distance of this Grail, the performance criteria can no longer be error rates on mostly fully-visible objects[1] present in biased test images. Instead, we need to design additional and alternative criteria. Also, if we wish to realistically benchmark state-of-the-art classifiers, we require test images which mimic the challenging conditions (e.g. local occlusion, insufficient global context) in which humans routinely recognize objects. To address these concerns, we make the following contributions:

- We systematically vary the level of object-part content, global visibility and spatial context in object images to create a PASCAL-based [8] benchmarking dataset named PPSS-12 (Sect. 4).
- We propose a novel semantic similarity measure called Contextual Dissimilarity Score (CDS). This measure has been designed to reflect a classifier's ability to predict the target category in a semantically meaningful manner across varying visibility and contextual settings (Sect. 5).
- We use our measure CDS and the PPSS-12 dataset to benchmark the top-performing object recognition classifiers trained on the ILSVRC-2012 dataset. The results (Sect. 6) show that our benchmarking procedure enables additional differentiation between the top-performers on the basis of their ability to handle missing content and incomplete object detail.

2   Overview of Our Approach

Figure 1 provides an overview of our approach. In the text that follows, circled numbers correspond to various data items and processing stages of our approach, as marked in Fig. 1.

We benchmark the top-performing [9] object classifiers trained on the 1000-class ILSVRC-2012 dataset – ALEXNET [10], VGG-19 [11], NIN [12], GOOGLENET [13]. For benchmarking purposes, we first create 'PASCAL Parts Simplified (PPS)-12' – a modified, 12-category image subset (②) of PASCAL-parts [14] which in turn is a database of object images with semantic-part annotations (Sect. 3).

For each image I (①) in PPS-12 containing a reference object, we systematically vary the object's level of global visibility and its spatial context in terms

---

[1] Anecdotally, this is the case in most object-recognition datasets.

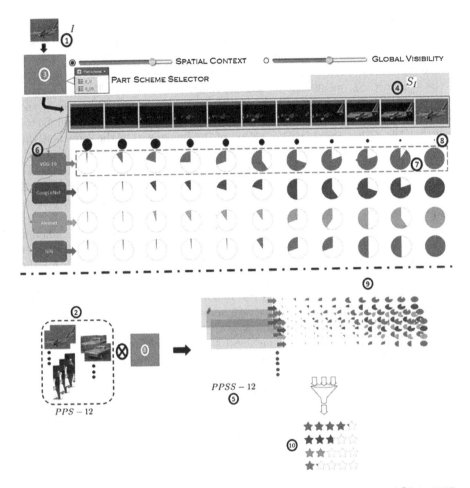

Fig. 1. A graphic overview of our approach (Sect. 2). For each image I (①) in PPS-12, we systematically vary the object's global visibility/spatial context in terms of its parts (③) to create sequence S_I of images (④). The main processing block is shown shaded in purple background above the black dash-dotted line. ⑤ refers to collection of such blocks, each of which contains the image sequences that form our benchmarking dataset $PPSS - 12$. The $\otimes$ in the lower half indicates that the various global visibility schemes/spatial context schemes are applied to the base dataset $PPS - 12$ to create the sequences which form $PPSS - 12$. For each sequence image, the degree of semantic similarity between its ground-truth label and that predicted by a classifier (⑥) is depicted as the proportion to which the corresponding circle underneath the image is filled (⑦). The similarity value of each sequence image is associated with a position-in-sequence based weighing factor depicted by the relative size of the black filled circle (⑧). The weighted similarities for the sequences in PPSS-12 are analyzed (⑨) to benchmark the classifiers (⑩). Best viewed in color.

of semantic object-parts (③) to create an associated sequence S_I of images (④) (Sect. 4). The collection of all such sequences comprise our benchmarking dataset 'PASCAL Parts Simplified Sequences (PPSS)-12' (⑤).

Having obtained the sequences, we first fix a classifier (⑥). For each image in the sequence S_I, we determine the normalized semantic similarity (Sect. 5) between the classifier-predicted label and ground-truth label[2]. We associate the similarity score of each image in the sequence with a normalized weight factor such that the earlier the relative location of the image within the sequence, the greater its weight[3]. We compute the weighted sum of similarity scores and normalize them by the sequence length to obtain a similarity-measuring score s. To obtain a measure similar in interpretation as error rate (i.e. lower the better), we subtract s from 1 to arrive at the final classifier-specific dissimilarity score which we term Contextual Dissimilarity Score CDS_I for image I (Sect. 5). By construction, early images of the sequence S_I contain relatively smaller amount of evidence for the reference object (see Figs. 2 and 3). Therefore, the larger the semantic similarity between classifier predictions and ground-truth for the initial images of sequence S_I, the greater the ability of the classifier to predict the target category in a semantically meaningful manner in challenging visibility and contextual settings and demonstrate human-like performance. This ability is numerically characterized by a low average CDS for the classifier. We gather statistics on CDS in the image sequences (⑨) on a per-image and per-classifier basis, across object categories. These statistics enable us to benchmark the classifiers as desired (⑩) (Sect. 6).

At this juncture, the reader might be inclined to question aspects of our semantic part-based benchmarking approach. We discuss the reasons and consequences of our choices in Sect. 7. For now, we move on to describe our preprocessing of the PASCAL-parts dataset.

3   Data Preprocessing

For the purpose of benchmarking, we start with a subset of PASCAL-parts [14], a 20-category object dataset containing named-part (semantic) annotations[4]. From these, we shortlist 12 categories (aeroplane, bicycle, bus, car, cat, cow, dog, horse, person, sheep, train) using the following criteria: (1) presence of at least two annotated parts (2) ease of annotating additional parts as required. We used object bounding box annotations from PASCAL-parts to obtain the cropped object images.

The part labeling scheme in PASCAL-parts contains labels on the basis of object orientation (left-facing, right-facing etc.) and intra-object location (leg-top, leg-bottom etc.). We simplified this scheme by ignoring such factors (i.e. orientation, location). In addition, certain crucial object parts have not been

---

[2] The degree of similarity is depicted as the proportion to which the circle underneath each sequence image is filled - see ⑦ in Fig. 1.

[3] The weighing factor is depicted by the relative size of the black filled circle - see ⑧.

[4] In this paper, we refer to named parts interchangeably with semantic parts.

annotated in PASCAL-parts dataset. Therefore, we modified the PASCAL-parts' labeling scheme by adding such parts to the annotation scheme. We used an in-house labeling tool to obtain annotations for these additional crucial parts.

In the end, we obtain an image set with 1850 object images across 12 object categories. We refer to this subset of PASCAL-parts with simplified/modified part annotations as PPS-12. Each object image in PPS-12 (See ① in Fig. 1) forms the basis for the construction of the image sequences in our benchmarking dataset PPSS-12[5]. Next, we describe how these image sequences are actually constructed.

4 Image Sequence Construction

For each object image I in PPS-12, we construct a sequence $\hat{S}_I$ of images (See ④ in Fig. 1). This sequence typically begins with only one semantic-part of the object in the image. The remaining object parts are successively added using a pre-defined 'part ordering scheme' (Sect. 4.1) to form the rest of the sequence. For each image in the sequence $\hat{S}_I$, a pre-defined 'content scheme' is applied to obtain the new sequence $\mathcal{S}_{\mathcal{I}}$. The content scheme controls either the amount of spatial context or level of object detail within the sequence (Sect. 4.2). Also, $\mathcal{S}_{\mathcal{I}}$ is constructed such that the final image in the sequence always coincides with the object image I which serves as the basis for the sequence construction in the first place. The collection of all such sequences constitutes our benchmarking dataset PPSS-12.

For the remainder of the section, we first describe the different object-part ordering schemes used to create image sequences. We subsequently describe the object content schemes (Sect. 4.2) activated during image sequence creation.

4.1 Part-Ordering Schemes

The part-ordering schemes essentially produce an ordered list of parts for each object category based on the scheme's criteria. During the image sequence creation (Sect. 4.2), this list forms the basis for incremental addition of object content.

In a recent work, Li et al. [15] augmented 850 images from PASCAL 2010 dataset with eye-fixation information to create their dataset PASCAL-S. In this dataset, each image is associated with a set of eye-fixation sequences. Each eye-fixation sequence, in turn, corresponds to spatial locations fixated upon by a human subject's eyes when shown the image. The PASCAL-parts derivative dataset we have created, PPS-12, is also derived from PASCAL 2010. We first identify images common to PASCAL-S and PPS-12. For these images, we analyze the density of fixation locations (from PASCAL-S) with respect to object part boundaries (from PPS-12). For each object category, we sort the parts in the decreasing order of fixation density to obtain the part-ordering scheme.

---

[5] Our dataset is available at http://val.serc.iisc.ernet.in/pbbm/.

In doing so, we implicitly make the assumption that fixation density is correlated with relative importance of image content, a phenomenon repeatedly observed in eye-fixation based image saliency studies [16]. We explored four variations in determining per-part fixation density resulting in four eye fixation-based part ordering schemes. We describe these schemes next.

Let $F^I(k) = \{f_j(k) = (x_j, y_j)\}, j = 1, 2, \ldots N_k$ denote the k-th eye fixation sequence (out of the N sequences from PASCAL-S) for an image $I \in$ PPS-12 containing an object from category $\mathbb{C}$. Here, (x_j, y_j) corresponds to the spatial location of the eye-fixation. Let $p_j(k), 1 \le j \le N_k$ denote the part within whose spatial boundary fixation $f_j(k)$ lies. N_k denotes the number of fixations in the k-th fixation sequence. Let $\mathcal{P}^\mathbb{C}$ denote the set of parts associated with category $\mathbb{C}$.

Unnormalized Sequence Position Scheme (E_{US}): Under this scheme, we assign part importance based on the total number of fixations within a part P's boundary. However, we also weigh each fixation by the relative position of the fixation within its original sequence. The part scheme factor for part P (of category $\mathbb{C}$) in image I is computed as:

$$E_{US}(P) = \frac{\sum\limits_{k=1}^{N} \sum\limits_{j=1}^{N_k} r_j(k) \mathbb{1}(P = p_j(k))}{\sum\limits_{Q \in \mathcal{P}^\mathbb{C}} \sum\limits_{k=1}^{N} \sum\limits_{j=1}^{N_k} r_j(k) \mathbb{1}(Q = p_j(k))} \tag{1}$$

$$\text{where } r_j(k) = \frac{N_k - j + 1}{N_k} \tag{2}$$

Here, the factor $r_j(k)$ captures the intuition that the earlier a fixation's location within its fixation sequence, the greater the prominence of the part within whose contour it falls. In the above equation and those that follow, $\mathbb{1}$ denotes the indicator function.

Unnormalized Part Count Scheme (E_U): This scheme is similar to E_{US} except that all fixations are considered equally important, i.e. $r_j(k) = 1$. Under this scheme, we simply count the number of fixations that lie within a part P's boundary as a measure of part importance.

Part Area Normalized Schemes $(E_A$ and $E_{AS})$: In the E_U scheme, a part can have a large importance score merely because it covers a larger portion of the image. Following Kiwon et al. [17], we normalize for part areas and construct part-area normalized versions of E_U and E_{US} as follows:

$$E_A(P) = \frac{E_U(P)}{A(P)} \tag{3}$$

$$\text{where } A(P) = \frac{\text{Area of part } P}{\text{Area of the object within } I} \tag{4}$$

$$E_{AS}(P) = \frac{E_{US}(P)}{A(P)} \tag{5}$$

In all the part schemes mentioned above, we sum the part-importance factor (Eqs. 1, 3, 5) for each part across all images of the category and sum-normalize to obtain the probability distribution of relative part importance. We obtain the final part scheme by listing the parts in the order of decreasing probability.

4.2 Content (Global Object Visibility and Object Context) Schemes

Having described the part schemes, we next describe how systematic variations in context and global object visibility are introduced into object images (See ③ in Fig. 1). We refer to the schemes which control image variations in these aspects (global object visibility, object context) as 'content schemes'. The content schemes essentially control the level of detail and the manner in which this detail is added as we progress across a part-image sequence. From a benchmarking point of view, these schemes are designed to evaluate the extent to which lack of content or the ability to exploit existing context affects a classifier's performance.

Object Context Scheme: Two variations exist within the object context scheme. In the first variation ('Intra-object context'), the images of the part-image sequence contain no contextual information other than that arising from incremental addition of object's parts. To ensure this, the object's parts are added to a completely black canvas. Figure 2 (top row) shows an example. In the second variation ('Intra-and-Neighborhood context'), the image content immediately surrounding the object is retained to provide neighborhood context. However, parts are incrementally added within a blacked-out bounding box enclosing the object. Figure 2 (bottom row) shows an example.

Fig. 2. Part sequences corresponding to change in 'object context' of a bus image. The top row corresponds to 'intra-object' context scheme while the bottom row corresponds to 'intra-and-neighborhood' context. Notice the road-like surroundings providing the context (surrounding the object) in the bottom row.

Global Object Visibility Scheme: Unlike the context-based schemes mentioned above, the visibility schemes additionally have access to global context from the entire image in a gist-like manner, including that from parts not yet added. Figure 3 shows example sequences for this content scheme. In this scheme, the entire image's data is present, albeit at a low level of detail to begin with.

Fig. 3. Part sequences corresponding to change in global object visibility of a `bicycle` image. We refer to the two variations of the visibility level as 'low level of detail' (top row) and 'higher level of detail' (bottom row).

As each part of the object is added, the part comes into focus. The net effect is a blurring of the image relative to already added parts. This scheme is inspired by the manner in which level of detail falls relative to the location fixated upon by a human eye. To achieve this fall-off effect, we utilize the visual-field simulation of Perry and Geisler [18]. By varying the parameters of the visual-field simulation, the level of detail in the immediate vicinity of each added part can be changed. For the purposes of our evaluation, we utilize two parameter settings which result in two variations of the global object visibility scheme which we refer to 'low level of detail' (top row of Fig. 3) and 'higher level of detail' (bottom row).

In the next section, we describe how the PPSS-12 sequences created by applying the content schemes to images in PPS-12 help determine the 'Contextual Dissimilarity Score' (CDS) for a fixed classifier. Later on (Sect. 6), we shall see how the process of determining CDS lets us benchmark the classifiers.

5   Determining Contextual Dissimilarity Score (CDS)

For a given image I from our PPS-12 dataset (①, Fig. 1), we first choose a part-ordering scheme (Sect. 4.1) and generate a sequence of images $\mathcal{S}_I$ according to this scheme. We then choose an object content scheme (Sect. 4.2) and apply it to each of the images in the sequence (③, Fig. 1). With the content-scheme applied sequence at hand (④, Fig. 1), we are ready to determine the Contextual Dissimilarity Score CDS_I for image I.

Let $\mathcal{S}_I = \{S_1, S_2, \ldots, S_N\}$ represent our aforementioned image sequence from PPSS-12. Note that by construction, sequence image S_N corresponds to given image I. Since our analysis is on a per-classifier basis, let us fix the classifier $\mathbb{C}$ (⑥, Fig. 1). Each image in sequence $S_j \in \mathcal{S}_I$ is input to the classifier to obtain the corresponding class prediction label c_j. Suppose the ground-truth label for S_j is g_j. In our case, c_j and g_j are drawn from two different label spaces (Imagenet-based and PASCAL-based) with varying levels of semantic granularity and therefore, an exact literal match may not be possible. Therefore, we utilize a semantic similarity measure $\mathcal{M}$ which provides a $[0,1]$-normalized score x_j reflecting the semantic similarity between c_j and g_j (i.e. $x_j = \mathcal{M}(c_j, g_j)$). Thus, we obtain a sequence of normalized scores $\mathcal{X}_I = \{x_1, x_2, \ldots, x_N\}$[6].

---

[6] The colored rows containing the pie-like shapes in Fig. 1 correspond to such similarity scores.

Also note that by construction, early images of the sequences contain relatively smaller amount of evidence for the reference object (see Figs. 2 and 3). Therefore, the higher the similarity score in the initial parts of the sequence, the greater the ability of the classifier to perform well in challenging conditions and demonstrate human-like performance. To characterize this notion, each similarity score x_j in the sequence in associated with a weight factor $w_j = \frac{N-j+1}{N}$ such that the earlier the location, the greater its weight ((8), Fig. 1)). We then compute $NWSS$ - the normalized weighted sum of similarity scores. To obtain a measure similar in interpretation as error rate (i.e. lower the better), we subtract $NWSS$ from 1 to arrive at the final Contextual Dissimilarity Score for image I (CDS_I).

$$CDS_I = 1 - \frac{\sum_{j=1}^{N} x_j w_j}{\sum_{j=1}^{N} w_j} \tag{6}$$

The resulting CDS_I is an indicator of the part-level and contextual content required by classifier $\mathbb{C}$ to recognize the object in image I. Therefore, obtaining a relatively smaller 'average CDS' when CDS_I are averaged across part-schemes, context-schemes and object categories indicates the ability of a classifier to perform well in spite of missing or poorly detailed object information.

6 Experimental Analysis

As a preliminary experiment, we computed the median top-1 error-rate of each classifier (ILSVRC-trained) on our PPS-12 (derived from PASCAL) images. Given the inherent dataset bias commonly present in recognition challenge datasets [3], the error-rates are higher unlike the low, barely distinguishable top-1 rates typically encountered on recognition challenge leaderboards (See Table 1). This result should not be surprising. Instead, it merely serves to reinforce the importance of cross-dataset validation in obtaining a fair assessment of classifiers' generalization capabilities [19].

Table 1. Cross-dataset error-rates: performance of the ILSVRC trained classifiers on our PASCAL-based PPS-12 dataset using manual mappings across the two datasets.

Classifier	GoogLeNet	VGG-19	NiN	AlexNet
Median error rate	0.24	0.25	0.32	0.35

As part of the main benchmarking process, we determine the CDS (Sect. 5) for all possible combinations of classifiers, part-ordering schemes (Sect. 4.1) and object content schemes (Sect. 4.2). This sets the stage for examining the effect of these schemes on the overall benchmarking process. For the similarity measure $\mathcal{M}$ between the category labels (Sect. 5), we used Wu-Palmer similarity measure [20]. This measure calculates relatedness of two words using a graph-distance based method applied to WordNet [21], a standard English lexical database containing groupings of cognitively similar concepts and their interrelationships.

6.1 Benchmarking Classifiers Across Object Content Schemes

In the discussion that follows, it is important to remember that smaller the CDS, better the classifier's performance.

'Intra-object' Context: For the first set of experiments, we analyze CDS for the 'intra-object' context scenario. This scenario consists of object images without any of the surrounding context except that arising out of the object's parts themselves (see top row of Fig. 2) and is perhaps the most challenging scenario for a classifier. On the other hand, it is also the most appropriate since the image content is precisely confined only to the object.

Fixing the content scheme to 'intra-object', for each classifier and for each category, we compute the median CDS. We do this initially for each part-ordering scheme and subsequently average the median scores over the schemes to obtain category-wise CDS. These category-wise scores are, in turn, averaged to obtain the CDS for each classifier. The results on a per-classifier basis can be seen in the first column ('Intra-object') of Table 2. As expected, the median scores are relatively high regardless of classifier.

'Intra-object and Neighborhood' Context: We repeat the previous experiment with the content-scheme now being 'intra-object and neighborhood'. In addition to object parts, contextual information from the immediate surroundings is additionally available in this scheme (see bottom row of Fig. 2).

Table 2. Benchmarking classifiers: Average median CDS across categories for different context schemes. The best CDS score for each content scheme is shown in bold. The bracketed percentages in column 2 indicate the improvement in CDS over column 1 with addition of context. The ones in column 4 indicate the improvement over column 3 when level of detail is increased. The best percentage improvement is also shown in bold. Note that smaller the CDS, better the performance.

Scheme	Context based		Global visibility based	
	Intra-object	Intra-object and neighborhood	Low level of detail	Higher level of detail
ALEXNET	0.4499	0.4470 (0.6 %)	0.4450	0.3803 (14.54 %)
GOOGLENET	0.5264	0.4319 (**17.95 %**)	0.4544	0.3490 (**30.20 %**)
NIN	0.4788	0.4492 (6.18 %)	0.4689	0.3882 (17.20 %)
VGG-19	**0.4136**	**0.4147** (−0.27 %)	**0.3628**	**0.2880** (20.62 %)

We hypothesized that such information would improve performance and that is indeed the case (see second column ('Intra-object and neighborhood') of Table 2).

Global Object Visibility: Next, we examine the impact of visibility-based content schemes (Sect. 4.2). As mentioned before, these schemes, unlike the intra-object and/or neighborhood context ones, have additional access to global context from the entire image in a gist-like manner, including that from parts not yet brought into focus (See Fig. 3). Therefore, the performance of the classifier for these schemes conveys the extent to which it utilizes the global context.

Keeping the classifier fixed and content-scheme as 'low-detail', for each category, we compute the median CDS for each part scheme and average them across part schemes to obtain category-wise CDS. These are averaged in turn to obtain the CDS for the classifier. As the results in Table 2 (third column) suggest, the presence of global information, even at a low level of detail and even with minimal object-specific information, is still powerful enough to improve performance, as evidenced by the lower CDS. Increase in the level of detail (i.e. lower level of blurring) causes the results to be on predictable lines, with the overall average median CDS trending downwards (See last column of Table 2).

6.2   Overall Performance and Additional Experiments

Examining the results in Table 2, it is evident that VGG-19 achieves the best performance (lowest average CDS) in general. More importantly, Table 2 also shows that our benchmarking procedure contrasts the performance of almost equally well-performing classifiers (GOOGLENET, VGG-19) better than the traditional accuracy-based counterparts—the CDS-based benchmarking values are generally further apart compared to the accuracy scores (compare Tables 1 and 2).

To determine which classifier exploits addition of object neighborhood-based context the most, we compute the percentage improvement in average CDS over the 'object only' (i.e. no neighborhood context) setting (Table 2, first column). As the bracketed numbers in second column of Table 2 show, GOOGLENET's performance improves the most. GOOGLENET also best exploits the increase in level of detail (fourth column of Table 2). We believe these results stem from the 'inception-style' mechanism GOOGLENET [11] uses to capture context.

To obtain a category-level perspective on the benchmarking performance, we determine the classifier that produces the lowest CDS most frequently across all combinations of part schemes and content schemes. The entries in Table 3 (top row) merely endorse the results seen earlier – VGG-19 is the best performer in general. At the other end, NIN and surprisingly (for a couple of categories), GOOGLENET have relatively higher CDS (bottom row of Table 3).

6.3   Relationship Between CDS and (Traditional) Error Measures

To verify that our CDS measure provides additional information beyond the traditional top-1 error measure, we computed the correlation between CDS and

Table 3. Category-wise best and worst performers (in terms of CDS) aggregated across part and content schemes.

Classifier	airplane	bicycle	bird	bus	car	cat	cow	dog	horse	person	sheep	train
Lowest-cds	GOOGLENET	VGG-19	VGG-19	VGG-19	VGG-19	VGG-19	VGG-19	VGG-19	VGG-19	VGG-19	ALEXNET	VGG-19
Highest-cds	NiN	NiN	NiN	NiN	NiN	NiN	NiN	GOOGLENET	NiN	GOOGLENET	GOOGLENET	GOOGLENET

the top-1 error rates across all the classifiers. For this, we determined the median error-rate and median CDS for each content scheme by averaging across the respective measures across part schemes and classifiers. Thus, we obtain two vectors, one for median error-rate and the other for median CDS. The correlation between these two vectors was found to be close to 0 (Pearson $\rho = 0.0227$, $p = 0.98$ and Spearman $\rho = 0, p = 1$). This result indicates that CDS measures an aspect of classifier performance distinct from the traditional top-1 measure.

7   Discussion and Related Work

Having presented the experiments and analysis, we now examine some of the design decisions and forces at play in our work.

Our benchmarking procedure relies crucially on semantic object part-based image sequences. Using 'named' semantic-parts ensures that all images of a category are treated *consistently*. This advantage is lost when we use purely statistically generated, unnamed, region-based part models[7] [22]. On a related note, Taylor and Likova [23] suggest that humans tap into generic concepts of objects, including linguistic propositions (e.g. named object-parts) while analyzing a scene. Furthermore, studies by Palmer [24] have shown that when parts correspond to a 'good' segmentation of a figure (e.g. object-part contours), the speed and accuracy of responses related to queries on figure attributes improves significantly. These observations further lend support for our use of semantic named object-parts. The burdensome aspect of semantic-part annotation does limit the number of categories benchmarked. However, recent trends seem to suggest the possibility of large, richly detailed datasets [25] and multi-task recognition frameworks [26] which can potentially offset this burden.

Our choice of part-importance order (Sect. 4.1) offers a future opportunity to explore connections between eye-fixation based saliency properties of a partial content image and its recognizability. We wish to point out that our part ordering schemes are not exhaustive – any other principled part-ordering scheme may also be utilized for additional hold-out style benchmarking. In this respect, it is interesting to note that Taylor and Likova [23] suggest a list specifying a Bayesian prior on possible object attributes (including semantic-parts) to characterize objects and related concepts.

---

[7] The lack of consistency also holds true for area-based approaches (e.g. systematically decreasing the percentage of object area occluded by a fixed percent).

The images from the sequences we used to compute the CDS are artificial in construction and one might argue that they are too structured and therefore, an imperfect representation of the object occlusion scenarios typically seen in real photos. An alternative could be to utilize realistic data wherein the extent and the manner in which the target object is occluded can be precisely quantified. This, in itself, is an extremely challenging task although newer datasets with depth ordering and occlusion level specified as part of annotations [27] may compensate to some extent[8]. The advantage of our constructs is that they let us quantify the global object visibility *consistently* – for a given location in the part importance order, the *same* part is missing in all the image sequences. Moreover, as the results indicate, state-of-the-art classifiers can still utilize available information effectively in spite of the artificial nature of sequence images.

Our benchmarking measure relies on a semantic measure of dissimilarity between the predicted label and ground-truth label. The deeper implication of our choice of similarity measure is that the median CDS for each classifier reflects the general ability of the classifier to utilize the semantics of the image to produce semantically meaningful predictions. We initially considered an alternative scheme: a more traditional 'hard' $0-1$ binary prediction in place of 'soft' semantic similarity. However, this approach requires a manual, subjectively grouped, many-to-one mapping between predicted-label set (Imagenet) and ground-truth label set (PASCAL).

On a deeper level, our overall approach reveals aspects of the object recognition task that each of the top-performers address better than the rest. As already pointed out, while VGG-19 is the top-performer in general, GOOGLENET is better (in percentage terms) at exploiting context from an object's immediate surroundings (Sect. 6.2). Therefore, while our benchmarking procedure is useful to differentiate classifiers, it can also be used to characterize the extent to which contextual and visibility factors are addressed by a classifier on a standalone basis. Such characterization can help classifier designers tweak their architectures and help improve the classifier's capabilities. In addition, as Table 2 shows, our benchmarking procedure contrasts the performance of almost equally well-performing classifiers (GOOGLENET, VGG-19) reasonably better than the traditional approach (Table 1). In addition, the moderately high CDS scores (Table 2) suggests that top of the line classifiers of current day are yet to perform well on images which mimic the challenging conditions (e.g. occlusion) in which humans routinely recognize objects. To confirm that humans recognize the objects much more robustly than machine classifiers, we performed a rudimentary user study in which we asked human subjects to recognize the PPSS-12 sequence images. We found that human CDS values were indeed disproportionately low compared to the classifiers[9]. Finally, we also wish to point out that

---

[8] The dataset was not publicly available at the time of our publication.

[9] In fact, the humans were able to correctly recognize the category at extremely early stages of the sequence – the highest median score across content schemes was $0.20(\pm 0.06)$ while the lowest was 0.

our benchmarking procedure is by no means complete - a gamut of additional transformations (e.g. rotation) and their combinations can be applied to create additional image sequences and benchmark them using the approach described in our work.

Typically, the state-of-the-art results reported on recognition leaderboards [1,2] and literature [6,28–30] correspond to ensemble models. However, the corresponding pre-trained models were not always available. To keep the benchmarking consistent, we utilized readily available, pre-trained, non-ensemble baseline models [9].

Related Work: One class of quantitative approaches which supplement the usual mAP/error-rate essentially use variations of the traditional measures or tend to be derived from them [31,32]. These additional measures (e.g. Area-Under-the-Curve(AUC), precision, recall) may provide additional differentiation between classifiers but unlike our work, do not provide insight into semantic aspects of data which affect classification. Somewhat similar to our approach, Aghazadeh and Carlsson [33] propose measures which quantify properties of training data (class bias, intra-class variation) and compare classifier performance on the basis of such measures for an object detection problem. However, their formulation involves comparison of features across image pairs whereas our measure is based on per-image statistics aggregated over a category. Hoiem et al. [34] characterize the effects of challenging extrinsic factors (e.g. occlusion, viewpoint) and intrinsic factors (e.g. aspect ratio, part visibility) for the object detection problem and suggest the factors most likely to impact performance. Analyzing user study data for downsampled versions of 256×256 images, Torralba [35] examine the effect of image resolution for scene and object recognition. However, their study is focused on human subject performance.

8   Conclusion

In this paper, we have demonstrated a semantic part-based procedure for benchmarking state-of-the-art classifiers. The benchmarking procedure relies on a semantic similarity measure that naturally addresses potential granularity differences between the category names in training and test datasets, thus eliminating laborious and subjective manual mapping. The measures we propose provide additional insights into the classifiers' ability to handle various degrees of object detail and missing object information *à la* humans. In our particular case, the benchmarking procedure enables performance evaluation of the ILSVRC-trained classifiers for test images sourced from an different dataset (PASCAL). Given this capability for additional differentiation, our benchmarking procedure can supplement existing procedures used in object recognition leaderboards. In addition, our benchmarking procedure and dataset are potentially useful for classifier designers on a standalone basis to analyze their classifier's ability to handle missing content and incomplete object detail.

The top performers in our benchmarking study do not explicitly consider object parts (semantic or otherwise) nor do they attempt to model occlusions.

However, architectures which are "part-aware" [36] and explicitly contain compensatory mechanisms for occlusion [37] hold great potential not only for our benchmarking procedure, but for the broader area of object recognition as well [7].

Please visit our project webpage http://val.serc.iisc.ernet.in/pbbm/ for additional and up-to-date information.

Acknowledgement. We wish to acknowledge NVIDIA for their generous grant of the GPU. This work was partially supported by Science and Engineering Research Board (SERB), Department of Science and Technology (DST), Govt. of India (Proj No. SB/S3/EECE/0127/2015) and Defence Research and Development Organization (DRDO), Govt. of India.

References

1. Everingham, M., Van Gool, L., Williams, C.K.I., Winn, J., Zisserman, A.: The PASCAL Visual Object Classes Challenge 2012 (VOC2012) Results (2012). http://www.pascal-network.org/challenges/VOC/voc2012/workshop/index.html
2. Russakovsky, O., Deng, J., Su, H., Krause, J., Satheesh, S., Ma, S., Huang, Z., Karpathy, A., Khosla, A., Bernstein, M., Berg, A.C., Fei-Fei, L.: Imagenet large scale visual recognition challenge. Int. J. Comput. Vis. (IJCV), pp. 1–42 (2015)
3. Torralba, A., Efros, A.: Unbiased look at dataset bias. In: 2011 IEEE Conference on Computer Vision and Pattern Recognition (CVPR), pp. 1521–1528. IEEE (2011)
4. Khosla, A., Zhou, T., Malisiewicz, T., Efros, A.A., Torralba, A.: Undoing the damage of dataset bias. In: Fitzgibbon, A., Lazebnik, S., Perona, P., Sato, Y., Schmid, C. (eds.) ECCV 2012. LNCS, vol. 7572, pp. 158–171. Springer, Heidelberg (2012). doi:10.1007/978-3-642-33718-5_12
5. Tommasi, T., Patricia, N., Caputo, B., Tuytelaars, T.: A deeper look at dataset bias. In: Gall, J., Gehler, P., Leibe, B. (eds.) GCPR 2015. LNCS, vol. 9358, pp. 504–516. Springer, Heidelberg (2015). doi:10.1007/978-3-319-24947-6_42
6. He, K., Zhang, X., Ren, S., Sun, J.: Delving deep into rectifiers: surpassing human-level performance on imagenet classification. CoRR abs/1502.01852 (2015)
7. Ullman, S., Assif, L., Fetaya, E., Harari, D.: Atoms of recognition in human and computer vision. In: Proceedings of the National Academy of Sciences (2016)
8. Everingham, M., Eslami, S.M.A., Van Gool, L., Williams, C.K.I., Winn, J., Zisserman, A.: The pascal visual object classes challenge: a retrospective. Int. J. Comput. Vis. **111**, 98–136 (2015)
9. Caffe: Models accuracy on ILSVRC 2012 data. (http://github.com/BVLC/caffe/wiki/Models-accuracy-on-ImageNet-2012-val) (2012)
10. Krizhevsky, A., Sutskever, I., Hinton, G.E.: Imagenet classification with deep convolutional neural networks. In: Advances in Neural Information Processing Systems, pp. 1097–1105 (2012)
11. Szegedy, C., Liu, W., Jia, Y., Sermanet, P., Reed, S., Anguelov, D., Erhan, D., Vanhoucke, V., Rabinovich, A.: Going deeper with convolutions. arXiv preprint. arXiv:1409.4842 (2014)
12. Simonyan, K., Zisserman, A.: Very deep convolutional networks for large-scale image recognition. arXiv preprint. arXiv:1409.1556 (2014)
13. Lin, M., Chen, Q., Yan, S.: Network in network. CoRR abs/1312.4400 (2013)

14. Chen, X., Mottaghi, R., Liu, X., Fidler, S., Urtasun, R., Yuille, A.: Detect what you can: detecting and representing objects using holistic models and body parts. In: IEEE Conference on Computer Vision and Pattern Recognition (CVPR) (2014)
15. Li, Y., Hou, X., Koch, C., Rehg, J.M., Yuille, A.L.: The secrets of salient object segmentation. In: 2014 IEEE Conference on Computer Vision and Pattern Recognition (CVPR), pp. 280–287. IEEE (2014)
16. Borji, A., Itti, L.: State-of-the-art in visual attention modeling. IEEE Trans. Pattern Anal. Mach. Intell. **35**, 185–207 (2013)
17. Yun, K., Peng, Y., Samaras, D., Zelinsky, G.J., Berg, T.: Studying relationships between human gaze, description, and computer vision. In: 2013 IEEE Conference on Computer Vision and Pattern Recognition (CVPR), pp. 739–746. IEEE (2013)
18. Geisler, W.S., Perry, J.S.: Real-time foveated multiresolution system for low-bandwidth video communication. In: Photonics West 1998 Electronic Imaging, pp. 294–305. International Society for Optics and Photonics (1998)
19. Farhadi, A., Endres, I., Hoiem, D.: Attribute-centric recognition for cross-category generalization. In: 2010 IEEE Conference on Computer Vision and Pattern Recognition (CVPR), pp. 2352–2359. IEEE (2010)
20. Wu, Z., Palmer, M.: Verbs semantics and lexical selection. In: Proceedings of the 32nd Annual Meeting on Association for Computational Linguistics, ACL 1994, pp. 133–138. Association for Computational Linguistics, Stroudsburg, PA, USA (1994)
21. Miller, G.A.: Wordnet: a lexical database for english. Commun. ACM **38**, 39–41 (1995)
22. Felzenszwalb, P.F., Girshick, R.B., McAllester, D., Ramanan, D.: Object detection with discriminatively trained part-based models. IEEE Trans. Pattern Anal. Mach. Intell. **32**, 1627–1645 (2010)
23. Tyler, C.W., Likova, L.T.: An algebra for the analysis of object encoding. NeuroImage **50**, 1243–1250 (2010)
24. Palmer, S.E.: Hierarchical structure in perceptual representation. Cogn. Psychol. **9**, 441–474 (1977)
25. Lin, T.-Y., Maire, M., Belongie, S., Hays, J., Perona, P., Ramanan, D., Dollár, P., Zitnick, C.L.: Microsoft COCO: common objects in context. In: Fleet, D., Pajdla, T., Schiele, B., Tuytelaars, T. (eds.) ECCV 2014. LNCS, vol. 8693, pp. 740–755. Springer, Heidelberg (2014). doi:10.1007/978-3-319-10602-1_48
26. Yao, J., Fidler, S., Urtasun, R.: Describing the scene as a whole: joint object detection, scene classification and semantic segmentation. In: 2012 IEEE Conference on Computer Vision and Pattern Recognition (CVPR), pp. 702–709. IEEE (2012)
27. Zhu, Y., Tian, Y., Mexatas, D., Dollár, P.: Semantic Amodal Segmentation. ArXiv e-prints (2015). http://arXiv.org/abs/1509.01329
28. Sermanet, P., Eigen, D., Zhang, X., Mathieu, M., Fergus, R., LeCun, Y.: Overfeat: integrated recognition, localization and detection using convolutional networks. CoRR abs/1312.6229 (2013)
29. Chatfield, K., Simonyan, K., Vedaldi, A., Zisserman, A.: Return of the devil in the details: delving deep into convolutional nets. CoRR abs/1405.3531 (2014)
30. Zeiler, M.D., Fergus, R.: Visualizing and understanding convolutional networks. In: Fleet, D., Pajdla, T., Schiele, B., Tuytelaars, T. (eds.) ECCV 2014. LNCS, vol. 8689, pp. 818–833. Springer, Heidelberg (2014). doi:10.1007/978-3-319-10590-1_53
31. Welinder, P., Welling, M., Perona, P.: A lazy man's approach to benchmarking: semisupervised classifier evaluation and recalibration. In: 2013 IEEE Conference on Computer Vision and Pattern Recognition (CVPR), pp. 3262–3269. IEEE (2013)

32. Xu, L., Li, J., Brenning, A.: A comparative study of different classification techniques for marine oil spill identification using radarsat-1 imagery. Remote Sens. Environ. **141**, 14–23 (2014)
33. Aghazadeh, O., Carlsson, S.: Properties of datasets predict the performance of classifiers. In: British Machine Vision Conference, pp. 22–31 (2013)
34. Hoiem, D., Chodpathumwan, Y., Dai, Q.: Diagnosing error in object detectors. In: Fitzgibbon, A., Lazebnik, S., Perona, P., Sato, Y., Schmid, C. (eds.) ECCV 2012. LNCS, vol. 7574, pp. 340–353. Springer, Heidelberg (2012). doi:10.1007/978-3-642-33712-3_25
35. Torralba, A.: How many pixels make an image? Vis. Neurosci. **26**, 123–131 (2009)
36. Parizi, S.N., Vedaldi, A., Zisserman, A., Felzenszwalb, P.F.: Automatic discovery and optimization of parts for image classification. CoRR abs/1412.6598 (2014)
37. Yilmaz, O.: Classification of occluded objects using fast recurrent processing. arXiv preprint. arXiv:1505.01350 (2015)

End-to-End Training of Object Class Detectors for Mean Average Precision

Paul Henderson[(⊠)] and Vittorio Ferrari

University of Edinburgh, Edinburgh, UK
p.m.henderson@ed.ac.uk

Abstract. We present a method for training CNN-based object class detectors directly using mean average precision (mAP) as the training loss, in a truly end-to-end fashion that includes non-maximum suppresion (NMS) at training time. This contrasts with the traditional approach of training a CNN for a window classification loss, then applying NMS only at test time, when mAP is used as the evaluation metric in place of classification accuracy. However, mAP following NMS forms a piecewise-constant structured loss over thousands of windows, with gradients that do not convey useful information for gradient descent. Hence, we define new, general gradient-like quantities for piecewise constant functions, which have wide applicability. We describe how to calculate these efficiently for mAP following NMS, enabling to train a detector based on Fast R-CNN [1] directly for mAP. This model achieves equivalent performance to the standard Fast R-CNN on the PASCAL VOC 2007 and 2012 datasets, while being conceptually more appealing as the very same model and loss are used at both training and test time.

1 Introduction

Object class detection is the task of localising all instances of a given set of object classes in an image. Modern techniques for object detection [1–4] use a convolutional neural network (CNN) classifier [5,6], operating on object proposal windows [7–9]. Given an image, they first generate a set of windows likely to include all objects, then apply a CNN classifier to each window independently. The CNN is trained to output one score for each possible object class on each window, and an additional one for 'background' or 'no object'. Such models are trained for window classification accuracy: the loss attempts to maximise the number of training windows for which the CNN gives the highest score to the correct class. At test time, the CNN is applied to every window in a test image, followed by a non-maximum suppression processing stage (NMS). This eliminates windows that are not locally the highest-scored for a class, yielding the output set of detections. Typically, the performance of the detector is evaluated using mean average precision (mAP) over classes, which is based on the ranking of detection scores for each class [10].

Thus, the traditional approach is to train object detectors with one measure, classification accuracy over all windows, but test with another, mAP over locally highest-scoring windows. While the training loss correlates somewhat with the test-time evaluation metric, they are not really the same, and furthermore, training ignores the effects of NMS. As such, the traditional approach is not true end-to-end training for the final *detection* task, but for the surrogate task of window *classification*.

In this work, we present a method for training object detectors directly using mAP computed after NMS as the loss. This is in accordance with the machine learning dictum that the loss we minimise at training time should correspond as closely as possible to the evaluation metric used at test time. It also fits with the recent trend towards training models end-to-end for their ultimate task, in vision [11–13] and other areas [14,15], rather than training individual components for engineered sub-tasks, and combining them by hand.

Directly optimising for mAP following NMS is very challenging for two main reasons: (i) mAP depends on the global ordering of class scores for all windows across all images, and as such is piecewise constant with respect to the scores; and, (ii) NMS has highly non-local effects within an image, as changing one window score can have a cascading effect on the retention of many other windows. In short, we have a structured loss over many thousands of windows, that is non-convex, discontinuous, and piecewise constant with respect to its inputs. Our main contribution is to overcome these difficulties by proposing new gradient-like quantities for piecewise constant functions, and showing how these can be computed efficiently for mAP following NMS. This allows us to train a detector based on Fast R-CNN [1] in a truly end-to-end fashion using stochastic gradient descent, but with NMS included at training time, and mAP as the loss.

Experiments on the PASCAL VOC 2007 and 2012 detection datasets [16] show that end-to-end training directly for mAP with NMS reaches equivalent performance to the traditional way of training for window classification accuracy and without NMS. It achieves this while being conceptually simpler and more appealing from a machine learning perspective, as exactly the same model and loss are used at both training and test time. Furthermore, our method is widely applicable on two levels: firstly, our loss is a simple drop-in layer that can be directly used in existing frameworks and models; secondly, our approach to defining gradient-like quantities of piecewise-constant functions is general and can be applied to other piecewise-constant losses and even internal layers. For example, using our method can enable training directly for other rank-based metrics used in information retrieval, such as discounted cumulative gain [17]. Moreover, we do not require a potentially expensive max-oracle to find the most-violating inputs with respect to the model and loss, as required by [2,18,19].

2   Background

We recap here how NMS is performed (Sect. 2.1) and mAP calculated (Sect. 2.2). Then, we describe Fast R-CNN [1] in more detail (Sect. 2.3), as it forms the basis for our proposed method.

2.1   Non-maximum Suppression (NMS)

Given a set of windows in an image, with scores for some object class, NMS removes those windows which are not locally the highest-scored, to yield a final set of detections [20]. Specifically, all the windows are marked as retained or suppressed by the following procedure: first, the highest-scored window is marked as retained, and all those overlapping with it by more than some threshold (*e.g.* 30% in [1,4]) intersection-over-union (IoU) are marked as suppressed; then, the highest-scored window neither retained nor suppressed is marked as retained, and again all others sufficiently-overlapping are marked as suppressed. This process is repeated until all windows are marked as either retained or suppressed. The retained windows then constitute the final set of detections.

2.2   Mean Average Precision (mAP)

The mAP [10,16,21] for a set of detections is the mean over classes, of the interpolated AP [22] for each class. This per-class AP is given by the area under the precision/recall (PR) curve for the detections (Fig. 1).

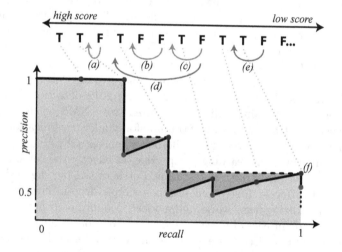

Fig. 1. Precision/recall curve (bottom) for a sequence of true-positive (TP) and false-positive (FP) detections ordered by score (top) for some object class with six ground-truth instances. Plotting the sequence of precision and recall values yields the black curve. The pink area shows the result of replacing each precision with the maximum at same or higher recall. AP is the total area of the pink and blue regions. The arrows (a–e) show the effect of positive perturbations to scores of FP detections. Blue arrows (a–c) show perturbations with no effect on AP: (a) the order of detections does not change; (b) the detection swaps places with another FP; (c) the detection swaps places with a TP, but a higher-recall TP (f) has higher precision so there is no change to area under the filled-in curve (pink shading). Orange arrows (d–e) show perturbations that do affect AP: (d) the same FP as (c) is moved beyond a TP that does appear on (hence affect) the filled in curve; (e) the FP moves past a single TP, altering the filled-in curve as far away as 0.5 recall. (Color figure online)

The PR curve is constructed by first mapping each detection to its most-overlapping ground-truth object instance, if any overlaps sufficiently—for PASCAL VOC, this is defined as overlapping with >50% IoU [16]. Then, the highest-scored detection mapped to each ground-truth instance is counted as a true-positive, and all other detections as false-positives. Next, we compute recall and precision values for increasingly large subsets of detections, starting with the highest-scored detection and adding the remainder in decreasing order of their score. Recall is defined as the ratio of true-positive detections to ground-truth instances, and precision as the ratio of true-positive detections to all detections. The PR curve is then given by plotting these recall-and-precision pairs as progressively lower-scored detections are included. Finally, dips in the curve are filled in (interpolated) by replacing each precision with the maximum of itself and all precisions occurring at higher recall levels (pink shading in Fig. 1) [10,22].

The area under the interpolated PR curve is the AP value for the class. For the PASCAL VOC 2007 dataset, this area is calculated by a rough quadrature approximation sampling at 11 uniformly spaced values of recall [10]; for the VOC 2012 dataset it is the true area under the curve [16].

2.3 Fast R-CNN

Model. Our model is based on Fast R-CNN [1] (Figs. 2a, b), without bounding-box regression. This model operates by classifying proposal windows of an image, as belonging to one of a set of object classes, or as 'background'. Whole images are processed by a sequence of convolutional layers; then, for each window, convolutional features with spatial support corresponding to that window are extracted and resampled to fixed dimension, before being passed through three fully-connected layers, the last of which yields a score for each object class and 'background'. The class scores for each window are then passed through a softmax function, to yield a distribution over classes.

Training. This network is trained with a window classification loss. If a window overlaps a ground-truth object with IoU > 0.5, its true class is defined as being that object class; otherwise, its true class is 'background'. For each window, the network outputs softmax probabilities for each class, and the negative log likelihood (NLL) of the true class is used as the loss for that window; the total loss over a minibatch is simply a sum of the losses over all windows in it. The network is trained by stochastic gradient descent (SGD) with momentum, operating on minibatches of two images at a time.

Testing. At test time, windows are scored by passing them forwards through the network, and recording the final softmax probabilities for each class. Then, for each class and image, NMS is applied to the scored windows (Sect. 2.1). Note that this NMS stage is not present at training time. Finally, the detections are evaluated using mAP over the full test set.

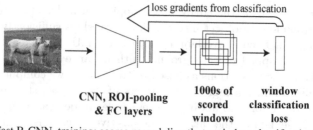

(a) Fast R-CNN, training: scores passed directly to window classification loss

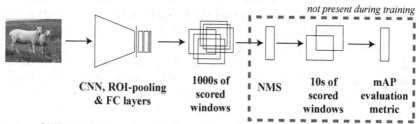

(b) Fast R-CNN, testing: NMS applied, and detections evaluated with mAP

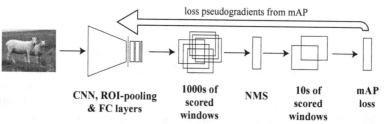

(c) Our method, both training and testing: exactly the same operations occur at train and test time , with identical model structure and the training loss matching the test-time evaluation metric

Fig. 2. Fast R-CNN [1] architecture during training (a) and testing (b) phases, and our architecture (c), which is the same in both phases.

3   Related Work

Nearly all works on object class detection train a window classifier, and ignore NMS and mAP at training time. Earlier approaches [20, 23–25] apply the classifier to all windows in a dense regular grid, while more recently, object proposal methods [7, 9] have been used to greatly reduce the number of windows [4, 8]. Below we review the few works that try to either train for AP or other structured losses, or include NMS at training time.

Blaschko and Lampert [26] formulate object detection as a structured prediction problem, outputing a binary indicator for object presence and a set of bounding-box coordinates. This is trained using a structured SVM, with a task loss that aims for correct classification and maximal IoU of predicted and ground-truth boxes in images containing the target class. Like our method, this is a structured loss involving IoU of detections and ground-truth objects; however, it does not correspond to maximising AP, and only a single detection is returned in each image, so there is no NMS. More recently, [2] uses the same structured SVM loss, but with a CNN in place of a kernelised linear model over SURF features [26]. This work directly optimises the structured SVM loss via gradient descent, allowing backpropagation to update the nonlinear CNN layers.

There exist works that train specifically for AP, but for classification problems, rather than for object detection with NMS. Yue et al. [18] optimizes AP in the structured SVM framework—with a linear model, trained using a hinge loss weighted according to AP. This requires solving a loss-augmented inference problem, i.e. finding the scores that maximise the sum of AP and the output of the current model. They present a dynamic programming algorithm to solve this, which has quadratic complexity in the number of training points. Extending this work, [19] presents a more general technique for training nonlinear structured models directly for non-differentiable losses, again assuming that loss-augmented inference can be performed efficiently. Using the same dynamic-programming approach as [18], they apply it to the case of single-class AP with a model based on R-CNN [4], without NMS at training time. While their method requires changes to the optimiser itself, ours does not. Instead, we simply define a new loss layer that can be easily dropped into existing frameworks, and do not require solving a loss-augmented inference problem. Furthermore, our approach can incorporate NMS and train simultaneously for multiple classes. Thus, while [19] trains for AP over binary window classification scores, ours trains directly for mAP over object detections.

Taylor et al. [27] discuss a different formulation for gradient-descent optimisation of certain losses based on ranking of scores (though not AP specifically). They define a smooth proxy loss for a non-differentiable, piecewise constant ranking loss. They treat the predicted score of each training point as a Gaussian random variable centered on the actual value, and hence compute the distribution of ranks for each score, by pairwise comparisons to all other scores. This distribution is used in place of the usual hard ranks when evaluating the loss, and the resulting quantity is differentiable with respect to the original scores. This method has cubic complexity in the number of training samples, making it intractable when there are tens of classes and thousands of windows (e.g. in PASCAL VOC).

Unlike most other approaches to object detection, [28] includes NMS at training time as well as test time. They use a deformable parts model over CNN features, that outputs scored windows derived from a continuous response map (in contrast to feeding fixed proposal windows through a CNN [1]). The windows are passed through a non-standard variant of NMS. Instead of training for mAP or window classification accuracy, the authors then introduce a new structured loss. This includes terms for detections retained by NMS, but also for suppressed windows, in a fashion requiring knowledge of which detection suppressed them. As such, it is deeply tied to the NMS implementation at training time, rather than being a generally-applicable loss such as mAP.

4 Proposed Method

We now describe our proposed method (Fig. 2c). We discuss how our model differs from Fast R-CNN (Sect. 4.1) and why it is challenging to train (Sect. 4.2). Then we introduce our general method for defining gradients of piecewise-constant functions (Sect. 4.3) and how we apply it to train our model (Sect. 4.4).

4.1 Detection Framework

Model. Our model is identical to Fast R-CNN as described above, up to the softmax layer: windows are still scored by passing through a sequence of convolutional and fully-connected layers. As in [1], we can use different convolutional network architectures pretrained for ILSVRC 2012 [21] classification, such as AlexNet [5] or VGG16 [6]. We omit the softmax layer, using the activations of the last fully-connected layer directly as window scores. In our experiment we found that the softmax has little effect on the final performance, but its tendency to saturate causes problems with propagating the loss gradients back through it. In contrast to Fast R-CNN, our model also includes an NMS layer immediately after the last fully-connected layer, which performs the same operation as used at test time for Fast R-CNN. We regard the NMS layer as part of the model itself, present at both training and test time.

Training. During training, we add a loss layer that computes mAP over the minibatch, after NMS. Thus, at training time, minibatches undergo exactly the same sequence of operations as at test time, and the training loss matches the test-time evaluation metric. The network is still trained using SGD with momentum. Section 4.2 describes how to define derivatives of the mAP and NMS layers, while Sect. 4.5 discusses some additional techniques used during training.

Testing. During testing, our method is identical to Fast R-CNN, except that the softmax layer is omitted.

4.2 Gradients of mAP and NMS Layers

In order to minimise our loss by gradient descent, we need to propagate derivatives back to the fully-convolutional layers of the CNN and beyond. However, mAP is a piecewise constant function of the detection scores, as it depends only on their ordering—each score can be perturbed slightly without changing the loss. The partial derivatives of such a loss function do not convey useful information for gradient descent (Fig. 3a) as they are almost everywhere zero (in the constant regions), and otherwise undefined (at the steps). The subgradient is also undefined, as the function is non-convex.

Furthermore, even if we could compute the derivatives of mAP with respect to the class scores, they still need to be propagated back through the NMS layer. This requires a definition of the Jacobian of NMS, which is again non-trivial.

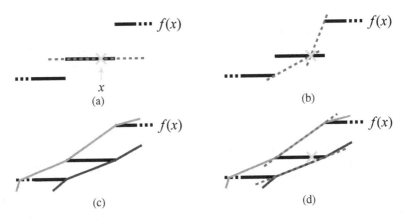

Fig. 3. A piecewise constant function $f(x)$ with steps at two points, and various definitions for gradients. (a) Conventional partial derivative (red dashed) at x, equal to zero, does not convey useful information for gradient descent. (b) Gradients at x given by positive-perturbing and negative-perturbing finite difference estimators. (c) Piecewise-linear upper (green) and lower (brown) envelopes of $f(x)$. (d) Gradients at x given by slope of upper/lower envelopes. When applied to our model, $f(x)$ is mAP, and the horizontal axis corresponds to the score of a single window with respect to which the partial derivative is being computed. (Color figure online)

Note that max-pooling layers are similarly non-differentiable, but good results are achieved by simply propagating the gradient back to the maximal input only. We could do similar for NMS: allow only the locally-maximal windows propagate gradients back; however, this loses valuable information. For example, if all detections overlapping some ground-truth object are suppressed, then there should be a gradient signal favouring increasing the score of those windows (or decreasing that of their suppressors). This does not occur if we naïvely copy gradients back through to maximal windows. In contrast, we require a Jacobian-like quantity for NMS that does capture this information.

We therefore develop general definitions for gradient-like quantities of piecewise-constant functions in Sect. 4.3, and then describe how to apply them efficiently to NMS and mAP in Sect. 4.4.

4.3   Pseudogradients of General Piecewise-Constant Functions

We consider how to define a general *pseudo partial derivative* (PPD) operation for piecewise-constant functions, that can be used to define quantities analogous to the gradient and the Jacobian. For any piecewise-constant function $f(\mathbf{x})$ with countably many discontinuities (steps), we denote the PPD with respect to x_i by $\widetilde{\partial}_{x_i} f$. When the PPD is non-zero we need to move some non-infinitesimal distance before any change in the function occurs (unlike a conventional partial derivative). However when there is a change, it will be in the direction indicated by the PPD, and in magnitude corresponding to the PPD (this is made more precise below). We then use our PPD to define an analogue to the gradient by $\widetilde{\nabla} f = (\widetilde{\partial}_{x_1} f, \ldots, \widetilde{\partial}_{x_N} f)$. Intuitively, this tells us locally what direction to move so that the function will decrease, if we move some non-infinitesimal distance in this direction. Similarly, for the Jacobian of vector-valued $\mathbf{f}$, we have $\widetilde{J}_{ij} = \widetilde{\partial}_{x_j} f_i$.

We now discuss two possible definitions for the PPD; these and the regular partial derivative are illustrated in Fig. 3 for a one-dimensional function, at a point lying in a constant region between two steps.

Finite Difference Estimators. Most simply, we can apply a traditional single-sided finite difference estimator, as used for computing numerical gradients of a differentiable function. Here, a small, fixed perturbation δx is added to x, the function evaluated at this point, and the resulting slope used to approximate the gradient, by $\widetilde{\partial}_x f = \frac{f(x+\delta x) - f(x)}{\delta x}$. The piecewise-constant functions we are interested in have finitely many steps, and so the probability of f being undefined at the perturbed point is zero. However, the constant regions of our function vary in size by several orders of magnitude, and so it is impossible to pre-select a suitable value for δx. Instead, we use an adaptive approach: given x, set δx to the smallest value such that $f(x + \delta x) \neq f(x)$, then compute $\widetilde{\partial}_x f$ as above (Fig. 3b). Note that this method is single-sided: it only takes account of the change due to perturbing x in one direction or the other. This is undesirable, as in general, it delivers different results for each direction, perhaps yielding complementary information. We address this issue by performing the same calculation independently with positive then negative perturbations δx^+ and δx^-, and taking a mean of the resulting pseudogradients. We refer to this mean pseudogradient as SDE, for symmetric difference estimator. This approach has the disadvantage that the magnitude of the gradient is sensitive to the exact location of x: if it is nearer to a step, the gradient will be larger, yet a correspondingly larger change to the network parameters may be undesirable.

Linear Envelope Estimators. An alternative approach to defining the PPD is to fit a piecewise-linear upper or lower envelope to the steps of the piecewise-constant function (Fig. 3c). The PPD $\widetilde{\partial}_x f$ is then given by the slope of the envelope segment at the point x (Fig. 3d). In practice, we take the average of the gradients of the upper and lower envelopes. Unlike SDE, this estimator does not become arbitrarily large as x approaches a step. If f has finitely many steps, then for all points before the first step and after the last, both linear envelopes have zero gradient; we find however that better results are achieved by using SDE in these regions, but with an empirically-tuned lower-bound on δx. We refer to this pseudogradient as MEE, for mean envelope estimator.

4.4 Application to mAP and NMS

To apply the above methods to mAP, we must compute the PPD of each class' AP with respect to each window score independently, holding the other scores constant. This raises two questions: (i) how to efficiently find the locations of the nearest step before and after a point, and (ii) how to efficiently evaluate the loss around those locations. We solve these problems by noting that changes to AP only occur when two scores change their relative ordering, and even then, only in certain cases. Specifically, AP changes value only when a window counted as a true-positive changes place with one counted as a false-positive. Also, the effective precision at a given recall is the maximum precision at that or any higher recall (Sect. 2.2 and Fig. 1). So we have further conditions, *e.g.* decreasing the score of a false-positive only affects AP when it drops below that of a true-positive at which precision is higher than any with even lower score. This effect and other perturbations are illustrated in Fig. 1 (blue and orange arrows).

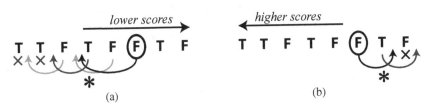

Fig. 4. Efficient calculation of smallest perturbations to detection scores to cause a step in AP. In each case the circled FP is currently being considered. (a) Iterating detections in decreasing order of score, finding the smallest increase to each score that causes a change in AP (higher for TPs, lower for FPs). Detections already considered have an arrow showing where they are perturbed to; a cross indicates no increase to that score affects AP. When considering the circled FP, the last-seen TP is shown by the orange asterisk; perturbing the score of the circled detection just beyond (left) of this is the minimal change to affect AP. (b) Similar but iterating in increasing order of score, and hence calculating minimal decreases in score to affect AP. (Color figure online)

Thus, for each class, we can find the nearest step before and after each point by making two linear passes over the detections, in descending then ascending order of score (Fig. 4). Assuming we have computed AP as described in Sect. 2.2, we know whether each detection is a true- or false-positive, and can keep track of the last-seen detection of each kind. In the descending pass, for each detection, we find the smallest increase to its score that would result in a change to AP, thus giving the location of the nearest step on the positive side. This score increase is that which moves it an infinitesimal amount higher than the score of the last-seen window of the other kind (true-positive vs. false-positive), subject to the additional conditions mentioned above. Similarly, in the ascending pass, we can find the required decreases in scores that would cause a change in AP. Once the step locations have been found, the new AP values resulting from perturbing the scores accordingly can be calculated by updating the relevant part of the PR curve, and then computing its area as normal. Given the step locations and AP values, it is then straightforward to use the methods of Sect. 4.3 to compute the SDE or MEE.

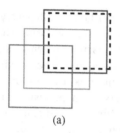
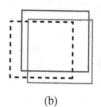

(a)   (b)

Fig. 5. Transitivity approximations for NMS. Dashed black box is a ground-truth object; coloured boxes are scored windows, red > green > blue. (a) Red overlaps green sufficiently for NMS inhibition, and green overlaps blue similarly, but red does not overlap enough with blue. However, whether red is retained indirectly affects whether blue is retained, as if red suppresses green, then green does not suppress blue. In our approximation, this long-distance interaction between red and blue is ignored; however the two local interactions (red-green and green-blue) are included. (b) Red and blue overlap each other sufficiently for NMS inhibition; given that red suppresses blue, our approximation assumes that blue overlaps the same ground-truth instance as red (if any). (Color figure online)

Incorporating NMS. We must also account for NMS when propagating gradients back. The PPDs of NMS can be used to define a Jacobian as described in Sect. 4.3, which may then be composed with the pseudogradient of mAP to define the gradient of mAP with respect to the pre-NMS scores. However, subject to a small approximation, it is both easier and more efficient to consider NMS simultaneously with AP when determining step locations and the resultant changes to the loss. Specifically, we introduce two transitivity approximations (Fig. 5): (i) we do not attempt to model cascaded long-distance interactions between windows through multiple steps of NMS; (ii) we assume in certain cases that windows suppressed by some detection overlap exactly the same ground-truth instances as the detection itself. Under these approximations, it is possible to compute the PPDs with respect to pre-NMS scores in linear time in the number of windows. This is achieved by: (i) adding gradient contributions due to windows suppressed by a true-positive or false-positive detection at the same time as that detection, as these suppressed windows need to have their scores perturbed to the same point as their suppressor did to cause a change in AP; (ii) including a third pass that adds gradient contributions from suppressed windows overlapping ground-truth instances that were missed entirely (*i.e.* no detection covers them); (iii) also adding gradient contributions from the detections that caused the suppressed-but-overlapping windows of (ii) to be suppressed.

4.5 Training Protocol

In order to train our model successfully, we make various changes to the training protocol used for Fast R-CNN in [1]. The impact of each of these changes is given in Sect. 5.

Minibatch Composition. We use larger minibatches than [1], as (i) object detection mAP has a much higher batch-to-batch variance than simple window classification accuracy, and (ii) including more windows increases the density of the gradient signal, as there are likely to be more false positives which score higher than some true positive (and vice versa). We also find that performance is improved by using proportionally fewer foreground windows (those overlapping a ground-truth instance as opposed to background) in each training minibatch. While Fast R-CNN uses 25% foreground windows, we use 5%, which roughly corresponds to the distribution of windows seen at test time, when 5% of all selective search proposals overlap a ground-truth instance.

Regularisation. Using our method, we found empirically that scores are prone to grow very large after several hundred iterations of training. This is effectively mitigated by introducing a regulariser on the window scores. We find that an L4 regulariser with very small weight performs best, as it gives greater freedom to smaller-magnitude scores while imposing a relatively hard constraint on magnitude, compared to the more common L1/L2 regularisation.

Log-Space. We find it is beneficial to follow gradients of $\log(\text{mAP} + \epsilon)$ instead of mAP itself, for some small, fixed constant ϵ. Early in training when mAP is low, scores of true-positive windows are uniformly distributed amongst those of false-positive windows, and so an increase in the score of a true-positive often yields only a very small gain in mAP. Using $\log(\text{mAP} + \epsilon)$ instead amplifies the effect of these changes, so training quickly escapes from the initial very low mAP.

Gradient Clipping. We find that numerical behaviour is improved (particularly at high learning rates) by clipping elements of the gradient to a fixed threshold.

5 Experiments

We now evaluate the performance of our approach on two datasets: PASCAL VOC 2007 and 2012 [16]. Both datasets have 20 object classes; for VOC 2007, we train on the trainval subset (5011 images) and test on the test subset (4952 images); for VOC 2012, we train on the train subset (5717 images) and test on the validation subset (5823 images). We also give results training on the union of VOC 2007 trainval and VOC 2012 trainval (total 16551 images), and testing on VOC 2007 test.

We compare our method to two others: (i) Fast R-CNN trained with the standard NLL loss for window classification, as described in [1] (bounding box regression is disabled, to give a fair comparison with our method); and (ii) [19], which also trains an R-CNN-like model for AP, but with a separate model for each class, no NMS at training time, and with a different way to compute parameter gradients. This is the closest work in spirit to ours.

Settings. We use Fast R-CNN as described in [1], built upon AlexNet [5] or VGG16 [6], with weights initialised on ILSVRC 2012 classification [21]. We then remove the softmax layers at both training and test time, as described in Sect. 4.1, and replace the training loss layer with our NMS layer and mAP loss.

Incorporating the techniques described in Sect. 4.5, the overall loss we min- imise by SGD is $L = -\log\{\sum_c \mathrm{AP}(\mathrm{NMS}(\mathbf{s}_c))/K\} + \lambda \sum_{c,b} |s_c^b|^4$, where $\mathbf{s}_c$ are the window scores for class c, K is the total number of classes, and b indexes over windows.

The AP calculation during training is always matched to that used for evalu- ation. When testing on VOC 2007, we train using the VOC 2007 approximation to AP (Sect. 2.2); when testing on VOC 2012, we train using the true AP. In order to compute pseudogradients for training, we try both SDE and MEE and compare their performance (Sect. 4.3). As our method works best with large minibatches, for the VGG16 experiments, we clamp the maximum image dimen- sion to 600 pixels, to conserve GPU memory (this does not have a significant impact on the baseline performance).

Table 1. Performance of our method measured by mAP on VOC 2007 test set, with different pseudogradients (MEE vs SDE), network architectures (AlexNet vs VGG16), and training sets (VOC 2007 trainval vs union of VOC 2007 trainval and VOC 2012 trainval). We also give results for Fast R-CNN trained using a traditional softmax loss, without bounding box regression.

Trained on...	2007 only		2007 + 2012	
	AlexNet	VGG16	AlexNet	VGG16
Ours, MEE	51.6	58.9	54.9	62.5
Ours, SDE	51.3	60.7	54.8	62.3
Fast R-CNN	52.0	62.4	53.8	63.5

Main Results on VOC 2007. Table 1 shows how our methods compare with Fast R-CNN, testing on the PASCAL VOC 2007 dataset. Overall, our method achieves comparable performance to Fast R-CNN. The results also show that using a larger training set (union of VOC 2007 and 2012 trainval subsets) increases performance by up to 3.6% mAP, compared to training from VOC 2007 trainval alone. This effect is significantly stronger for our method than for Fast R-CNN: for AlexNet, we gain 3.3% mAP compared with 1.8% for Fast R-CNN; for VGG16, we gain 3.6% compared with 1.1% for Fast R-CNN. This indicates that our approach particularly benefits from more training data, possi- bly because optimising for mAP implies many comparisons between windows. Of our two pseudo-gradient estimators, MEE slightly outperforms SDE, in all cases apart from VGG16 training on VOC 2007 trainval only. This is likely because MEE is insensitive to the distances from points to nearest steps, in contrast to SDE (Sect. 4.3); hence, MEE is a more robust estimator of the impact of a score change, whereas SDE may introduce very large derivatives for a particular window. In all cases, VGG16 significantly outperforms AlexNet, confirming previous studies [1,6].

Ablation Study. In Sect. 4.5, we noted that certain modifications to the original training procedure of Fast R-CNN were necessary to achieve these results. Ablating away these modifications reduces our mAP, as follows (all using AlexNet on VOC 2007 with the MEE gradient estimator): (i) minibatch composition: increasing foreground fraction to 25% (as used in Fast-RCNN): -6.1 mAP (ii) minibatch size: halving batch size but doubling iteration count (so the same amount of data is seen): -0.8 mAP (iii) score regularisation: with L2 regularisation instead of L4 and the constant adjusted appropriately: -1.0 mAP. With no regularisation, training fails after <100 iterations as the magnitude of the classification scores explode. (iv) gradient clipping: with this disabled, training fails after <100 iterations due to numerical issues caused by large gradients.

Comparison to [19] on VOC 2012. The only previous work that attempts to train a CNN-based object detector directly for AP is [19]. Table 2 compares this method to ours; we use the PASCAL VOC 2012 dataset (testing on the validation subset) as this is what [19] reports results on. Our method achieves comparable performance to [19], with the MEE estimator again being slightly better than SDE.

Unlike our method, [19] trains a separate model for each class; their dynamic-programming solution to the loss-augmented inference problem is for single-class AP only (not mAP over all classes). Moreover, their training procedure does not take into account NMS.

Discussion. We hypothesise that our methods do not significantly outperform Fast R-CNN overall for three reasons. (i) Our gradients are sparser than those of a softmax loss: not every window propagates information back for every class, as changing scores of certain windows has no effect on mAP (*e.g.* low-scored background windows suppressed by NMS). For example, for VOC 2007, around 20% of scores have a non-zero gradient — compared with 100% when using a softmax loss. (ii) mAP is a more rapidly changing function than the softmax loss: an estimate over a minibatch is a much higher-variance estimator of loss over the full set. (iii) It can be shown numerically that mAP over a minibatch of

Table 2. Performance of our method compared with [19] (which trains for single-class AP, with a technique very different from ours). All models were trained on VOC 2012 train subset, tested on VOC 2012 validation subset, and use AlexNet. Bounding box regression was not used in any of the models.

Ours, MEE	Ours, SDE	Song et al. [19]
48.2	48.0	48.5

images is a biased estimator of mAP over the population of images from which that minibatch was drawn.

The real advantage of our method over the standard training procedure of Fast R-CNN is being more principled by respecting the theoretical need for having the same evaluation during training and test.

6   Conclusions

We have presented two definitions of pseudo partial derivatives of piecewise-constant functions. Using these, we have trained a Fast R-CNN detector directly using mAP as the loss, with identical model structure at training and test time, including NMS during training. This ensures that training is truly end-to-end for the final detection task, as opposed to window classification. Our method achieves equivalent performance to Fast R-CNN. It is easily integrated with standard frameworks for SGD, such as Caffe [29], as our NMS and mAP loss layers can be dropped in without affecting the minimisation algorithm or other elements of the model. Our definitions of pseudogradients open up the possibility of training for other piecewise-constant losses. In particular, ranking-based metrics are common in information retrieval, including simple AP on document scores, and discounted cumulative gain [17]. Our method is very general as it does not require definition of an efficient max-oracle, in contrast to [19] and structured SVM methods. Indeed, our approach can also be applied to piecewise-constant internal layers of a network, allowing back-propagation of gradients through such layers.

References

1. Girshick, R.: Fast R-CNN. In: ICCV (2015)
2. Zhang, Y., Sohn, K., Villegas, R., Pan, G., Lee, H.: Improving object detection with deep convolutional networks via Bayesian optimization and structured prediction. In: CVPR (2015)
3. Sermanet, P., Eigen, D., Zhang, X., Mathieu, M., Fergus, R., LeCun, Y.: OverFeat: integrated recognition, localization and detection using convolutional networks. In: ICLR (2014)
4. Girshick, R., Donahue, J., Darrell, T., Malik, J.: Rich feature hierarchies for accurate object detection and semantic segmentation. In: CVPR (2014)
5. Krizhevsky, A., Sutskever, I., Hinton, G.E.: Imagenet classification with deep convolutional neural networks. In: NIPS (2012)
6. Simonyan, K., Zisserman, A.: Very deep convolutional networks for large-scale image recognition. In: ICLR (2015)
7. Alexe, B., Deselaers, T., Ferrari, V.: What is an object? In: CVPR (2010)
8. Uijlings, J.R.R., van de Sande, K.E.A., Gevers, T., Smeulders, A.W.M.: Selective search for object recognition. IJCV **104**, 154–171 (2013)
9. Zitnick, C.L., Dollár, P.: Edge boxes: locating object proposals from edges. In: Fleet, D., Pajdla, T., Schiele, B., Tuytelaars, T. (eds.) ECCV 2014. LNCS, vol. 8693, pp. 391–405. Springer, Heidelberg (2014). doi:10.1007/978-3-319-10602-1_26

10. Everingham, M., Van Gool, L., Williams, C.K.I., Winn, J., Zisserman, A.: The PASCAL Visual Object Classes (VOC) challenge. IJCV **88**, 303–338 (2010)
11. Long, J., Shelhamer, E., Darrell, T.: Fully convolutional networks for semantic segmentation. In: CVPR (2015)
12. Vinyals, O., Toshev, A., Bengio, S., Erhan, D.: Show and tell: a neural image caption generator. In: CVPR (2015)
13. Pfister, T., Charles, J., Zisserman, A.: Flowing ConvNets for human pose estimation in videos. In: ICCV (2015)
14. Sutskever, I., Vinyals, O., Le, Q.V.: Sequence to sequence learning with neural networks. In: NIPS (2014)
15. Levine, S., Finn, C., Darrell, T., Abbeel, P.: End-to-end training of deep visuomotor policies. JMLR **17**, 1–40 (2016)
16. Everingham, M., Eslami, S., van Gool, L., Williams, C., Winn, J., Zisserman, A.: The PASCAL visual object classes challenge: a retrospective. IJCV **111**, 98–136 (2015)
17. Järvelin, K., Kekäläinen, J.: IR evaluation methods for retrieving highly relevant documents. In: SIGIR (2000)
18. Yue, Y., Finley, T., Radlinski, F., Joachims, T.: A support vector method for optimizing average precision. In: SIGIR (2007)
19. Song, Y., Schwing, A.G., Zemel, R.S., Urtasun, R.: Training deep neural networks via direct loss minimization. In: ICML, pp. 2169–2177 (2016)
20. Felzenszwalb, P., Girshick, R., McAllester, D., Ramanan, D.: Object detection with discriminatively trained part based models. IEEE Trans. PAMI **32**, 1627–1645 (2010)
21. Russakovsky, O., Deng, J., Su, H., Krause, J., Satheesh, S., Ma, S., Huang, Z., Karpathy, A., Khosla, A., Bernstein, M., Berg, A., Fei-Fei, L.: ImageNet large scale visual recognition challenge. IJCV **115**, 211–252 (2015)
22. Salton, G., McGill, M.J.: Introduction to Modern Information Retrieval. McGraw-Hill, New York (1986)
23. Harzallah, H., Jurie, F., Schmid, C.: Combining efficient object localization and image classification. In: ICCV (2009)
24. Dalal, N., Triggs, B.: Histogram of oriented gradients for human detection. In: CVPR (2005)
25. Viola, P., Jones, M.: Rapid object detection using a boosted cascade of simple features. In: CVPR, pp. 511–518 (2001)
26. Blaschko, M.B., Lampert, C.H.: Learning to localize objects with structured output regression. In: Forsyth, D., Torr, P., Zisserman, A. (eds.) ECCV 2008. LNCS, vol. 5302, pp. 2–15. Springer, Heidelberg (2008). doi:10.1007/978-3-540-88682-2_2
27. Taylor, M., Guiver, J., Robertson, S., Minka, T.: SoftRank: optimising non-smooth rank metrics. In: WSDM (2008)
28. Wan, L., Eigen, D., Fergus, R.: End-to-end integration of a convolution network, deformable parts model and non-maximum suppression. In: CVPR (2015)
29. Jia, Y.: Caffe: an open source convolutional architecture for fast feature embedding (2013). http://caffe.berkeleyvision.org/

R-CNN for Small Object Detection

Chenyi Chen[1(✉)], Ming-Yu Liu[2], Oncel Tuzel[2], and Jianxiong Xiao[1]

[1] Princeton University, Princeton, NJ, USA
chency05thu@gmail.com
[2] Mitsubishi Electric Research Labs (MERL), Cambridge, MA, USA

Abstract. Existing object detection literature focuses on detecting a big object covering a large part of an image. The problem of detecting a small object covering a small part of an image is largely ignored. As a result, the state-of-the-art object detection algorithm renders unsatisfactory performance as applied to detect small objects in images. In this paper, we dedicate an effort to bridge the gap. We first compose a benchmark dataset tailored for the small object detection problem to better evaluate the small object detection performance. We then augment the state-of-the-art R-CNN algorithm with a context model and a small region proposal generator to improve the small object detection performance. We conduct extensive experimental validations for studying various design choices. Experiment results show that the augmented R-CNN algorithm improves the mean average precision by 29.8% over the original R-CNN algorithm on detecting small objects.

1 Introduction

We have witnessed several breakthroughs in the field of visual object detection in the past decade, demonstrated by the ever-increasing performance improvement on the PASCAL VOC [1]. However, the object detection problem still remains largely unsolved as none of the state-of-the-art object detectors is close to perfect. Moreover, the performance on the PASCAL VOC can be misleading due to the dataset bias as pointed out by Torralba et al. [2]. It is expected that when the application domain has a very different bias to the one in the PASCAL VOC, the performance of the state-of-the-art detectors for the PASCAL VOC would degrade significantly.

In this paper, we study the small object detection problem. By small objects, we refer to objects with smaller physical sizes in the real world. We also limit our interest to the small objects that each occupies a small part of an image. This means that comparing to the PASCAL VOC where the majority of objects are big in the real world and each occupies a large portion of an image, we are considering an application domain with a selection bias toward small objects as shown in Fig. 1.

Electronic supplementary material The online version of this chapter (doi:10. 1007/978-3-319-54193-8_14) contains supplementary material, which is available to authorized users.

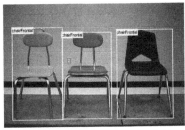

(a) Typical objects in PASCAL   (b) Small objects in the real world

Fig. 1. Detecting small objects with low-resolution inputs.

It is true that one can always have a higher resolution image or take a closer snapshot of a small object in order to detect it. But the low-resolution inputs for small objects is deeply embedded in the nature of visual perception, and a robust vision system should be able to deal with it. For example, the physical size of a typical desk and monitor is many times bigger than a mouse. As a human, when we see a desk with a monitor and a mouse, we recognize all of them in one shot. We do not look particularly closer to the mouse to put a large image at the center of our retina. It is desirable that a computer vision system possesses a similar capability.

Moreover, detecting small objects is itself an intriguing problem due to several unique challenges. First, there are much more possibilities for the locations of small objects. The precision requirement for accurate localization is several magnitudes higher than that for typical PASCAL VOC objects. Second, there are much fewer pixels available for small objects, which means much weaker signal for the detector to utilize. Third, there are only limited prior knowledge and experiences in this area since most of the prior works are tuned for the big object detection problem. Practically, there is no benchmark dedicated to such a task[1]. In fact, we do not have much understanding on how difficult the small object detection task is or how well existing object detectors work. In order to better assess the performance of an algorithm for the small object detection problem, we establish a small object detection benchmark.

The R-CNN algorithm [3,4], which extracts discriminative features using deep convolutional neural network from region proposals, has been established as the state-of-the-art approach for object detection as supported by the achieved impressive performance on the PASCAL VOC benchmark. In this paper, we extend the R-CNN algorithm to deal with the small object detection problem. Specifically, we propose a region proposal network tailored for capturing the "objectness" for small objects in order to obtain a small set of proposals while still keeping a high recall rate. We also propose a way to encode the context information from the surrounding areas of an object proposal. We show that the

---

[1] Although standard datasets such as the Microsoft COCO contains several "small" object categories, many of the instances of the objects in the "small" object categories occupy a large part of an image.

extended R-CNN algorithm achieves a mean Average Precision (mAP) of 23.5% on the benchmark dataset, which significantly outperforms a mAP of 18.1% achieved by the original R-CNN algorithm. We also present extensive experimental evaluations on various design choices for understanding their impacts to the small object detection performance.

1.1 Related Work

Earlier work on small object detection is mostly about detecting vehicles using hand-crafted features and shallow classifiers in aerial images [5,6]. In this paper, we cover a diverse set of small objects in the daily life and augment the state-of-the-art R-CNN algorithm for detecting them. [7] analyzes the influences of object characteristics on the performance of multiple detectors, with "object size" among the characteristics being studied. The results reveal that the detection accuracy drops as the object size becomes smaller, which provides some initial insight into the small object detection problem.

The PASCAL VOC [1] is the most widely used benchmark dataset for general object detection. It contains 20 object categories including "cow", "vehicle", and "dog". The object instances in the PASCAL VOC are usually large. Many of them occupy a major portion of the image. Our focus is on small objects where the object instance should only occupy a small portion of the image. In this sense, directly using the PASCAL VOC dataset is inappropriate. Microsoft COCO dataset [8] is proposed to advance the object detection techniques by placing it in the context of scene understanding, and the dataset contains many categories of small objects. To better represent the problem, we compose our small object detection dataset by using a subset of images from both the COCO dataset and the Scene UNderstanding database (SUN) [9], which also contains a large amount of small objects in various scenes.

[3,4] propose the R-CNN algorithm, which combines convolutional neural networks with bottom-up region proposals [10] for object detection. R-CNN significantly outperforms conventional approaches on the PASCAL VOC dataset and establishes the new state-of-the-art in object detection research. Recently, some work improves the region proposal generation part of R-CNN and obtain faster computation speed and more accurate detection performance. [11] generates region proposals using edge cues. [12] computes "objectness" of region proposals based on a convolutional neural network. The MultiBox method [13] directly predicts a set of class-agnostic bounding boxes along with a single object-ness score for each box, the method is not translation-invariant. [14] propose a translation-invariant Region Proposal Network (RPN) that shares convolutional layers with the detection network and achieve faster computation speed and better performance. The above algorithms are designed for detecting large objects in the PASCAL VOC. We focus on the small object detection problem and systematically study the applicability of the R-CNN style algorithms for detecting small object in the image.

Generally, context is useful for improving the object detection performance in natural scenes [15,16]. Based on R-CNN, [17] proposes a pipeline for action recognition using more than one regions. [18] proposes a multi-region object detection system that can steering the ConvNet to focus on different regions of the object. [19] use both segmentation and context to improve object detection accuracy. [20] studies the role of context in existing object detection approaches and further proposed a model that exploits both the local and global context. In this work, we also leverage the context information to get better performance.

Many researches have been shown to improve the localization accuracy of object detectors. [21] introduces a Bayesian optimization-based algorithm that iteratively searches for better bounding boxes for object detection. [22] casts object detection as an iterative classification problem and proposed Attention-Net which achieves more accurate localization. [23,24] propose object detection pipelines that completely eliminate region proposal generation stage by predicting category scores and bounding box locations altogether from feature maps. [25] shows the overall performance of object detection can also be improved by using image renderings for data augmentation.

1.2 Contributions

This paper makes the following contributions:

1. We propose a dataset containing diverse small objects to facilitate the study of the applicability of state-of-the-art deep learning-based object detectors for detecting small objects in the image.
2. From systematic experiment design and performance comparison, we augment the R-CNN algorithm, which boosts the small object detection performance by 29.8% on the benchmark dataset.

2   Small Object Dataset

We compose our dataset for the small object detection problem by using a subset of images from both the Microsoft COCO and SUN datasets. We call the dataset the "small object dataset". We manually select ten small object categories where the largest physical dimension of instances in the categories are smaller than 30 cm. The selected object categories are "mouse", "telephone", "switch", "outlet", "clock", "toilet paper", "tissue box", "faucet", "plate", and "jar". A small object is not necessarily small in the image. For instance, the "tissue box" may occupy a large portion of an image. We use the ground truth bounding box locations in the COCO and SUN datasets to prune out big object instances and compose a dataset containing purely small objects with small bounding boxes.

The statistics of the small object dataset is shown in Table 1. It contains about 8,393 object instances in 4,925 images. The "mouse" category has the largest number of object instances: 2,137 instances in 1,739 images. The "tissue box" category has the smallest number of instances: 103 instances in 100 images. All the object instances in our dataset are small. Median of relative areas (the ratio of the bounding box area over the image area) of all the object instances

Table 1. Statistics of our small object dataset. Relative area (%) of each instance is computed as the ratio of the bounding box area over the image area.

Category	mouse	telephone	switch	outlet	clock	toilet paper	tissue box	faucet	plate	jar
Number of images	1739	345	425	916	746	157	100	1094	419	252
Number of instances	2137	363	487	1210	814	175	103	1388	1005	711
Median relative area	0.35	0.38	0.08	0.08	0.25	0.40	0.58	0.43	0.37	0.29
Median top-10% area	2.76	1.99	0.33	0.37	1.92	1.43	1.94	2.02	2.40	1.57

Table 2. Median relative area (%) of the object categories in the PASCAL VOC.

Category	cat	sofa	train	dog	table	mbike	horse	bus	aero	bicycle
Median relative area	46.40	33.87	32.33	30.96	23.73	23.69	23.15	23.04	22.83	14.38

Category	person	bird	cow	chair	tv	boat	sheep	plant	car	bottle
Median relative area	8.14	8.03	6.68	6.09	5.96	3.82	3.34	2.92	2.79	1.38

in the same category is between 0.08% to 0.58%. This corresponds to 16×16 to 42×42 pixel^2 areas in a VGA image. As a comparison, median of relative areas of object categories in the PASCAL VOC dataset is between 1.38% to 46.40%, as shown in Table 2. Even the smallest object category is much larger than the biggest object category in our dataset.

Our small object dataset is considered more challenging than the PASCAL VOC in at least two ways: First, the appearance cue available for distinguishing a small object from background clutters is much less due to the small size. Second, the number of bounding box hypotheses for a small object in an image is much larger than that for a big object in the PASCAL VOC.

During evaluation, the small object dataset is split into two subsets: one for training and the other for testing. The number of object instances per category in the training set is roughly two times the corresponding number in the testing set. There are no common images between the two sets.

Performance Metric: We use the standard performance metric for comparing various object detection algorithms. An object bounding box hypothesis is considered as a true detection if its overlap ratio with the ground truth bounding box is greater than 0.5, where the overlapping ratio is measured using the Intersection over Union (IoU) measure. The detection algorithm returns a confidence score for each object bounding box hypothesis. We vary the threshold and compute the precision recall curve for each object. We then use the average precision of the curve to report the performance of the detector for an object category. The performance of the detector for the entire dataset is measured using the mean Average Precision (mAP) score.

3   R-CNN for Small Object Detection

The R-CNN algorithm [3] has been established as the de facto algorithm for deep learning-based object detection. It significantly outperforms conventional approaches in the PASCAL VOC by capitalizing the following two insights: First, it uses object proposals rather than sliding windows. Before the R-CNN, most object detectors such as DPM adopt a image pyramid plus sliding window approach [26] to generate potential object locations and handle various scales. In the R-CNN pipeline, a fixed number (e.g. 2000) of boxes are proposed per image which most likely contain the target objects. The problem of various scales is also handled automatically by the proposal generation. Fewer but better proposals contribute a lot to the good performance of the R-CNN. Second, it leverages ImageNet pre-trained deep neural network models, which is then fine-tuned using the PASCAL VOC. The pre-training process is proven to be crucial to the performance. Without the pre-training process, the R-CNN works poorly.

Given the region proposals, training an R-CNN object detector generally composing two major steps: supervised pre-training and domain-specific fine-tuning. During supervised pre-training, ImageNet data are used to train the entire network from scratch. In the domain-specific fine-tuning, the weights of the network are initialized by the pre-trained model and trained by the domain-specific data (for example, PASCAL VOC). Training images for the ConvNet are region proposal patches being resized and warped to the required resolution (e.g. 227×227). Both the positive and negative patches are sampled from the region proposals according to certain overlap thresholds.

In the following sections, we investigate into various necessary changes for successfully extending the R-CNN algorithm for small object detection. We follow the same procedure to train our small object detection networks, but based on the nature of the problem, in the domain-specific fine-tuning stage, we only sample the negative patches from the region proposals. The positive patches are generated by randomly deviating from the ground truth box. We also try to balance the positive patches of each category by sampling complementary number of positive patches per category per instance.

The Fast R-CNN algorithm [4] simplifies the R-CNN pipeline by proposing a *ROIPooling* layer that crops the proposals from the feature map instead of the input image. Although the Fast R-CNN reduces the time cost and further improves the performance on PASCAL VOC, the core idea of R-CNN is intact. Adding the *ROIPooling* leads to the primary difference between the two methods: in R-CNN, all the proposal boxes (even small ones) are resized to a canonical size, this means that full feature map is generated for each proposal box at the last pooling layer. However, in Fast R-CNN, a small proposal box gets mapped to only a small map (sometimes 1*1*n) at the last pooling layer. Such a small feature map may lack necessary information for the classification step, adding unnecessary uncertainty into the study. Thus, we feel that the R-CNN is more suitable than the Fast R-CNN algorithm in this case. Moreover, as we do not have much knowledge about how the deep learning-based method works on small objects, the original R-CNN pipeline provides a more convenient way to better understand the problem. For example, it is more convenient to visualize the neuron responses of the R-CNN than the Fast R-CNN. By working with proposal patch input, analyzing the effects of up-sampling and context is also easier. Thus in this paper, we choose to follow the original R-CNN pipeline.

Moreover, in our work, we do not implement bounding box regression. Although bounding box regression is proven as an effective way to increase the localization accuracy, it is not a major issue for small object detection. We believe the challenges come from the region proposal generation and classification, while bounding box regression will be less useful on poor proposal and classification results. So in this paper, we will only focus on generating better region proposals and searching for stronger classifiers.

For all the experiments, our training pipeline consists of two stages: in the first stage, the weights of the ConvNets are initialized with corresponding ImageNet pre-trained models. We then fix the convolutional layers and only update fully connected layers for 8000 iterations with a learning rate of 0.0005. In the second stage, all the layers are updated with a learning rate of 0.00005. We use stochastic gradient descent with momentum of 0.9 for optimization, the batch size is 100. The training is terminated after 80000 iterations.

3.1   Small Proposal Generation

Selective search and edge box are two popular choices for object proposal generation. They use mid-level image cues, such as segments and contours and are object category-agnostic. While the selective search and edge box work well for generating proposals for big objects in the PASCAL VOC. We empirically find them rendering unsatisfactory results for generating small object proposals even after an exhaustive search of the algorithm parameter space. With 2000 object proposals per image, the typical recall rate is lower than 60%, leading to poor performance for detecting small objects using R-CNN. Further investigation shows that both of the algorithms favor salient objects with closed contours and distinctive colors. Since the nature of the small objects are non-prominent, they are non-ideal for small object proposal generation.

The Region Proposal Network (RPN) [14] is the current state-of-the-art method for proposal generation. It attaches nine anchor boxes - derived from three different aspect ratios at three different scales - to each spatial dimension of the feature map from the $conv5\_3$ layer of the VGG16 network [27] for region proposal classification and bounding box regression. The three aspect ratios used are 0.5 (landscape), 1 (square), and 2 (portrait), and the areas of the square shape bounding boxes at the three scales are 128^2, 256^2, and 512^2 pixel^2, respectively. The RPN achieves good performance for big object proposal generation. But we find that directly applying the RPN to the small object proposal generation results in poor performance. Several modifications are necessary as described below.

We first notice that the RPN anchor boxes are too large. Even the smallest anchor box is much bigger than most instances in our small object dataset. Based on the statistics of the small object size in the dataset, we choose 16^2,

Table 3. Recall rate (%) of the region proposal generation methods.

Method	mouse	tel.	switch	outlet	clock	t. paper	t. box	faucet	plate	jar	Average
DPM, 300 prop. per category	70.9	58.0	70.5	80.9	79.1	**86.6**	76.2	69.3	58.0	**63.4**	71.3
RPN original, 300 prop	85.0	63.4	78.7	73.1	66.0	76.1	50.0	76.0	58.6	31.8	65.9
RPN modified, 300 prop	**88.4**	**82.4**	**80.9**	**83.1**	**86.9**	83.6	**88.1**	**86.4**	**71.9**	58.9	**81.1**
DPM, 500 prop. per category	73.2	61.8	74.3	82.2	82.5	86.6	78.6	73.9	62.2	**72.9**	74.8
RPN original, 500 prop	85.7	64.9	79.2	74.7	68.4	77.6	57.1	78.0	61.4	38.2	68.5
RPN modified, 500 prop	**89.9**	**86.3**	**82.0**	**84.2**	**88.9**	**91.0**	**90.5**	**89.8**	**76.4**	67.1	**84.6**
DPM, 1000 prop. per category	76.5	67.2	78.7	84.2	86.9	89.6	81.0	79.7	68.3	**81.7**	79.4
RPN original, 1000 prop	87.0	70.2	79.8	75.6	71.7	82.1	66.7	80.9	66.4	46.2	72.7
RPN modified, 1000 prop	**92.4**	**93.1**	**83.6**	**86.0**	**90.2**	**97.0**	**92.9**	**93.3**	**82.5**	76.4	**88.7**
DPM, 2000 prop. per category	80.2	72.5	82.0	86.2	89.9	92.5	83.3	83.3	73.9	**87.8**	83.2
RPN original, 2000 prop	87.7	75.6	80.3	76.0	75.1	89.6	76.2	84.0	69.4	54.6	76.9
RPN modified, 2000 prop	**94.1**	**94.7**	**85.3**	**87.1**	**90.9**	**97.0**	**97.6**	**95.3**	**86.1**	85.2	**91.3**

40^2, and 100^2 pixel^2 for the square shape anchor box sizes. For the aspect ratios, we keep the original values used in the original paper. We further notice that the stride length of the *conv5_3* feature map, which is 16 pixels, is too large. It is larger than most of the "switch" and "outlet" objects in our dataset. The other candidate feature maps for attaching the anchor boxes are *conv2_2*, *conv3_3* and *conv4_3*. We empirically compare the performance and find that *conv4_3* renders the best performance for small object proposal generation. The *conv4_3* feature map has a theoretical receptive field of 92×92 pixel^2, which appears to be more appropriate than 196×196 pixel^2 from the *conv5_3* feature map.

For benchmarking the performance of deep learning for small object detection, we also apply the Deformable Part Model (DPM) [28] detector to detect the small object. The DPM detector was the state-of-the-art algorithm on the PASCAL VOC dataset before the R-CNN algorithm. The DPM detector is based on the Histogram of Oriented Gradient (HOG) features and latent support vector machine. To accommodate the small object size, we down-sample the root and part template sizes of the DPM detector by half. The DPM is a category-specific object detector. We train a DPM detector for each class.

Evaluation: In Table 3, we compare the recall rate of the proposal generation methods for the small object detection problem. Specifically, we compare the recall performance of using the DPM detector, the original RPN, and the proposed modification of RPN. We vary the number of proposals per image and show the recall numbers. The DPM is category-specific. We use the top scored bounding boxes from all the classes for computing the recall rate. The effective number of bounding boxes are 10 times the number of the RPN. As discussed, the modified RPN renders the best recall performance. For 2000 proposals, the recall rate for the "tissue box" is about 97.6%. The recall rate for the "jar" is the worst. It is 85.2% with 2000 proposals. However, this is still much better than 54.6% achieved by the original RPN method. From the table, we also find that the original RPN algorithm renders worse performance than the DPM algorithm. The proposed modification of the RPN algorithm considers the nature of small object and largely improve the performance. Overall, the proposed modification

Table 4. Up-sampling effects. Both networks are trained and tested with DPM proposals, 500 per image per category.

	mouse	tel.	switch	outlet	clock	t. paper	t. box	faucet	plate	jar	Average
Partial AlexNet	29.8	3.1	5.3	18.0	19.6	15.5	1.9	6.7	**5.4**	2.0	10.7
Full AlexNet	**42.9**	**7.7**	**9.4**	**22.7**	**28.2**	**26.7**	**15.7**	**18.6**	**5.4**	**3.4**	**18.1**
Median size	32.4	54.0	25.5	25.8	38.5	73.1	90.0	50.8	39.2	29.4	45.9
Up-sampling ratio	7.0	4.2	8.9	8.8	5.9	3.1	2.5	4.5	5.8	7.7	5.8

achieves an average recall rate of 91.3%, which is relatively 19% better than the original RPN method.

3.2   Up-Sampling

The first question encountered as extending the R-CNN algorithm to the small object detection is whether to aggressively up-sample the image or not. Unlike the objects in the PASCAL VOC, the bounding boxes of the small objects in our dataset are very small. In Table 4, we show the median bounding box size (square root of the box area) of the objects per category and the corresponding up-sampling ratios required to match the input size (227×227 in this case) of the deep convolutional neural networks. We find that, generally, 6 to 7 times up-sampling is required, which will introduce a large amount of up-sampling artifacts. One way to reduce the artifacts is to use low resolution small input patches with a ConvNet deviated from the standard pre-trained models. For example, we can exclude the pre-trained weights in the last few fully connected layers and only use the convolution layers. However, using small patches as input may create other disadvantages:

1. The receptive field over small patch is larger than the same receptive field over large patch. This means given a small patch, the network can only look at the object in a coarse scale, thus possibly loses useful information regarding the parts of the object.
2. Small input patch produces lower dimensional feature vector, thus the size of the vector may not be large enough to accommodate all the crucial information.
3. Since all the fully connected layers need to be trained from scratch, we only utilize the partial strength of the pre-trained models.

To answer this question. We design an experiment comparing the two solutions using the following two networks:

1. Partial AlexNet [29]: Using *conv1* to *pool5* layers from the AlexNet. The object proposals are re-scaled to 67×67. The *pool5* layer produces a $1 \times 1 \times 256$ feature vector, which is used to get the final classification scores.
2. Full AlexNet: Using the entire AlexNet structure. The object proposals are up-sampled to 227×227 and contains a large amount of artifacts.

The results are shown in Table 4. From the table, we found that although with the up-sampling artifacts. The full AlexNet still renders much better performance. So in our following experiments, we will only use the aggressively up-sampled proposal patches as input.

3.3   Context

Context is an important cue for object detection. We expect that it will be even more important for small object detection, since small objects are simple in shape and usually only cover a small image region. The feature extracted from the proposal region is less discriminative, so when only given the proposal region, it can be very difficult to recognize, even for human beings.

We investigate into several methods for incorporating context information to boost small object detection performance, and based on the R-CNN algorithm, we propose a simple method that works quite well. When given an object proposal in an image, in addition to cropping the proposal region, we crop the corresponding context region enclosing the proposal region, with the center coinciding with the center of the proposal region. The context region is set to be several times larger than the proposal region. We then feed both regions into a neural network. The neural network consists of three sub-networks where the first one takes the proposal region as input, the second one takes the context region as input, and the last one takes the concatenation of the outputs of the others as input and computes the final classification score. We call this neural network ContextNet, and the structure is shown in Fig. 2.

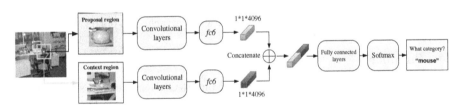

Fig. 2. ContextNet: the neural network for integrating context information. The two front-end sub-networks take proposal region patches and context region patches as input respectively, the back-end sub-network takes in the concatenation of the two feature vectors and computes the final classification score.

The two front-end sub-networks have identical structure. Each consists of a few convolutional layers followed by one fully connected layer, which are derived from the first six layers of the AlexNet (or the equivalent layers of VGG16). Input image regions to the two sub-networks are resized to 227×227 (224×224 for VGG16) patches. Each of the front-end sub-networks outputs a 4096 dimensional feature vector. The back-end sub-network consists of two fully connected layers and outputs the predicted object category label. During training, the front-end sub-networks are initialized using the ImageNet pre-trained model. However, the weights of the two sub-networks evolve separately - the weights are not shared.

Table 5. Results of ContextNet. All the networks are trained (2000 per image) and tested (500 per image) with modified RPN proposals.

Method	mouse	tel.	switch	outlet	clock	t. paper	t. box	faucet	plate	jar	mAP
Baseline AlexNet	48.2	10.6	8.9	21.4	32.3	**34.1**	**23.0**	25.1	6.7	3.6	21.4
ContextNet (AlexNet, 3x)	54.8	9.1	12.8	**30.7**	28.5	28.4	18.6	**30.8**	**10.6**	**6.4**	23.1
ContextNet (AlexNet, 7x)	**56.4**	**12.2**	**12.9**	26.3	**32.7**	34.0	18.7	26.8	9.9	4.6	**23.5**

Evaluation: We evaluate the performance of the AlexNet-based ContextNet with two variants: the 3x and 7x models. The context region of the 3x model is three times larger than the proposal region in both height and width dimension. The 7x model is defined in a similar way and it uses a very larger context region. We also include the AlexNet R-CNN model as the baseline.

The performance is shown in Table 5. We find that the neural network with context integration achieves better performance than the baseline model. The improvement with the 7x model is slightly better than that with the 3x model. Overall, the relative mAP improvement over the baseline are 7.9% and 9.8% for the 3x and 7x models, respectively. We also investigate a ConvNet-based co-occurrence model, which leverages the detection of big objects to better localize the small objects. The spatial relation between the big and small objects are posed as learnable parameter integrated into an end-to-end training framework. However, we find this method is only effective when attached to the Baseline AlexNet, it does not make any improvement when attached to both ContextNets.

3.4   Summary

In Table 6, we list the average precision of our R-CNN models on small object dataset, we also list the DPM as a baseline. Not surprising at all, DPM is significantly outperformed by all the deep learning-based models. And deeper network (VGG16) has superior performance over shallow network (AlexNet).

To demonstrate the influence of region proposal quality on the final average precision, we compare two AlexNet models: one using the DPM detection outputs as proposals, and the other use the modified RPN proposals. From Table 3, we know the modified RPN proposals have much higher recall rate than the DPM proposals, and consequently, the AlexNet trained on modified RPN proposals performs much better (Table 6).

Fewer Proposals: In Table 7, we show the average precision of the ContextNet using 7x context region on different number of proposals per image. We find it achieves higher average precision on a smaller number of proposals. Small object detection is very vulnerable to false positives. Using a smaller number of proposals eliminates a large amount of potential false positives and improves the average precision. 300 proposals per image produces the best performance.

Table 6. Results of DPM, AlexNet R-CNN, and VGG16 R-CNN. The AlexNet in row 2 is trained and tested with DPM proposals, 500 per image per category. The AlexNet in row 3 and the VGG16 in row 4 are trained (2000 per image) and tested (500 per image) with modified RPN proposals.

Method	mouse	tel.	switch	outlet	clock	t. paper	t. box	faucet	plate	jar	mAP
DPM	18.9	0.3	1.9	23.0	9.1	18.3	2.0	5.7	2.4	0.4	8.2
DPM prop. + AlexNet	42.9	7.7	9.4	22.7	28.2	26.7	15.7	18.6	5.4	3.4	18.1
RPN prop. + AlexNet	48.2	10.6	8.9	21.4	**32.3**	**34.1**	23.0	25.1	6.7	3.6	21.4
RPN prop. + VGG16	**56.8**	**16.4**	**14.2**	**31.1**	31.9	29.4	**23.4**	**31.3**	**9.3**	**4.2**	**24.8**

Table 7. Results of ContextNet. Both networks are trained (2000 per image) and tested (various) with modified RPN proposals.

Method	mouse	tel.	switch	outlet	clock	t. paper	t. box	faucet	plate	jar	mAP
AlexNet, 7x, 300 prop	**56.9**	**12.4**	**13.6**	**28.0**	32.4	35.6	17.9	**27.2**	9.8	**5.1**	**23.9**
AlexNet, 7x, 500 prop	56.4	12.2	12.9	26.3	**32.7**	34.0	18.7	26.8	**9.9**	4.6	23.5
AlexNet, 7x, 1000 prop	55.4	11.2	11.4	25.7	29.5	**37.6**	18.5	25.7	9.1	4.2	22.8
AlexNet, 7x, 2000 prop	54.9	10.9	10.9	24.6	29.8	35.0	**19.5**	24.8	8.4	3.9	22.3
VGG16, 7x, 300 prop	**60.6**	13.7	**21.5**	**41.5**	**37.7**	33.3	**22.0**	30.3	15.8	7.2	**28.4**
VGG16, 7x, 500 prop	60.2	14.0	20.0	40.7	36.4	**35.7**	20.4	**31.4**	**16.0**	**7.7**	28.3
VGG16, 7x, 1000 prop	59.6	**14.6**	18.9	39.9	36.2	34.9	18.7	30.9	15.3	7.4	27.6
VGG16, 7x, 2000 prop	58.4	13.7	18.1	38.2	33.6	33.0	18.5	30.1	14.0	7.1	26.5

Stronger Pre-trained Model: We also experiment with replacing the AlexNet with the VGG16 net to verify if the performance boost in the big object detection due to the stronger pre-trained model is also true for small object detection. The results are shown in Table 7. From the table, we find that the stronger pre-trained model leads to improved performance for all the proposal numbers.

In Fig. 3, we show the detection results of the ContextNet (AlexNet, 7x) model on several images in the testing set. We use a fix threshold and show the output bounding boxes after non-maximum suppression. Since the target objects are too small for visualization, we put a zoom-in window to highlight the output bounding boxes. From the figure, one can see that the small object detector works well on many categories. It can detect object instances with very low resolution.

Fig. 3. Examples of the detection results on some testing images.

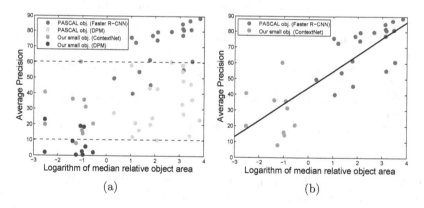

(a)   (b)

Fig. 4. Comparison of methods on small objects and PASCAL. In both (a) and (b), a marker represents the mAP of a detector on an object category. Specifically, red represents Faster R-CNN on PASCAL objects, green represents our ContextNet on our small objects, light and dark blue represent DPM on PASCAL objects and our small objects, respectively. (Color figure online)

As one of the major purposes of this paper is to study the applicability of the state-of-the-art object detection algorithms to the small object detection problem, by summarizing the findings, we now can answer this question. Our answer is based on two observations: (1) before the R-CNN algorithm, the state-of-the-art object detector on PASCAL VOC was DPM. Since our work is a preliminary stage of small object detection, we think it is comparable to DPM on PASCAL. Shown in Fig. 4a, the average precision of our best model, e.g. ContextNet (VGG16, 7x), on small object categories is distributed in the same range (indicated by the black dashed lines) as that of DPM on PASCAL. Numerically, on the small object dataset, our deep learning-based algorithm (mAP 28.3) has close performance to the DPM on PASCAL (mAP 33.7). (2) on PASCAL VOC, the R-CNN style algorithm improves the mAP of DPM from 33.7 to 70.4. While on the small object dataset, our best model improves the mAP of DPM from 8.2 to 28.3, which indicates the deep learning models are still very effective on small objects. Thus, we think they are applicable to small object detection problem.

4 Visualization

We visualize the neurons in our ContextNet (AlexNet, 7x) model to better understand what the network learns as learning to detect small objects. We plot the training patches that excite each neuron in the $fc6$ layer most for both the proposal and context front-end sub-networks.

In Fig. 5, we display the top 20 image patches with the highest response to several neurons in the proposal sub-network. We find that the patches are dominated by mouse and round shape objects (e.g. row 1 to row 5). This partially explains why the network performs better for the "mouse" and "clock" categories. We also find the neurons in row 2 fire when seeing Apple mouses or

Fig. 5. The proposal patches that have the largest excitation to the neurons in $fc6$ of proposal sub-network. Please refer to the main text for further discussions.

similar shapes, while those in row 9 response to oval pattern. In row 10, we can see outlet patches are mixed with speaker and clock patches. The neurons in row 11 and row 12 correspond to a monitor detector and a toilet detector. This is surprising since our dataset does not contain these two object category labels. The figure also suggest that there is not much high-level features to distinguish small objects. Hence, the network relies on basic shape patterns to detect small objects (e.g. row 6 to row 8).

In Fig. 6, we display the top 8 image patches with the highest response to several neurons in the context sub-network. Since the 7x context region covers a large image area, the context patches fire for the same neuron have more diverse patterns. As expected, strong scene-specific patterns exist on many neurons. The neuron in row 1 looks at computers, and the neuron in row 2 evolves for bedroom scene. The neurons in row 3, 4, and 5 respond to tables, toilets and sinks, respectively. The neuron in row 6 activates on kitchen scene. These neurons provide context information to resolve the ambiguity in the proposal patches.

Fig. 6. The context patches that have the largest excitation to the neurons in *fc6* of context sub-network. Please refer to the main text for further discussions

5 Conclusion

We extended the state-of-the-art R-CNN algorithm to deal with the small object detection problem. We composed a small object dataset to facilitate the study. Through detailed experimental validation and analysis, we found that, with a carefully designed region proposal network and context modeling, the deep learning-based object detection algorithm achieves similar performance improvement over the conventional approach for small object detection as it did for big object detection.

References

1. Everingham, M., Van Gool, L., Williams, C.K., Winn, J., Zisserman, A.: The PASCAL Visual Object Classes (VOC) challenge. Int. J. Comput. Vis. **88**, 303–338 (2010)
2. Torralba, A., Efros, A., et al.: Unbiased look at dataset bias. In: 2011 IEEE Conference on Computer Vision and Pattern Recognition (CVPR), pp. 1521–1528 (2011)
3. Girshick, R., Donahue, J., Darrell, T., Malik, J.: Rich feature hierarchies for accurate object detection and semantic segmentation. In: 2014 IEEE Conference on Computer Vision and Pattern Recognition (CVPR), pp. 580–587 (2014)
4. Girshick, R.: Fast R-CNN. In: Proceedings of the IEEE International Conference on Computer Vision, pp. 1440–1448 (2015)
5. Kembhavi, A., Harwood, D., Davis, L.S.: Vehicle detection using partial least squares. IEEE Trans. Pattern Anal. Mach. Intell. **33**, 1250–1265 (2011)
6. Morariu, V., Ahmed, E., Santhanam, V., Harwood, D., Davis, L.S., et al.: Composite discriminant factor analysis. In: 2014 IEEE Winter Conference on Applications of Computer Vision (WACV), pp. 564–571 (2014)
7. Hoiem, D., Chodpathumwan, Y., Dai, Q.: Diagnosing error in object detectors. In: Fitzgibbon, A., Lazebnik, S., Perona, P., Sato, Y., Schmid, C. (eds.) ECCV 2012. LNCS, vol. 7574, pp. 340–353. Springer, Heidelberg (2012). doi:10.1007/978-3-642-33712-3_25
8. Lin, T.-Y., Maire, M., Belongie, S., Hays, J., Perona, P., Ramanan, D., Dollár, P., Zitnick, C.L.: Microsoft COCO: common objects in context. In: Fleet, D., Pajdla, T., Schiele, B., Tuytelaars, T. (eds.) ECCV 2014. LNCS, vol. 8693, pp. 740–755. Springer, Heidelberg (2014). doi:10.1007/978-3-319-10602-1_48
9. Xiao, J., Ehinger, K.A., Hays, J., Torralba, A., Oliva, A.: Sun database: exploring a large collection of scene categories. Int. J. Comput. Vis. **119**, 1–20 (2014)
10. Uijlings, J.R., van de Sande, K.E., Gevers, T., Smeulders, A.W.: Selective search for object recognition. Int. J. Comput. Vis. **104**, 154–171 (2013)
11. Zitnick, C.L., Dollár, P.: Edge boxes: locating object proposals from edges. In: Fleet, D., Pajdla, T., Schiele, B., Tuytelaars, T. (eds.) ECCV 2014. LNCS, vol. 8693, pp. 391–405. Springer, Heidelberg (2014). doi:10.1007/978-3-319-10602-1_26
12. Kuo, W., Hariharan, B., Malik, J.: DeepBox: learning objectness with convolutional networks. arXiv preprint arXiv:1505.02146 (2015)
13. Erhan, D., Szegedy, C., Toshev, A., Anguelov, D.: Scalable object detection using deep neural networks. In: 2014 IEEE Conference on Computer Vision and Pattern Recognition (CVPR), pp. 2155–2162 (2014)
14. Ren, S., He, K., Girshick, R., Sun, J.: Faster R-CNN: towards real-time object detection with region proposal networks. In: Advances in Neural Information Processing Systems, pp. 91–99 (2015)
15. Divvala, S.K., Hoiem, D., Hays, J.H., Efros, A., Hebert, M., et al.: An empirical study of context in object detection. In: 2009 IEEE Conference on Computer Vision and Pattern Recognition (CVPR), pp. 1271–1278 (2009)

16. Torralba, A., Murphy, K.P., Freeman, W.T., Rubin, M., et al.: Context-based vision system for place and object recognition. In: Proceedings of the IEEE International Conference on Computer Vision, pp. 273–280 (2003)
17. Gkioxari, G., Girshick, R., Malik, J.: Contextual action recognition with R*CNN. arXiv preprint arXiv:1505.01197 (2015)
18. Gidaris, S., Komodakis, N.: Object detection via a multi-region and semantic segmentation-aware CNN model. arXiv preprint arXiv:1505.01749 (2015)
19. Zhu, Y., Urtasun, R., Salakhutdinov, R., Fidler, S.: segDeepM: exploiting segmentation and context in deep neural networks for object detection. arXiv preprint arXiv:1502.04275 (2015)
20. Mottaghi, R., Chen, X., Liu, X., Cho, N.G., Lee, S.W., Fidler, S., Urtasun, R., et al.: The role of context for object detection and semantic segmentation in the wild. In: 2014 IEEE Conference on Computer Vision and Pattern Recognition (CVPR), pp. 891–898 (2014)
21. Zhang, Y., Sohn, K., Villegas, R., Pan, G., Lee, H.: Improving object detection with deep convolutional networks via Bayesian optimization and structured prediction. arXiv preprint arXiv:1504.03293 (2015)
22. Yoo, D., Park, S., Lee, J.Y., Paek, A., Kweon, I.S.: AttentionNet: aggregating weak directions for accurate object detection. arXiv preprint arXiv:1506.07704 (2015)
23. Redmon, J., Divvala, S., Girshick, R., Farhadi, A.: You only look once: unified, real-time object detection. arXiv preprint arXiv:1506.02640 (2015)
24. Liu, W., Anguelov, D., Erhan, D., Szegedy, C., Reed, S.: SSD: single shot multibox detector. arXiv preprint arXiv:1512.02325 (2015)
25. Pepik, B., Benenson, R., Ritschel, T., Schiele, B.: What is holding back convnets for detection? In: Gall, J., Gehler, P., Leibe, B. (eds.) GCPR 2015. LNCS, vol. 9358, pp. 517–528. Springer, Heidelberg (2015). doi:10.1007/978-3-319-24947-6_43
26. Liu, M.Y., Mallya, A., Tuzel, O., Chen, X.: Unsupervised network pretraining via encoding human design. In: 2016 IEEE Winter Conference on Applications of Computer Vision (WACV), pp. 1–9. IEEE (2016)
27. Simonyan, K., Zisserman, A.: Very deep convolutional networks for large-scale image recognition. arXiv preprint arXiv:1409.1556 (2014)
28. Felzenszwalb, P.F., Girshick, R.B., McAllester, D., Ramanan, D.: Object detection with discriminatively trained part-based models. IEEE Trans. Pattern Anal. Mach. Intell. 32, 1627–1645 (2010)
29. Krizhevsky, A., Sutskever, I., Hinton, G.E.: ImageNet classification with deep convolutional neural networks. In: Advances in Neural Information Processing Systems, pp. 1097–1105 (2012)

Image Set Classification via Template Triplets and Context-Aware Similarity Embedding

Feng-Ju Chang$^{(\boxtimes)}$ and Ram Nevatia

Institute for Robotics and Intelligent Systems,
Univeristy of Southern California, Los Angeles, USA
fengjuch@usc.edu

Abstract. We present a template-triplet-based embedding approach to optimize the ensemble SoftMax similarity between templates (sets) for improved image set classification. More specifically, a triplet is created among "three" whole templates or subtemplates of images to incorporate the (sub)template structure into metric learning. To further account for intra-class variations of images, we introduce a factorization technique to integrate image-specific context for learning sample-specific embedding. We evaluate our approach on several benchmark datasets, and demonstrate its effectiveness for image set classification.

1 Introduction

With the growth of video data and camera networks, image *set* classification problem has attracted significant attention recently in computer vision and pattern recognition. One representative application is set-based face recognition, where a single training or testing "unit" is a set of face images or a video rather than one image. Given multiple images to describe different aspects of a person of interest, the face recognition accuracy can be improved with high potential.

There has been an increasing number of methods addressing the fundamental component in image set classification; i.e., the set-to-set similarity. According to how such a similarity is computed, existing methods can be divided into two paradigms: (1) "Set fusion followed by set matching" [1–18], which first constructs a single representation for each set and compares two sets according to such representations. (2) "Set matching followed by set score fusion" [19,20], which first computes matching scores between images of two sets and then pools those scores by late fusion schemes such as the average or max fusion.

The methods in paradigm (1) usually assume a certain distribution on the sets and/or require a large amount of samples in a set for the robust set modeling. Unfortunately, it is often not the case in practice that a set may have few or even a single image, or contain a *mixture* of images and video frames. This kind of set is called *template* in the recently available IARPA Janus Benchmark (IJB-A) dataset for unconstrained face recognition [21]. Because of the existence of

Electronic supplementary material The online version of this chapter (doi:10.1007/978-3-319-54193-8_15) contains supplementary material, which is available to authorized users.

© Springer International Publishing AG 2017
S.-H. Lai et al. (Eds.): ACCV 2016, Part V, LNCS 10115, pp. 231–247, 2017.
DOI: 10.1007/978-3-319-54193-8_15

multi-media in a template, the intra-class variation becomes much larger than sets involving only a single media such as Labeled Face in the wild [22] (multiple images as a set) and YouTube face dataset [23] (a video as a set). Thus, in this paper we focus on paradigm (2) for its wide applicability. *In the following, the word "template" is used interchangeably with "set" in the paper because of their common usage in literature.*

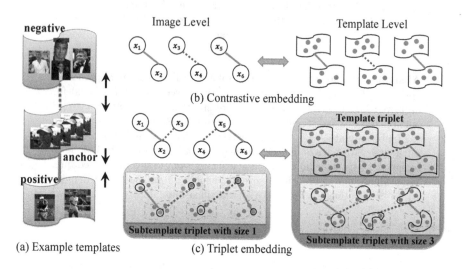

Fig. 1. Example of templates and template triplet embedding: (a) shows an example of anchor, positive, and negative templates. The goal is to enlarge similarity of positive pairs (solid-blue edges) while reducing that of negative ones (dashed-red edges). Template triplet embedding, right hand side of (c), is compared to the conventional contrastive embedding (b), and image triplet embedding, left hand side of (c); the circle denotes an image sample of the template. Previous metric learning works for template-based classification are usually based on the contrastive embedding or image-triplet embedding. Our approach, on the other hand, introduces the (sub)template triplets to take the template structure into account. Note that the image-triple embedding can be seen as a special case when the subtemplate size equals 1. Better view in color. (Color figure online)

To obtain more discriminative similarity among templates, a commonly used technique is distance or similarity metric learning, with either *contrastive* embedding [3,6,7,9,12,18,24] or *triplet* embedding [11,25,26] as illustrated in Fig. 1. Triplet embedding, which simultaneously considers a positive and negative pair of samples w.r.t. an anchor sample, has been demonstrated to outperform contrastive embedding [25–27]. Previous methods, however, either work on the image level or generate a single representation for each template by average pooling (i.e., paradigm (1)) before constructing triplets. The structure within templates, which is likely to convey significant discriminative information, is thus ignored. Besides, due to the large intra-class variations, it is hard to enforce all the image samples in a template (of the same class) to be close in the embedding space by a

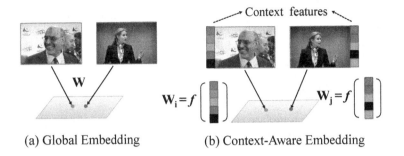

(a) Global Embedding   (b) Context-Aware Embedding

Fig. 2. Global and context-aware embedding: A global embedding is learned generally in previous work. We propose integrating the context features to achieve sample-specific embedding, where $f(.)$ consists of three factorized matrices as described in Sect. 3.3.

global embedding (shared for all samples) or by the class-specific embedding [11] (shared for samples of the same class). This work distinguishes itself with the following two contributions to address the above problems.

First, we introduce a *template*-triplet-based embedding approach to optimize template-to-template similarity computed by the ensemble SoftMax function [19, 20] — the ensemble SoftMax function fuses image-to-image scores at multiple scales (i.e., paradigm (2)) and has been shown effective in [19,20] as well as in our experiments. Different from *image*-triplet embedding [25,27], our triplet can be created not only on the entire template, but also on the *subtemplate* of samples with any reasonable size. Note that a template triplet becomes an image triplet when the *subtemplate* size equals to 1. Therefore, the proposed approach generalizes the triplet embedding to subtemplates of a predefined size. To our best knowledge, we are the first to apply triplet based metric learning on the "set matching followed by set score fusion" paradigm for the image set classification. An illustration of this approach is presented in Fig. 1.

Second, inspired by the scene-specific image captioning task [28,29], we propose a "context-aware" metric learning approach by integrating image-specific context into metric learning so as to achieve distinct embedding for every image, as illustrated in Fig. 2. The image-specific context, in face recognition task, can be defined as the capturing condition and the measurable attributes of a face (e.g., poses and expressions). Both of them may be provided by metadata of the dataset or by the existing face attribute detectors. The metadata, however, are usually inaccessible. Attributes could constitute useful context but the state-of-art attribute detectors seem to be too noisy, which hinders the performance improvements. Hence, we exploit the image features themselves as context. Although these features are also used to generate matching scores, the ways of using them on the context (via matrix factorization) and matching (with ensemble softmax similarity) are quite different. Of course, all features, including, attributes, derive from the image so are correlated to some extent. Furthermore, a mechanism to alleviate the influence of noisy context is also introduced. Experiments on four benchmark datasets for face template classification [21,23,30,31] demonstrate the efficacy of the proposed approach.

The rest of this paper is organized as follows. Related work is introduced in Sect. 2, and Sect. 3 presents the proposed approach. The experimental results and conclusion come in Sects. 4 and 5.

2   Related Work

The literature of image template classification and metric learning is quite extensive, so our survey focuses on the work most relevant to the establishment of the proposed method.

2.1   Image Template-to-Template Similarity Computation

Set Fusion Followed by Set Matching: In [12,25], the sample mean is used to represent a template before matching by the inner product. It is risky to take the average in the feature level because some discriminative information maybe lost. Besides, averaging over different medias may destroy the intrinsic structure within each media. [1–3] represent a template by the convex or affine combination of samples in a template and [4,5] use the subspace representation by applying PCA on the templates. The templates are then matched by Eulidean distance or principle angles.

In [6,7,32], geometric structures such as Riemannian manifolds are exploited, which assume a single or multivariate Gaussian distributions on templates [18]; this assumption could be easily violated in practice. To handle arbitrary distribution in a template, [18] proposed to model the image template as probability distribution functions using kernel density estimations and the Log-Euclidean distance or K-L divergence are employed for matching. To capture the nonlinear variations, nonlinear manifold modeling methods and the associated manifold-to-manifold distance are introduced in [8,9,11]. In the above approaches, a large amount of samples is usually needed to model a distribution or manifold well, which hinders them to be applicable on the dataset with templates of extremely small sizes. Instead of using a single statistics to model a set, [33] represents an image set with the sample mean, sample covariance, and the Gaussian mixture model (GMM) where they bridge the gap between the Euclidean space and Riemannian manifold.

Set Matching Followed by Set Score Fusion: Different from the approaches involving the template modeling before matching, image-to-image matching are first performed among templates in [19,20], the ensemble softmax fusion (which is the ensemble of average, max, and weighted average score fusions) is then applied to obtain the template-to-template similarity. This method does not assume any distributions for the template and can handle the template with extreme size. Besides, it could be more robust to the noise due to the "soft" average over the similarity scores on image level.

2.2 Metric Learning for Template-to-Template Matching

There are several methods to handle image-to-image matching such as LDML [24], ITML [34,35]. LDML exploits the linear logistic dicriminant model to estimate the probability of the two images belonging to the same object, while ITML and [35] impose prior knowledge on the metric so that the learned one can be closed to the known prior and invariant to the rigid transformation respectively. KISSME [36], on the other hand, models the genuine and imposter as multivariate Gaussians and the distance is defined by the likelihood ratio test.

For template-to-template matching, the template-to-template distance (similarity) is first defined by either of the above-mentioned strategies, and then the image-to-image metric learning is employed to enhance the discriminative power of the defined measures. There are two typical types of losses exploited in metric learning: *contrastive* loss [3,6,7,9,12,18,24] or *triplet* loss [11,25,26]. The goal of *contrastive* loss is to minimize the intra-class distance but maximize the inter-class distance, while the *triplet* loss ensures the distance between an *anchor* and a *positive* sample, both of which have the same class label, is minimized and the one between the *anchor* and a *negative* sample of different classes is maximized. Note the distance can be replaced by the similarity and the minimization (maximization) becomes maximization (minimization).

The *triplet* loss was previously exploited in image level when a template is represented as a single image or features [25]. Instead, we introduce a new type of triplet, called *template* triplet, which directly works on the templates rather than on images. Unlike [11] where the triplets are created only on k nearest neighbors, the farthest positive template from the anchor template is selected to be in the template triplet.

Our context-aware metric learning method is most related to [37,38]. They learn a specific metric for a group of images, while ours achieves truly image-specific metric with the help of context. The proposed method to incorporate context into metric learning is based on the matrix factorization, which has been shown effective in image captioning [28,29] and image modeling [39,40] especially for the integration of the sample-specific clues.

3 Proposed Approach

In this section, we introduce a new triplet creation approach for similarity metric learning by creating the triplets on the *templates*. We also propose to integrate the context features to achieve a specific embedding for every image.

First, the similarity between templates is defined in Sect. 3.1. Then, we describe in Sects. 3.2 and 3.3 how to learn discriminative embedding via the template triplet loss and the context-aware metric learning respectively.

3.1 Template-to-Template Similarity

Given two input images with features x_i and x_j, a common similarity measure is inner product, defined as $s_{inn}(x_i, x_j) = x_i^T x_j$. In template based recognition tasks, it is often the case to compare two given templates instead of two images since a subject or visual class usually contains more than one image, and the images are grouped into *templates* (as in the Janus benchmark [21]). Therefore, how to measure the similarity between two templates becomes very important.

Witnessing the success of the ensemble SoftMax fusion [19,20] for summarizing the image-level matching scores among two templates in face recognition, we exploit it as the similarity measure in this paper. Given two templates $D_1 = \{x_1, \ldots, x_M\}$ and $D_2 = \{z_1, \ldots, z_N\}$, their ensemble SoftMax similarity is defined as:

$$s(D_1, D_2) = \sum_\alpha s_\alpha(D_1, D_2), \tag{1}$$

where

$$s_\alpha(D_1, D_2) = \frac{\sum_{x_i \in D_1} \sum_{z_j \in D_2} exp(\alpha s_{inn}(x_i, z_j)) \times s_{inn}(x_i, z_j)}{\sum_{x_r \in D_1} \sum_{z_q \in D_2} exp(\alpha s_{inn}(x_r, z_q))}. \tag{2}$$

This similarity can be regarded as an ensemble of multiple fusion schemes including the min fusion ($\alpha \to -\infty$), average fusion ($\alpha \to 0$), max fusion ($\alpha \to \infty$), and weighted average fusion ($0 < \alpha < \infty$) of a set of image-to-image similarity scores[1]. Thus, the ensemble SoftMax similarity can not only handle varying template size (number of samples in a template), arbitrary template distribution, but also robust to the noise via the "soft" average. In the experiments (cf. Table 1), we compare it to other common types of template-based similarities to demonstrate its effectiveness.

3.2 Ensemble SoftMax Similarity Embedding via Template Triplets

Inspired by the effectiveness of the triplet based metric learning on the recognition tasks [25,27], we propose a new type of triplet for similarity metric learning, called *template* triplets, based on the ensemble SoftMax similarity. A diagram for our approach is shown in Fig. 1 where the triplets are generated among templates instead of among image samples. The *anchor* template and *positivie* template share the same class label, and the *negative* template is of the different class label.

Our aim is to learn a discriminative embedding W, where the ensemble SoftMax similarity of the *anchor* template $A = \{a_1, \ldots, a_M\}$ to the *positive* template $P = \{p_1, \ldots, p_Q\}$ is larger than the one to the *negative* template $N = \{n_1, \ldots, n_R\}$ as described by:

$$s_W(A, P) > s_W(A, N) \tag{3}$$

---

[1] In this work, we use totally 21 values of α in $\{0, 1, \cdots, 20\}$ to combine the advantages of multiple fusion schemes, following [19,20].

where

$$s_{\boldsymbol{W}}(A, X) = \sum_{\alpha} \frac{\sum_{\boldsymbol{a}_i \in A} \sum_{\boldsymbol{x}_j \in X} exp(\alpha (\boldsymbol{W}\boldsymbol{a}_i)^T (\boldsymbol{W}\boldsymbol{x}_j)) \times (\boldsymbol{W}\boldsymbol{a}_i)^T (\boldsymbol{W}\boldsymbol{x}_j)}{\sum_{\boldsymbol{a}_r \in A} \sum_{\boldsymbol{x}_q \in X} exp(\alpha (\boldsymbol{W}\boldsymbol{a}_r)^T (\boldsymbol{W}\boldsymbol{x}_q)))} \quad (4)$$

and $X \in P, N, \boldsymbol{x} \in \boldsymbol{p}, \boldsymbol{n}$

Given a set of labeled templates, the optimal embedding $\hat{\boldsymbol{W}}$ can be solved by the following optimization problem:

$$\hat{\boldsymbol{W}} = \arg \min_{\boldsymbol{W}} \sum_{A,P,N \in \mathcal{T}_A} \max(s_{\boldsymbol{W}}(A, N) - s_{\boldsymbol{W}}(A, P) + \beta, 0) \quad (5)$$

Here, β is the margin of the hinge loss, and $\mathcal{T}_A$ represents all possible template triplets (A, P, N).

We apply the stochastic gradient descent (SGD) with early stopping technique for optimization. In each iteration of SGD, a template triplet is generated for updating $\boldsymbol{W}$ w.r.t. the hinge loss in Eq. (5) (Please refer to the supplementary material for derivations of the gradients). Our template triplet based metric learning approach is summarized in Algorithm 1, where an *epoch* means a process going through all the labeled templates, and T determines how many times we repeat this process. Besides, the two subscripts of $\boldsymbol{W}$ correspond to the current epoch and template indices.

Template Triplet Creation: It is infeasible to consider the exponentially large set of triplets in Eq. (5). Therefore, we attempt to find out the most violating (A, P, N) w.r.t. Eqs. (5) and (3) such that the size of total triplets is linear in the number of labeled templates. The creation of the template triplet is described step be step:

1. An *anchor* template A is randomly selected first. (Actually, we go through all the labeled templates in the training process in a random order, as can be seen in Algorithm 1)
2. We then pick the most dissimilar positive template P to the currently considered anchor template A as described below:

$$P = \arg \min_{\hat{P}} \quad s_{\boldsymbol{W}}(A, \hat{P}) \quad (6)$$

3. Finally, the *negative* template most close to A is picked:

$$N = \arg \max_{\hat{N}} \quad s_{\boldsymbol{W}}(A, \hat{N}) \quad (7)$$

3.3   On Incorporation of Context for Sample-Specific Embedding

Learning a good embedding for template-to-template matching helps enlarge the inter-class variations (better discrimination ability); the intra-class variations, however, may still be hardly narrowed since only a single embedding is learned

---

Algorithm 1: *Ensemble SoftMax Similarity Embedding based on Template Triplets*

Input: J labeled templates $\{(\mathcal{L}_j = \{\boldsymbol{x}_1, ...\boldsymbol{x}_M\}, \boldsymbol{x}_i \in \mathcal{R}^D, y_j \in \{1, \cdots, C\})\}_{j=1}^J$ of C classes, the dimension $d \leq D$ of the output embedding, the number of epochs T, the margin β, the step size η, and subtemplate size m. ;

Output: Learned similarity metric or embedding $\hat{\boldsymbol{W}}_{T,J} \in \mathcal{R}^{d \times D}$;

Initialize: $\boldsymbol{W}_{0,J} \leftarrow I_D$ for $d = D$, where I denotes an identity matrix. Otherwise, derive $\boldsymbol{W}_{0,J}$ such that $\boldsymbol{W}_{0,J}^T \boldsymbol{W}_{0,J}$ is a low rank approximation of I_D;

for $t \leftarrow 1, 2, \ldots, T$ **do**

   a. $\boldsymbol{W}_{t,0} \leftarrow \boldsymbol{W}_{t-1,J}$.;

   b. Order the templates $\{\mathcal{L}_j\}_{j=1}^J$ by a random permutation $\pi : j \rightarrow \pi(j)$.;

   for $\ell \leftarrow \pi(1), \pi(2), \ldots, \pi(J)$ **do**

      1. Let ℓ be the *anchor* template A in eq. (5).;

      2. Pick a template $\tilde{l}$, $\tilde{l} \neq \ell$ and $y(\tilde{l}) = y(\ell)$ as the *positive* template P via eq. (6). ;

      3. Pick a template $\tilde{l}$, $y(\tilde{l}) \neq y(\ell)$ as the *negative* template N via eq. (7).;

      4. If $m \neq \infty$, subsample A, P, N to be subtemplates of size at most m.;

      5. Compute the stochastic gradient $\boldsymbol{H}$ w.r.t. $\boldsymbol{W}$, according to eq. (5) on the chosen (A, P, N).;

$$\boldsymbol{W}_{t,\ell} \leftarrow \boldsymbol{W}_{t,\ell-1} - \eta \times \boldsymbol{H} \ ;$$

   c. Evaluate $\boldsymbol{W}_{t,J}$ on the held-out validation set. If the performance does not improve for K epochs, we stop the iteration on t and return $\hat{\boldsymbol{W}}_{T,J} = \boldsymbol{W}_{t,J}$. In our experiments, we set $K = 5$.

---

generally and shared by all the image samples in previous work. To better address intra-class variations, we propose to learn a specific embedding for each image by factorizing the metric to be learned with its own context, the same as image features in this work.

Given the context $\boldsymbol{c}_i$ of the i-th image, to inject $\boldsymbol{c}_i \in \mathcal{R}^{d_c}$ and thus adapt the template based metric learning to be image-specific, we factorize the embedding $\boldsymbol{W}_i$ as follows:

$$\boldsymbol{W}_i = \boldsymbol{F} \operatorname{diag}(\boldsymbol{Lc}_i) \boldsymbol{G} = \sum_{k \in d_c} (\boldsymbol{Lc}_i)_k \boldsymbol{F}_{*,k} \boldsymbol{G}_{k,*} \tag{8}$$

where $\operatorname{diag}(.)$ means the diagonal matrix, $(\boldsymbol{Lc}_i)_k$ is the k-th element of the vector, and $\boldsymbol{F}_{*,k}$ as well as $\boldsymbol{G}_{k,*}$ indicate the k-th column and row of the two (suitably sized) matrices $\boldsymbol{F}$ and $\boldsymbol{G}$, shared for all images, and $\boldsymbol{L}$ is another matrix that linearly transforms the context vector.

Because of the potential human annotation error or the ambiguity in how the context is defined, the context information may be noisy and would be harmful for context-aware metric learning. To alleviate this problem, we concatenate $\boldsymbol{c}_i$ with a value 1 as a "pseudo" *non-context* feature for all images:

$$\boldsymbol{W}_i = \boldsymbol{F} \operatorname{diag}(\boldsymbol{L}[\boldsymbol{c}_i^T, 1]^T) \boldsymbol{G} \tag{9}$$

Note that when c_i is a zero vector for all images or when the first d_c columns of L are all 0, W_i degenerates to a globally-shared W. With the above decomposition in Eq. (9), the same algorithm as Algorithm 1 can be applied. The difference is now we are to learn the three matrics F, L, and G rather than W.

Initialization of F, L, G: It is critical to initialize F, L, and G. If we start from $W = I_D$ (D is the input feature dimension), F and G are both initialized as I_D, and $L \in \mathcal{R}^{D \times (d_c+1)}$ is set by all zeros except for the last column as all 1's.

On the other hand, if we have $\hat{W}$ learned from Algorithm 1, then F, L, and G can be initialized by the singular value decomposition of W. Given $W = U \Sigma V^T$, F and G can be set as U and V^T respectively, and the singular values of Σ are placed in the last column of L with all other entries as 0's.

4   Experimental Results

This section presents the experiments and results of our proposed methods, (1) Template Triplet based Ensemble SoftMax Similarity Embedding (denoted as TT-ESSE in the following) and (2) Context-Aware sample specific embedding (denoted as TT-ESSE + context in the following) on four image set classification datasets. We describe in turn the adopted datasets, features and context, and the experimental protocols, commonly used evaluation measures along the results.

4.1   Datasets

To evaluate the performance of the proposed method, we conduct experiments on four publicly available datasets, including YouTube Celebrity (YTC) [30], YouTube Face (YTF) [23], and IARPA Janus Benchmark A (IJB-A) [21] for face recognition, and UCSD Traffic dataset (Traffic) [31] for scene classification.

YouTube Celebrity (YTC) [30]: This dataset contains 1,910 video clips of 47 subjects collected from YouTube. Each subject consists of $\sim$40 templates in average and $\sim$170 images/frames are in each template.

YouTube Face (YTF) [23]: 3,425 videos of 1,595 different people are downloaded from YouTube. An average of 2 videos or templates are available for each subject, and the average length of a video clip is 181 frames. We downsampled every video about ten times because of the large redundancies.

IARPA Janus Benchmark A (IJB-A) [21]: IJB-A contains totally 5,712 images and 2,085 videos for 500 subjects. Each subject consists of $\sim$11 images and $\sim$4 videos. A template can be of a mixture of images and video frames.

UCSD Traffic dataset (Traffic) [31]: The traffic video database is collected over two days from the highway traffic in Seattle with a single stationary traffic camera. It consists of 254 video sequences, and are manually labeled in terms of the amount of traffic congestion: heavy, medium, and light traffics.

The distributions of number of images per template on the above datasets are shown in Fig. 3.

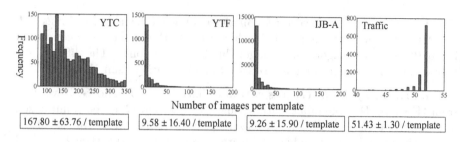

| 167.80 ± 63.76 / template | 9.58 ± 16.40 / template | 9.26 ± 15.90 / template | 51.43 ± 1.30 / template |

Fig. 3. The distribution of the number of image samples per template on four used datasets: YTC and Traffic datasets have at least 167 frames and 50 images per template respectively, while less than 10 images/ videos are in IJB-A and YTF datasets. Furthermore, about 50.14% templates on IJB-A are of a single image.

4.2 Features and Context

For the YTC dataset, we follow [18] and describe each region with a histogram of Local Binary Patterns (LBP). For the YTF and IJB-A datasets, we follow [20] to use the very deep VGGNet [41] CNN with 19 layers, trained on the large scale image recognition benchmark (ILSVRC) [42]; this CNN is then finetuned on the CASIA WebFace dataset [43]. Different from [20] where synthesized images of various poses, shapes and expressions are included for CNN finetuning, we only exploit the real images and the images rendered to the closet poses for the finetuning (Please refer to supplementary material for more details). For the Traffic dataset, HoG features [44] are exploited to describe each frame. Note that all the features are ℓ_2 normalized before metric learning in all the experiments. In the contex-aware metric learning approach, we employ the input features as context since they contain not only identity but also context information the such as poses, shapes, expressions, illuminations. Experiments using more sophisticated context or attribute detectors are remained for future work.

4.3 Protocols and Evaluation Measures

Following the standard practice [12], we split the YTC dataset into 5 folds, each of which contains 3 and 6 randomly selected videos from each person as the training/gallery templates and probe templates respectively. The average recognition rate is reported for YTC. The experiments performed on the Traffic dataset use the 4 splits provided with dataset [31], each of which contains 75% training/gallery templates, and 25% probe templates. The average recognition rate is also used for this dataset.

For the YTF dataset, 5,000 video pairs are randomly collected from the database, where half of them are of the same person, and half are different people. These pairs are divided into 10 splits and ensure the splits are subject-mutually exclusive (Please refer to [23] for more details). We consider the *Unrestricted* protocol and the average of true acceptance rates (**TARs**) at 10% and 1% false acceptance rates (**FARs**), verification accuracy (under a threshold selected on the validation set), equal error rate (**EER**) over 10 splits are reported.

In IJB-A dataset, there are 10 random training and testing splits, where 333 subjects are randomly sampled and placed in the training split, and the other 167 are placed in the testing split. We follow the *compare* (verification) protocol (for face verification) as defined in [21], and evaluate the verification performance by the average of TARs at 1%, 0.1%, and 0.01% FARs over 10 splits.

Table 1. Face Recognition Accuracies (in terms of TARs (%) at different FARs) on IARPA Janus Benchmark A (Verification Protocol) [21] with different kinds of template-to-template similarity measures. Note Paradigm (1) is "set fusion followed by set matching", and Paradigm (2) is "set matching followed by set score fusion".

Paradigm	Similarity measure	@10%FAR	@1%FAR	@0.1%FAR	@0.01%FAR
(1)	Avg pooling + inner prod	94.49	76.28	50.31	19.76
	Avg pooling + cos sim	93.84	79.98	58.89	**29.42**
	KDE [18]	92.68	81.32	**61.93**	23.42
(2)	Min fusion	44.88	17.64	8.02	3.44
	Average fusion	94.49	76.28	50.31	19.76
	Max fusion	92.23	73.47	41.61	13.29
(2)	Ensemble SoftMax fusion	**95.49**	**84.30**	61.15	20.45

Table 2. Average recognition rate (ARR) (%) on the YTC dataset

Method	inn+ESS [19]	ITSE [25]	TT-ESSE-3	TT-ESSE-5	TT-ESSE-whole
ARR	51.99	63.85	**65.33**	**64.87**	63.22
Method		ITSE + context	TT-ESSE-3 + contexts	TT-ESSE-5 + contexts	TT-ESSE-whole + contexts
ARR	-	**67.46**	**66.29**	**67.14**	**65.38**

4.4 Experimental Settings

In Algorithm 1, the number of epochs T is set to 10. The margin β and the step size η are selected in the range of $\{0.1, 0.2, \cdots, 1\}$ and $10^{\{-3,-2,\cdots,0\}}$ respectively on the held-out validation set. Besides, we set $d = D$ in all the experiments.

In addition to considering the entire template in the metric learning (denoted as TT-ESSE-whole), we also set subtemplate size to be 3 and 5 in the experiments and denote them as TT-ESSE-3 and TT-ESSE-5).

4.5   Comparisons of Ensemble SoftMax Similarity to the Other Template Based Similarity Measures

We compare the ensemble SoftMax similarity on IJB-A [21] to the other commonly used measures, Paradigm (1) and Paradigm (2), as introduced in Sect. 1. The average feature pooling with inner product or cosine similarity are denoted as Avg pooling + inner prod. and Avg pooling + cos sim. in Table 1. The kernel density estimation method (KDE) [18] is also compared. Both of them belong to Paradigm (1). Paradigm (2) contains several special cases of the ensemble Soft-Max fusion with the cosine similarity for image-to-image matching. As can be seen in Table 1, the ensemble SoftMax similarity (ESS) outperforms the others in most cases.

4.6   Comparisons of the Proposed Approaches to the Existing Image Template Classification Methods

We compare our method to the ensemble SoftMax similarity (ESS) [19,20] with the inner product as image-to-image matching score computation (denoted as inn+ESS in Tables 2, 3 and 4). We also compare to the image-triplet similarity embedding [25] with ESS (denoted as ITSE in Tables 2, 3 and 4). In ITSE, we consider the image triplets in metric learning but apply ESS in testing to compute the similarity[2]. This method then is equivalent to TT-ESSE-1 of our approach with a subtemplate size 1.

Table 3. Average verification performances (%) on the YTF dataset

Method	TAR@10%FAR	TAR@1%FAR	Verification accuracy	EER
inn+ESS [19]	85.64	55.96	68.18	12.12
ITSE [25]	86.92	61.04	72.10	11.60
TT-ESSE-3	86.96	**64.40**	72.22	11.56
TT-ESSE-5	**88.28**	**65.00**	71.82	10.92
TT-ESSE-whole	**87.60**	**62.04**	72.58	11.40
ITSE + context	**88.80**	**65.36**	**79.78**	**10.64**
TT-ESSE-3 + context	**88.04**	**65.04**	72.44	11.04
TT-ESSE-5 + contexts	87.84	64.16	**77.42**	11.24
TT-ESSE-whole + context	**88.16**	**63.72**	**77.56**	10.96

---

[2] Note that [25] performs average pooling + inner product in testing. Here we apply ESS becuase of its superior performance as shown in Table 1.

Table 4. Average recognition rate (ARR) (%) on the Traffic dataset

Method	inn+ESS [19]	ITSE [25]	TT-ESSE-3	TT-ESSE-5	TT-ESSE-whole
ARR	91.36	92.94	**93.34**	**93.73**	**94.12**
Method		ITSE + context	TT-ESSE-3 + context	TT-ESSE-5 + context	TT-ESSE-whole + context
ARR	-	**94.11**	**93.71**	91.74	92.94

Tables 2 and 3 illustrate the performances of TT-ESSE with and without context on YTC and YTF datasets. It can be seen that on the YTC dataset ITSE significantly improves over inn+ESS, and creating template triplets can further boost the performances ($\sim$2% relative improvements) with the subtemplate size 3 (TT-ESSE-3). Adding context leads to further $\sim$5% relative improvements when the subtemplate size is 1 or 5 (TT-ESSE-1 + context and TT-ESSE-5 + context). As for the YTF dataset, TT-ESSE-5 outperforms ITSE by $\sim$6% in terms of TAR@1%FAR, and the efficacy of context is shown with $\sim$9% relative improvements based on the verification accuracy.

Table 5. Average verification performances (%) on the IJB-A dataset

Method	TAR@1%FAR	TAR@0.1%FAR	TAR@0.01%FAR
GOTS [21]	40.6	19.8	-
OpenBR [45]	23.6	10.4	-
Wang *et al.* [46]	73.3	51.4	-
Deep Multi-Pose [47]	78.7	-	-
VGG-FACE [48]	80.5	60.4	-
inn+ESS [19][†]	84.3	61.2	20.5
Chen *et al.* [49]	78.7	-	-
[7]	68.8	28.6	16.8
KISSME [36]	65.4	39.4	15.2
ITSE [25][†]	84.5	65.1	35.2
TT-ESSE-3	**84.8**	**65.4**	**35.6**
TT-ESSE-5	**84.8**	**66.3**	**36.5**
TT-ESSE-whole	**85.0**	65.1	34.5
ITSE + context	**85.5**	**65.9**	36.2
TT-ESSE-3 + context	**85.3**	**66.4**	**36.5**
TT-ESSE-5 + context	**85.3**	66.2	**36.4**
TT-ESSE-whole + context	**85.4**	**66.5**	**36.0**

[†]Our reimplementation.

The benefit of template triplets (TT-ESSE) is also demonstrated on the Traffic dataset in Table 4. Note that the context integration works well for the subtemplate size 1 (TT-ESSE-1 + context) but not for larger subtemplate sizes (TT-ESSE-5 + context and TT-ESSE + whole). We believe that more elaborated context rather than HoG features is required for further improvements.

The results on IJB-A dataset are shown in Table 5. The improvements of TT-ESSE are limited since the average number of images per template is quite small as can be seen in Fig. 3. In average, the templates are of less than 10 images, and half of them are only of a single image, which makes the generation of good template triplets hard. According to these observations, we hypothesize that our approach would work better if the templates are of sufficient number of images such as at least 10 images. Table 6 presents the performances on the template pairs with at least 10 images. As can be seen, the improvements of TT-ESSE over the baselines are much more significant ($\sim$6% relative improvements) in terms of TAR@0.1%FAR compared to the ones in Table 5.

Table 6. Average verification performances (%) of the templates with at least 10 samples on the IJB-A dataset

Method	TAR@1%FAR	TAR@0.1%FAR
ITSE [25]	92.97	76.09
TT-ESSE-3	93.67	**79.82**
TT-ESSE-5	**95.16**	**81.38**
TT-ESSE-whole	**95.18**	**81.41**

Finally, we compare our method to the widely used metric learning approaches, [7,18], and KISSME [36]. The results on YTC, YTF, and Traffic datasets are shown in Table 7 based on different evaluation measures (The most significant one used in the literature), and the results on the IJB-A dataset is in Table 5. All the hyper-parameters were selected following [7,18,36]. It can be seen that our approach performs better than these methods where strong assumptions are made for a template's distribution.

Table 7. Average verification performances (%) of three baselines and the proposed method on the three datasets with the evaluation measure shown in the bracket

Method	YTC (ARR)	YTF (EER)	Traffic (ARR)
[7]	45.6	24.3	83.5
[18]	58.0	20.9	94.1
KISSME [36]	53.1	12.8	18.9
Ours	67.4	10.6	94.1

In summary, "Template Triplet" and "Context-Aware" are orthogonal methods to improve the image-based metric learning (ITSE) [25]. In all cases, we observe consistent gains from ITSE to TT-ESSE (by template triplet), and from ITSE to ITSE+context (by context). We also see consistent gains from TT-ESSE to TT-ESSE+context on YTC and YTF datasets, suggesting that the two methods can complement each other. Our best combination overall achieves $\sim2.5\%$ gain (averaged over datasets and measures) over ITSE.

5   Conclusion

In this paper, we propose a template triplet embedding approach to address image set classification, and further incorporate the image-specific context to learn the sample-specific metrics. Experiments on four image set classification datasets demonstrate the effectiveness of our approach.

Acknowledgement. This research is based upon work supported in part by the Office of the Director of National Intelligence (ODNI), Intelligence Advanced Research Projects Activity (IARPA), via IARPA 2014-14071600010. The views and conclusions contained herein are those of the authors and should not be interpreted as necessarily representing the official policies or endorsements, either expressed or implied, of ODNI, IARPA, or the U.S. Government. The U.S. Government is authorized to reproduce and distribute reprints for Governmental purpose notwithstanding any copyright annotation thereon. Moreover, we gratefully acknowledge USC HPC for hyper-computing.

References

1. Cevikalp, H., Triggs, B.: Face recognition based on image sets. In: CVPR, pp. 2567–2573 (2010)
2. Hu, Y., Mian, A.S., Owens, R.: Sparse approximated nearest points for image set classification. In: CVPR, pp. 121–128 (2011)
3. Zhu, P., Zhang, L., Zuo, W., Zhang, D.: From point to set: extend the learning of distance metrics. In: ICCV, pp. 2664–2671 (2013)
4. Yamaguchi, O., Fukui, K., Maeda, K.I.: Face recognition using temporal image sequence. In: FG, pp. 318–323 (1998)
5. Kim, T.K., Kittler, J., Cipolla, R.: Discriminative learning and recognition of image set classes using canonical correlations. Pattern Anal. Mach. Intell. **29**, 1005–1018 (2007)
6. Hamm, J., Lee, D.D.: Grassmann discriminant analysis: a unifying view on subspace-based learning. In: ICML, pp. 376–383 (2008)
7. Huang, Z., Wang, R., Shan, S., Chen, X.: Projection metric learning on Grassmann manifold with application to video based face recognition. In: CVPR, pp. 140–149 (2015)
8. Wang, R., Shan, S., Chen, X., Gao, W.: Manifold-manifold distance with application to face recognition based on image set. In: CVPR, pp. 1–8 (2008)
9. Wang, R., Chen, X.: Manifold discriminant analysis. In: CVPR, pp. 429–436 (2009)
10. Chen, S., Sanderson, C., Harandi, M., Lovell, B.: Improved image set classification via joint sparse approximated nearest subspaces. In: CVPR, pp. 452–459 (2013)

11. Lu, J., Wang, G., Deng, W., Moulin, P., Zhou, J.: Multi-manifold deep metric learning for image set classification. In: CVPR, pp. 1137–1145 (2015)

12. Lu, J., Wang, G., Moulin, P.: Image set classification using holistic multiple order statistics features and localized multi-kernel metric learning. In: ICCV, pp. 329–336 (2013)

13. Wang, R., Guo, H., Davis, L.S., Dai, Q.: Covariance discriminative learning: a natural and efficient approach to image set classification. In: CVPR, pp. 2496–2503 (2012)

14. Huang, Z., Wang, R., Shan, S., Li, X., Chen, X.: Log-Euclidean metric learning on symmetric positive definite manifold with application to image set classification. In: ICML, pp. 720–729 (2015)

15. Shakhnarovich, G., Fisher, J.W., Darrell, T.: Face recognition from long-term observations. In: Heyden, A., Sparr, G., Nielsen, M., Johansen, P. (eds.) ECCV 2002. LNCS, vol. 2352, pp. 851–865. Springer, Heidelberg (2002). doi:10.1007/3-540-47977-5_56

16. Arandjelović, O., Shakhnarovich, G., Fisher, J., Cipolla, R., Darrell, T.: Face recognition with image sets using manifold density divergence. In: CVPR, pp. 581–588 (2005)

17. Wang, W., Wang, R., Huang, Z., Shan, S., Chen, X.: Discriminant analysis on Riemannian manifold of Gaussian distributions for face recognition with image sets. In: CVPR, pp. 2048–2057 (2015)

18. Harandi, M., Salzmann, M., Baktashmotlagh, M.: Beyond Gauss: image-set matching on the Riemannian manifold of PDFs. In: ICCV, pp. 4112–4120 (2015)

19. Masi, I., Rawls, S., Medioni, G., Prem, N.: Pose-aware face recognition in the wild. In: CVPR (2016)

20. Masi, I., Tran, A.T., Leksut, J.T., Hassner, T., Medioni, G.: Do we really need to collect millions of faces for effective face recognition? arXiv preprint arXiv:1603.07057 (2016)

21. Klare, B.F., Klein, B., Taborsky, E., Blanton, A., Cheney, J., Allen, K., Grother, P., Mah, A., Burge, M., Jain, A.K.: Pushing the Frontiers of unconstrained face detection and recognition: IARPA Janus Benchmark A. In: CVPR, pp. 1931–1939 (2015)

22. Huang, G.B., Ramesh, M., Berg, T., Learned-Miller, E.: Labeled faces in the wild: a database for studying face recognition in unconstrained environments. Technical report 07–49, University of Massachusetts, Amherst (2007)

23. Wolf, L., Hassner, T., Maoz, I.: Face recognition in unconstrained videos with matched background similarity. In: CVPR, pp. 529–534 (2011)

24. Guillaumin, M., Verbeek, J., Schmid, C.: Is that you? Metric learning approaches for face identification. In: ICCV, pp. 498–505 (2009)

25. Sankaranarayanan, S., Alavi, A., Chellappa, R.: Triplet similarity embedding for face verification. arXiv preprint arXiv:1602.03418 (2016)

26. Van Der Maaten, L., Weinberger, K.: Stochastic triplet embedding. In: IEEE International Workshop on Machine Learning for Signal Processing, pp. 1–6 (2012)

27. Schroff, F., Kalenichenko, D., Philbin, J.: FaceNet: a unified embedding for face recognition and clustering. In: CVPR, pp. 815–823 (2015)

28. Jin, J., Fu, K., Cui, R., Sha, F., Zhang, C.: Aligning where to see and what to tell: image caption with region-based attention and scene factorization. arXiv preprint arXiv:1506.06272 (2015)

29. Mao, J., Xu, W., Yang, Y., Wang, J., Huang, Z., Yuille, A.: Deep captioning with multimodal recurrent neural networks (m-RNN). arXiv preprint arXiv:1412.6632 (2014)

30. Kim, M., Kumar, S., Pavlovic, V., Rowley, H.: Face tracking and recognition with visual constraints in real-world videos. In: CVPR, pp. 1–8 (2008)

31. Chan, A.B., Vasconcelos, N.: Probabilistic kernels for the classification of auto-regressive visual processes. In: CVPR, pp. 846–851 (2005)

32. Harandi, M.T., Salzmann, M., Hartley, R.: From manifold to manifold: geometry-aware dimensionality reduction for SPD matrices. In: Fleet, D., Pajdla, T., Schiele, B., Tuytelaars, T. (eds.) ECCV 2014. LNCS, vol. 8690, pp. 17–32. Springer, Heidelberg (2014). doi:10.1007/978-3-319-10605-2_2

33. Huang, Z., Wang, R., Shan, S., Chen, X.: Face recognition on large-scale video in the wild with hybrid Euclidean-and-Riemannian metric learning. Pattern Recogn. **48**, 3113–3124 (2015)

34. Davis, J.V., Kulis, B., Jain, P., Sra, S., Dhillon, I.S.: Information-theoretic metric learning. In: ICML, pp. 209–216 (2007)

35. Bosveld, J., Mahmood, A., Huynh, D.Q., Noakes, L.: Constrained metric learning by permutation inducing isometries. IEEE Trans. Image Process. **25**, 92–103 (2016)

36. Koestinger, M., Hirzer, M., Wohlhart, P., Roth, P.M., Bischof, H.: Large scale metric learning from equivalence constraints. In: CVPR (2012)

37. Sharma, G., Pérez, P.: Latent max-margin metric learning for comparing video face tubes. In: CVPR Workshops, pp. 65–74 (2015)

38. Cinbis, R.G., Verbeek, J., Schmid, C.: Unsupervised metric learning for face iden-tification in TV video. In: ICCV, pp. 1559–1566 (2011)

39. Memisevic, R., Hinton, G.: Unsupervised learning of image transformations. In: CVPR, pp. 1–8 (2007)

40. Salakhutdinov, R., Hinton, G.E.: Deep Boltzmann machines. In: AISTATS, vol. 1, p. 3 (2009)

41. Simonyan, K., Zisserman, A.: Very deep convolutional networks for large-scale image recognition. arXiv preprint arXiv:1409.1556 (2014)

42. Russakovsky, O., Deng, J., Su, H., Krause, J., Satheesh, S., Ma, S., Huang, Z., Karpathy, A., Khosla, A., Bernstein, M., et al.: ImageNet large scale visual recog-nition challenge. Int. J. Comput. Vis. **115**, 211–252 (2015)

43. Yi, D., Lei, Z., Liao, S., Li, S.Z.: Learning face representation from scratch. arXiv preprint arXiv:1411.7923 (2014)

44. Dalal, N., Triggs, B.: Histograms of oriented gradients for human detection. In: CVPR, pp. 886–893 (2005)

45. Klontz, J.C., Klare, B.F., Klum, S., Jain, A.K., Burge, M.J.: Open source biometric recognition. In: BTAS, pp. 1–8 (2013)

46. Wang, D., Otto, C., Jain, A.K.: Face search at scale: 80 million gallery. arXiv preprint arXiv:1507.07242 (2015)

47. AbdAlmageed, W., Wua, Y., Rawlsa, S., Harel, S., Hassner, T., Masi, I., Choi, J., Leksut, J.T., Kim, J., Natarajan, P., et al.: Face recognition using deep multi-pose representations. arXiv preprint arXiv:1603.07388 (2016)

48. Parkhi, O.M., Vedaldi, A., Zisserman, A.: Deep face recognition. In: Proceedings of the British Machine Vision, vol. 1, p. 6 (2015)

49. Chen, J.C., Patel, V.M., Chellappa, R.: Unconstrained face verification using deep CNN features. arXiv preprint arXiv:1508.01722 (2015)

Object-Centric Representation Learning from Unlabeled Videos

Ruohan Gao[✉], Dinesh Jayaraman, and Kristen Grauman

University of Texas at Austin, Austin, USA
rhgao@cs.utexas.edu

Abstract. Supervised (pre-)training currently yields state-of-the-art performance for representation learning for visual recognition, yet it comes at the cost of (1) intensive manual annotations and (2) an inherent restriction in the scope of data relevant for learning. In this work, we explore unsupervised feature learning from unlabeled video. We introduce a novel *object-centric* approach to temporal coherence that encourages similar representations to be learned for object-like regions segmented from nearby frames. Our framework relies on a Siamese-triplet network to train a deep convolutional neural network (CNN) representation. Compared to existing temporal coherence methods, our idea has the advantage of lightweight preprocessing of the unlabeled video (no tracking required) while still being able to extract object-level regions from which to learn invariances. Furthermore, as we show in results on several standard datasets, our method typically achieves substantial accuracy gains over competing unsupervised methods for image classification and retrieval tasks.

1 Introduction

The emergence of large-scale datasets of millions of labeled examples such as ImageNet has led to major successes for supervised visual representation learning. Indeed, visual feature learning with deep neural networks has swept the field of computer vision in recent years [1–3]. If learned from labeled data with broad enough coverage, these learnt features can even be transferred or repurposed to other domains or new tasks via "pretraining" [4,5]. However, all such methods heavily rely on ample manually provided image labels. Despite advances in crowdsourcing, massive labeled image datasets remain quite expensive to collect. Furthermore, even putting cost aside, it is likely that restricting visual representation learning to a bag of unrelated images (like web photos) may prevent algorithms from learning critical properties that simply are not observable in such data—such as certain invariances, dynamics, or patterns rarely photographed in web images.

Due to these restrictions, *unsupervised visual feature learning* from unlabeled images or videos has therefore increasingly drawn researchers' attention. An unsupervised approach has several potential advantages. First, it is in principle much more scalable, because unlabeled images and videos can be obtained

© Springer International Publishing AG 2017
S.-H. Lai et al. (Eds.): ACCV 2016, Part V, LNCS 10115, pp. 248–263, 2017.
DOI: 10.1007/978-3-319-54193-8_16

essentially for free. Moreover, features learnt in an unsupervised way, particularly from diverse videos, may prove even more effective as a generalizable base representation by being unencumbered by the "closed world" restriction of categorically labeled data. Recent years have seen a number of exciting ideas in unsupervised visual feature learning, particularly using temporal coherence of unlabeled video [6–10], self-supervision from image context [11,12] or ego-motion [13,14], as well as earlier attempts based on autoencoders [15–18].

In this work, we focus on learning from unlabeled videos and build upon the idea of *temporal coherence* as a form of "free" supervision to learn image representations invariant to small transformations. Temporal coherence, a form of *slow feature analysis* [19], is based on the observation that high-level signals cannot change too quickly from frame to frame; therefore, temporally close frames should also be close in the learned feature space. Most prior work in this space produces a holistic image embedding, and attempts to learn feature representations for video frames as a whole [6,8–10,20]. Two temporally close video frames, although similar, usually have multiple layers of changes across different regions of the frames. This may confuse the deep neural network, which tries to learn good feature representations and embed these two frames in the deep feature space as a whole. An alternative is to track local patches and learn a localized representation based on the tracked patches [7,21,22]. In particular, a recent approach [7] uses sophisticated visual motion tracking to connect "start" and "end" patches for training pairs. However, such tracking is biased towards moving objects, which may limit the invariances that are possible to learn. Furthermore, processing massive unlabeled video collections with tracking algorithms is computationally intensive, and errors in tracking may influence the quality of the patches used for learning.

With these limitations in mind, we propose a new way to learn visual features from unlabeled video. Similar to existing methods, we exploit the general principle of temporal coherence. Unlike existing methods, however, we neither learn from whole-frames of video nor rely on tracking fragments in the video. Instead, we propose to focus temporal coherence on *object-centric* regions discovered with object proposal regions. In particular, we first generate object-like region proposals on temporally adjacent video frames using Selective Search [23]. Then we perform feature learning using a ranking-based objective that maps spatio-temporally adjacent region proposals closer in the embedding space than non-neighbors. The idea is that two spatio-temporally close region proposals should be embedded close in the deep feature space since they likely belong to the same object or object part, in spite of superficial differences in their pose, lighting, or appearance. See Fig. 1.

Why might such an object-centric approach have an advantage? How might it produce features better equipped for object recognition, image classification, or related tasks? First, unlike patches found with tracking, patches generated by region proposals can capture static objects as well as moving objects. Static objects are also informative in the sense that, beyond object motion, there might be other slight changes such as illumination changes or camera viewpoint changes across video frames. Therefore, static objects should not be neglected in the

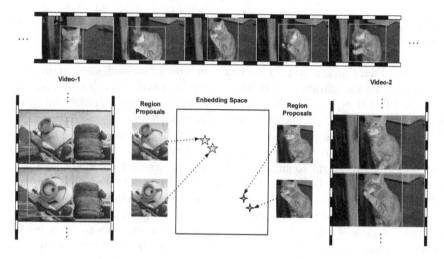

Fig. 1. Video frames tend to change coherently across the video, as shown in the two example videos above. These changes usually come from certain objects in the frames. In our framework, we first generate region proposals independently on whole video frames and find spatio-temporally close region proposal pairs. Instead of embedding video frames directly, we embed these region proposals. Spatio-temporally close region proposal pairs should be close in the deep embedding space, since they are likely to belong to the same object or object part.

learning process. Secondly, our framework can also help capture the object-level regions of interest in cases where there is motion only on a part of the object. For example, in Fig. 1, the cat is moving its paw but otherwise staying similarly posed. Visual motion tracking will have difficulty catching such subtle changes (or will catch only the small moving paw), while region proposals used as we propose can easily capture the entire object with its part-level change. Thirdly, our method is much more efficient to generate training samples since no tracking is needed.

In results on three challenging datasets, we show the impact of our approach for unsupervised convolutional neural network (CNN) feature learning from video. Compared to the alternative whole-frame and tracking-based paradigms discussed above, our idea shows consistent advantages. Furthermore, it often outperforms an array of state-of-the-art unsupervised feature learning methods [7,12,13] for image classification and retrieval tasks. In particular, we observe relative gains of 10 to 30% in most cases compared to existing pre-trained models. Overall, our simple but effective approach is an encouraging new path to explore learning from unlabeled video.

2 Related Work

Unsupervised feature learning has a rich history and can be traced to seminal work for learning visual representations which are sparse and reconstructive [24]. More recent advances include training a deep belief network by stacking layer-by-layer RBMs [18] and injecting autoencoders [16]. Building on this concept, multi-layer autoencoders are scaled up to large-scale unlabeled data [15], where it is shown that neurons in high layers of an unsupervised network can have high responses on semantic objects or object parts. Recently, some approaches explore the use of spatial context of images as a source of a (self-)supervisory signal for learning visual representations [11,12]. In [11], the learning is driven by position prediction of context patches, while in [12], the algorithm is driven by context-based pixel prediction in images.

Most existing work for learning representations from unlabeled video exploits the concept of *temporal coherence*. The underlying idea can be traced to the concept of slow feature analysis (SFA) [19,25], which proposes to use temporal coherence in a sequential signal as "free" supervision as discussed above. Some methods attempt to learn feature representations of video frames as a whole [6,9,10,20], while others track local patches to learn a localized representation [7,21,22]. Our approach builds on the concept of temporal coherence, with the new twist of learning from localized object-centric regions in video, and without requiring tracking.

Another way to learn a feature embedding from video is by means of a "proxy" task, solving which entails learning a good feature embedding. For example, the reconstruction and prediction of a sequence of video frames can serve as the proxy task [8,26]. The idea is that in order to reconstruct past video frames or predict future frames, good feature representations must be learnt along the way. Ego-motion [13,14] is another interesting proxy that is recently adopted to learn feature embeddings. Learning the type of ego-motion that corresponds to video frame transformations entails learning good visual features, and thus proprioceptive motor signals can also act as a supervisory signal for feature learning. We offer empirical comparisons to recent such methods, and show our method surpasses them on three challenging datasets.

3   Our Framework

Given a large collection of unlabeled videos, our goal is to learn image representations that are useful for generic recognition tasks. Specifically, we focus on learning representations for *object-like* regions. Our key idea is to learn a feature space where representations of object-like regions in video vary slowly over time. Intuitively, this property induces invariances (such as to pose, lighting etc.) in the feature space that are useful for high-level tasks. Towards this goal, we start by generating object-like region proposals independently on each frame of hundreds of thousands of unlabeled web videos (details below). We observe that objects in images vary slowly over time, so that correct region proposals in one frame tend to have corresponding proposals in adjacent frames (1 s apart in our setting). Conversely, region proposals that are spatio-temporally adjacent usually correspond to the same object or object part. For example, in Fig. 1, semantically meaningful objects like the cartoon character minions and the cat are proposed in the video frames by Selective Search. Based on this observation, we train deep neural networks to learn feature spaces where spatio-temporally adjacent region proposals in video are embedded close to each other. Next, we describe these two stages of our approach in detail: (1) region proposal pair selection (Sect. 3.1), and (2) slow representation learning (Sect. 3.2).

3.1   Selecting Region Proposal Pairs from Videos

We start by downloading hundreds of thousands of YouTube videos (details in Sect. 4). Among these, we are interested in mining region proposal pairs that (1) correspond to the same object and (2) encode useful invariances.

Whole Frame Pair Selection. We start by selecting frame pairs likely to yield useful region proposals, based on two factors:

Pixel Correlation. We extract video frames at a frame rate of fps = 1, and every two adjacent frames form a candidate video frame pair. We compute pixel space correlations for all frame pairs. Very low correlations usually correspond to scene cuts, and very high correlations to near-static scenes, where the object-like regions do not change in appearance, and are therefore trivially mapped to the same feature representation. Thus, neither of these cases yields region pairs useful for slow representation learning. Therefore, at this stage, we only select frame pairs whose correlation score $\in (0.3, 0.8)$.

Mean Intensity. Next, we discard video frame pairs where either one of the frames has a mean intensity value lower than 50 or larger than 200. These are often "junk" frames which do not contain meaningful region proposals, such as the prologue or epilogue of a movie trailer.

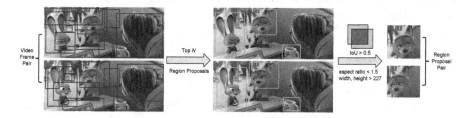

Fig. 2. A large quantity of region proposals can be generated in each video frame. We only keep the top-N scored region proposals (N = 100 in our experiments, N = 3 in the figure above for demonstration). For one region proposal in the first frame, we find the region proposal in the other frame with the largest IoU score. If the two spatio-temporally adjacent region proposals satisfy the thresholds on IoU, width, height and aspect ratio, two 227 × 227 regions are then cropped from them to form a region proposal pair.

Proposal Pair Selection. We now generate region proposal pairs from these selected video frame pairs using a standard object propsal generation method, as illustrated in Fig. 2. We use Selective Search [23], which generates hundreds of region proposals from each frame. Starting from the large candidate pool of all region proposal pairs from adjacent frames, we use the following filtering process to guarantee the *quality, congruity,* and *diversity* of our region proposal pairs:

Quality. We only keep the top-100 scored region proposals from Selective Search for each frame. These region proposals are of higher quality and tend to be more object-like. Furthermore, we only keep region proposals of width and height both larger than 227 and aspect-ratio smaller than 1.5. We then crop 227×227 regions to get the final region proposal pairs.

Congruity. As mentioned above, regions corresponding to the same object are likely to be spatially close in adjacent frames, and conversely, spatially close high quality proposals in adjacent frames usually correspond to the same object. With this in mind, at this stage, we only retain those region proposals from neighboring frames that have a spatial overlap score (intersection over union) exceeding 0.5. Where there are multiple candidate pairings for a single region proposal, we only retain the pairing with the largest IoU score.

Diversity. To increase the diversity of our region proposal pairs and avoid redundant pairs, we process each video sequentially, and compute the pixel-space correlation of each candidate pair with the last selected pair from the same video. We save the current pair only if this correlation with the last selected pair is <0.7. For computational efficiency, we first downsample the region patches to 33×33 and then calculate the correlation. This step reduces redundancy and ensures that each selected pair is more likely to be informative.

Fig. 3. Examples of region proposal pairs extracted from unlabeled videos. Note that both moving objects (e.g. dog, horse, cat, human, etc.) and static objects (e.g. flower, tree, chair, sofa, etc.) can be captured by region proposals. Moving objects or object parts (like moving animals and human beings) are informative. Changes like illumination variations and camera viewpoint changes across static objects (from both natural scenes and indoor scenes) also form informative pairs from which to learn.

Our method of generating patch pairs is much more efficient than tracking [7]. Our lightweight approach takes only a minute to generate hundreds of region proposal pairs, which is more than 100 times faster than tracking (our implementation). Figure 3 shows some examples of the region proposal pairs selected by our approach. We can see that these regions tend to belong to objects or object parts. Region pairs corresponding to both static and moving object regions are selected. The difference between the two region proposals in a pair comes from various sources: movements of the captured objects, partial movement of object-level regions, illumination changes across frames, viewpoint changes, etc. A full coverage of different types of changes among region proposal pairs can help to guarantee useful invariances in features trained in the next stage.

3.2   Learning from Region Proposal Pairs

After generating millions of region proposal pairs from those unlabeled videos, we then train a convolutional neural network based on these generated pairs. The trained CNN is expected to map region proposal pairs from image space to the feature space. Specifically, given a region proposal R as an input for the network, we aim to learn a feature mapping $f : R \Rightarrow f(R)$ in the final layer. Let (R_1, R_2) denote a region proposal pair (spatio-temporally close region proposals) from one video, and R^- denote a randomly sampled region proposal from another video[1]. In our target feature space, R_1 and R_2 should be closer to each other compared to R^-, i.e. $D(R_1, R^-) > D(R_1, R_2)$, where $D(\cdot)$ is the cosine distance in the embedding space:

$$D(R_1, R_2) = 1 - \frac{f(R_1) \cdot f(R_2)}{\| f(R_1) \| \| f(R_2) \|}.$$

We use a "Siamese-triplet" network to learn our feature space (see Fig. 4). A Siamese-triplet network consists of three base networks with shared parameters, to process R_1, R_2 and R^- in parallel. Based on the distance inequality we desire (discussed above), we use the loss function for triplets to enforce that region proposals that are not spatio-temporally close are further away than spatio-temporally close ones in the feature space. The idea of using triplets for learning embeddings has renewed interest lately [27–29] and can be traced to [30], which uses triplets to learn an image-to-image distance function that satisfies the property that the distance between images from the same category should be less than the distance between images from different categories.

Specifically, given a triplet (R_1, R_2, R^-), where R_1 and R_2 are two spatio-temporally close region proposals and R^- is a random region proposal from another video, the loss function in the feature space is defined as follows:

$$L(R_1, R_2, R^-) = \max\{0, D(R_1, R_2) - D(R_1, R^-) + M\},$$

where $D(R_1, R_2)$ is the cosine distance of region proposals R_1 and R_2 in the feature space, and M represents the margin between the two distances. This hinge loss forces the CNN to learn feature representations such that $D(R_1, R_2) <$

---

[1] 25,000 videos are used to generate training samples. The chance that the object proposal from one video and a random proposal from another video are similar (or of the exact same object instance) is negligible.

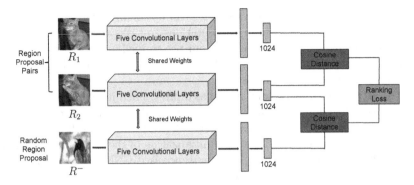

Fig. 4. The Siamese-triplet network consists of three base networks with shared parameters. The five convolutional layers of the base network are the same as the architecture of AlexNet, and two fully connected layers are stacked on the pool5 outputs. The final ranking loss is defined based on the last fully connected layers. The whole Siamese-triplet network is trained from scratch, using only our region proposals as input data.

$D(R_1, R^-) - M$. In other words, the distance between the embeddings of spatio-temporally adjacent region proposals should be smaller than that between a query region proposal and a random region proposal, by a margin M. The overall objective function for training is as follows:

$$\min_{W} \frac{\lambda}{2} \|W\|_2^2 + \sum_{i=1}^{N} \max\{0, D(R_{i_1}, R_{i_2}) - D(R_{i_1}, R_i-) + M\},$$

where W contains the parameter weights of the network, N is the number of triplet samples and, λ denotes the weight decay.

In our experiments, the convolutional layers of the base network follow the AlexNet architecture [1], and two fully connected layers are stacked on the pool5 outputs, whose neuron numbers are 4096 and 1024 respectively. Mini-batch Stochastic Gradient Descent (SGD) is used during training. Initially, for each region proposal pair (R_1, R_2), K negative region proposal patches R^- are randomly sampled in the same batch to form K triplets. After training for several epochs, hard negative mining is used to enhance training. More specifically, for each pair (R_1, R_2), the ranking losses of all other negative region proposals in the same batch are calculated, and the top K ones with highest losses are selected to form K hard triplets of samples. The idea is analogous to the hard-negative mining procedure in SVM. For details, please refer to [7].

3.3 Transferring Learnt Features for Supervised Tasks

Until this point, our CNN has been trained in a purely unsupervised manner, using only unlabeled videos from YouTube. We evaluate how well features directly extracted from our unsupervised pre-trained models can benefit recognition tasks in Sect. 4.2. To evaluate whether these features are useful for generic supervised recognition tasks, we can optionally adapt and specialize these purely unsupervised visual representations to tasks with labeled data. In our experiments in Sect. 4.3, we fine-tune our pre-trained model on the PASCAL VOC multi-label classification task (VOC 2007 and VOC 2012) and MIT Indoor Scene single-label classification task.

We directly adapt our ranking model as a pre-trained network for the classification task. The adaptation method we use is similar to the approach applied in RCNN [5]. However, RCNN uses the network pre-trained with ImageNet classification data (with semantic labels) as initialization of their supervised task. In our case, the pre-trained network is the *unsupervised* CNN trained using unlabeled videos. More specifically, we use the weights of the convolutional layers in the base network of our Siamese-triplet architecture to initialize corresponding layers for the classification task. For weights of the fully connected layers, we initialize them randomly.

4   Experiments

We present results on three datasets, with comparisons to several existing unsupervised methods [7,12,13] plus multiple informative baselines. We consider both the purely unsupervised case where all learning stems from unlabeled videos (Sects. 4.1 and 4.2) as well as the fine-tuning case where our method initializes a network trained with relatively few labeled images (Sect. 4.3).

Implementation Details. We use 25,000 unlabeled videos from YouTube downloaded from the first 25,000 URLs provided by Liang et al. [31], which used thousands of keywords based on VOC to query YouTube. These unlabeled videos are of various categories, including movie trailers, animal/human activities, etc. Most video clips are dedicated to several objects, while some can contain hundreds of objects. We do not use any label information associated with each video. We extract video frames at the frame rate of 1 fps. Using Selective Search and our filtering process, we can easily obtain millions of region proposal pairs from these unlabeled videos. For efficiency, evaluation is throughout based on training our model with 1M region proposal pairs, which requires about 1 week to train. We can expect even better results if more data is used, though for greater computational expense.

We closely follow the Siamese-triplet network implementation from [7] to learn our feature space. We set the margin parameter $M = 0.5$, weight decay $\lambda = 0.0005$, number of triplets per region proposal pair $K = 4$, and the batch size to be 100 in all experiments. The training is completed with Caffe [32] based on 1M region proposal pairs (2M region proposal patches). We first train our model without hard negative mining at a constant learning rate $\epsilon = 0.001$ for 150K iterations. Then we apply hard negative mining with hard ratio 0.5 to continue training with initial learning rate $\epsilon_0 = 0.001$. We reduce the learning rate by a factor of 10 at every 100K iterations and train for another 300K iterations.

Baselines. We compare to several existing unsupervised feature learning methods [7,12–14]. In all cases, we use the authors' publicly available pre-trained models. To most directly analyze the benefits of learning from region proposals, we implement three other baselines: *full-frame*, *square-region*, and *visual-tracking*. For fair comparison, we use 1M frame/patch pairs as training data for each of these three baselines. Note that all compared methods use unlabeled videos for pre-training, except for [12], which uses unlabeled images.

Full-Frame. In every video, we take every two temporally adjacent video frames (1 s apart) to be a positive pair. Namely, for a full-frame triplet (R_1, R_2, R^-), R_1 and R_2 are two temporally adjacent video frames and R^- is a random full frame from another video. We take 1M full-frame pairs (2M full frames) as training data and follow the same procedures to train the baseline model.

Square-Region. Same as our proposed framework, we only generate square regions on selected video frame pairs using the proposed initial filtering process. For a selected temporally adjacent video frame pair (1 s apart), we generate a random 227×227 patch in one frame and get the patch at the same position in the other frame. These two patches form a positive pair. We repeat this process 10 times and generate 10 square-region pairs for every video frame pair. Namely, for a square-region triplet (R_1, R_2, R^-), R_1 and R_2 are two 227×227 patches from two temporally adjacent video frames at the same position, and R^- is a random patch of the same size from another different video. We expect these square-region patches to be less object-like compared to patches obtained using our framework, because they are random regions from the frame. But note that square-region pairs also benefit from the same well-defined pruning steps we propose for our method. We take 1M square-region pairs (2M patches) as training data and follow the same procedures to train the baseline model.

Visual-Tracking. In [7], Wang and Gupta extract patches with motion and track these patches to create training instances. Specifically, they first obtain SURF [33] interest points and use Improved Dense Trajectories (IDT) [34] to obtain the motion of each SURF point. Then they find the best bounding box such that it contains most of the moving SURF points. After obtaining the initial bounding box (the first patch), they perform tracking using the KCF tracker [35] and track along 30 frames in the video to obtain the second patch. These two patches form a positive pair. We take 1M visual-tracking pairs (2M patches) from their publicly available dataset of collected patches as our training data and follow the same procedures to train the baseline model.

Random-Gaussian. As a sanity check, we also provide a random initialization baseline. Specifically, we construct a CNN using the AlexNet architecture, and initialize all layers with weights drawn from a Gaussian distribution. This randomly initialized AlexNet serves as a pre-trained model.

4.1 Qualitative Results of Unsupervised Feature Learning

Visualization of Filters. We first analyze our learnt features qualitatively. We visualize the conv1 features learnt in our unsupervised CNN as well as the first three baselines above. The visualization is shown in Fig. 5. We observe that

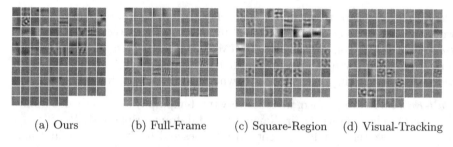

(a) Ours   (b) Full-Frame   (c) Square-Region   (d) Visual-Tracking

Fig. 5. Visualization of Conv1 features

Fig. 6. Top response images from PASCAL VOC 2012 for 6 of the conv5 neurons of our unsupervised CNN. Each seems to correspond to one semantically meaningful category, despite no labeled data during learning.

the conv1 features learnt using our framework are much more distinctive than the three baselines, suggesting a more powerful basis.

To better understand the internal representations of the units learnt by our unsupervised CNN, in Fig. 6, we get the top response images for units in the conv5 layer. We use all images in PASCAL VOC 2012 as the database. Although here we do *not* provide any semantic information during training, we can see that units in conv5 layers nonetheless often fire on semantically meaningful categories. For example, unit 33 fires on buses, unit 57 fires on cats, etc.

Neighbors in Unsupervised Learned Feature Space. Next we analyze the learned features via a retrieval task. We perform Nearest Neighbors (NN) using ground-truth (GT) windows in the PASCAL VOC 2012 val set as queries, and a retrieval database consisting of all selective search windows (having more than 0.5 overlap with GT windows) in the PASCAL VOC 2012 train set.

Figure 7 shows example results. We can see that our unsupervised CNN is far superior to AlexNet initialized with random weights. Furthermore, despite having access to zero labeled data, our features (qualitatively) appear to produce neighbors comparable to AlexNet trained on ImageNet with over 1 million semantic labels. For example, in the first row, given an image of a dog lying

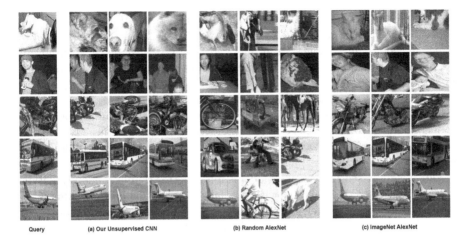

Query   (a) Our Unsupervised CNN   (b) Random AlexNet   (c) ImageNet AlexNet

Fig. 7. Nearest neighbor results. We compare three models: (a) Our unsupervised CNN pre-trained using 1M region proposal pairs; (b) AlexNet initialized with random parameters; (3) AlexNet trained with labeled ImageNet. Our unsupervised CNN is far superior to random AlexNet, and even learns some concepts similar to ImageNet AlexNet trained with semantic labels.

down as a query, our method successfully retrieves images of dogs of different postures. It even outperforms ImageNet AlexNet, which retrieves two cats as nearest neighbors. In comparison, AlexNet initialized with random weights mostly retrieves unrelated images. Note that there is no class supervision or fine-tuning being used here in retrieval for our unsupervised CNN, and *all the gains for our pre-trained model come from unlabeled videos.*

Quantitatively, we can measure the retrieval rate by counting the number of correct retrievals in the top-20 neighbors. Given a query image, a retrieval is considered correct if the semantic class for the retrieved image and the query image are the same. Using pool5 features extracted from our unsupervised CNN and cosine distance, we obtain 32% retrieval rate, which outperforms 17% by random AlexNet (our method's initialization) by a large margin. ImageNet AlexNet achieves 62% retrieval rate, but note that it is provided with substantial labeled data from which to directly learn semantics.

4.2   Unsupervised Feature Learning Recognition Results

Next we evaluate how well our unsupervised approach can benefit recognition tasks—*without* any network fine-tuning with labeled data. We test on three datasets: MIT Indoor 67 [36], PASCAL VOC 2007 and 2012.

For MIT Indoor 67, the database contains 67 indoor categories, and a total of 15,620 images. We use the subset defined in [36]. The training set contains 5360 (67×80) images and the test set contains 1340 (67×20) images. For PASCAL VOC 2007, we use all single-labeled images, namely 3103 single-labeled images from the trainval set as training data, and 3192 single-labeled images in the test set as testing data. Similarly, for PASCAL VOC 2012, this amounts to 3730 single-labeled training images (train set) and 3837 test images (val set). For all our baselines, we extract pool5 features from the network and train a linear classifier using softmax loss.

Table 1 shows the results. The third row shows the results of four unsupervised pre-trained models using other approaches (using their publicly available models), and the fourth row shows the results of our four baselines. Overall, our method obtains gains averaging 20% over the existing methods, and 30% over the additional baselines.

Our unsupervised pre-trained model is much better than a randomly initialized AlexNet. Note that we do not use any label information or fine-tuning here; our results are obtained using our unsupervised CNN trained purely on unlabeled videos. We also out-perform pre-trained models from Pathak et al. [12], Jayaraman and Grauman [14] and Agrawal et al. [13] by quite a large margin (around 20%). The pre-trained model from Wang and Gupta [7] has better performance, but their model is trained using substantially more data (4M visual-tracking pairs compared to our 1M pairs).

For fair and efficient comparison with our method and baselines, we take 1M of the collected visual-tracking pairs from [7] and implement their method. This is our "visual-tracking" baseline. Our approach outperforms all our baselines, including "visual tracking" on all three datasets. Surprisingly, our square-region baseline also has impressive performance. We attribute its competitive performance to our well-defined filtering process. Although the patches used for square-region baseline may not correspond to a certain object or object part, the two patches that form the square-region pair are guaranteed to be relevant. Moreover, square-region patches are taken randomly from the whole frame, and therefore it has no bias towards either objects or scenes.

Table 1. Quantitative comparisons for image classification on MIT Indoor 67, PASCAL VOC 2007 and PASCAL VOC 2012. The third outlined row shows the results of four unsupervised pre-trained models using other approaches. The fourth outlined row shows the results of our four baselines. The visual-tracking baseline is the same approach as Wang and Gupta [7], but uses the same amount of data as ours to train the model for fairest comparison.

Method	Supervision	MIT Indoor 67	VOC 2007	VOC 2012
ImageNet	1.2M labeled images	54%	71%	72%
Wang and Gupta [7]	4M visual tracking pairs	38%	47%	48%
Jayaraman and Grauman [14]	Egomotion	26%	40%	39%
Agrawal et al. [13]	Egomotion	25%	38%	37%
Pathak et al. [12]	Spatial context	23%	36%	36%
Full-frame	1M video frame pairs	27%	40%	40%
Square-region	1M square region pairs	32%	42%	42%
Visual-tracking [7]	1M visual tracking pairs	31%	42%	42%
Random Gaussian	-	16%	30%	28%
Ours	1M region proposal pairs	34%	46%	47%

4.3 Fine-Tuned Feature Learning Recognition Results

Finally, we evaluate our learnt features on image classification tasks after fine-tuning on the three datasets. For PASCAL VOC 2007, we fine-tune using VOC 2007 trainval set (5011 images) and test on VOC 2007 test set (4952 images). For PASCAL VOC 2012, we fine-tune using VOC 2012 train set (5717 images) and test on VOC 2012 val set (5823 images). For MIT Indoor 67, we fine-tune using the training set (5360 images) and test on the test set (1340 images). We use a simple horizontal flip as data augmentation and fine-tune for the same number of iterations for all methods. For PASCAL VOC multi-label classification tasks, we use the standard mean Average Precision (mAP) to evaluate the predictions. For MIT Indoor Scene single-label classification task, we report the classification accuracy.

Table 2 shows the results. Note that for the fine-tuning task, most of the learning for the network comes from labeled images in the fine-tuning dataset. The pre-trained models serve only as an initialization for fine-tuning. Our method consistently outperforms all our baseline methods and the pre-trained models from Pathak et al. [12], Jayaraman and Grauman [14] and Agrawal et al. [13]. The pre-trained model from [7] has better performance if using four times the training data as our method. However, for the comparable setting with the same amount of input video and identical fine-tuning procedures, our method is superior to the tracking-based approach (see Visual-Tracking [7] row). This comparison is the most direct and speaks favorably for the core proposed idea.

Table 2. Classification results on MIT Indoor 67, PASCAL VOC 2007 and PASCAL VOC 2012. Accuracy is used for MIT Indoor 67 and mean average precision (mAP) is used for PASCAL VOC to compare the models. The third row shows the results of four unsupervised pre-trained models using other approaches. The fourth row shows the results of our four baselines.

Pretraining method	Supervision	MIT Indoor 67	VOC 2007	VOC 2012
ImageNet	1.2M labeled images	61.6%	75.1%	73.9%
Wang and Gupta [7]	4M visual tracking pairs	41.6%	47.8%	47.4%
Jayaraman and Grauman [14]	Egomotion	31.9%	41.7%	40.7%
Agrawal et al. [13]	Egomotion	32.7%	42.4%	40.2%
Pathak et al. [12]	Spatial context	34.2%	42.7%	41.4%
Full-frame	1M video frame pairs	33.4%	41.9%	40.3%
Square-region	1M square region pairs	35.4%	43.2%	42.3%
Visual-tracking [7]	1M visual tracking pairs	36.6%	43.6%	42.1%
Random Gaussian	-	28.9%	41.3%	39.1%
Ours	1M region proposal pairs	38.1%	45.6%	44.1%

5 Conclusion and Future Work

We proposed a framework to learn visual representations from unlabeled videos. Our approach exploits object-like region proposals to generate associated regions across video frames, yielding localized patches for training an invariant embedding without explicit tracking. Through various experiments on image retrieval and image classification, we have shown that our method provides useful feature representations despite the absence of strong supervision. Our method outperforms multiple existing approaches for unsupervised pre-training, providing a new promising tool for the feature learning toolbox. In future work we plan to consider how spatio-temporal video object segmentation methods could enhance the proposals employed to create training data for our framework.

Acknowledgements. This research is supported in part by ONR PECASE N00014-15-1-2291. We also thank Texas Advanced Computing Center for their generous support and the anonymous reviewers for their comments.

References

1. Krizhevsky, A., Sutskever, I., Hinton, G.E.: ImageNet classification with deep convolutional neural networks. In: NIPS (2012)
2. Simonyan, K., Zisserman, A.: Very deep convolutional networks for large-scale image recognition. In: ICLR (2015)
3. He, K., Zhang, X., Ren, S., Sun, J.: Deep residual learning for image recognition. In: CVPR (2016)
4. Donahue, J., Jia, Y., Vinyals, O., Hoffman, J., Zhang, N., Tzeng, E., Darrell, T.: DeCAF: a deep convolutional activation feature for generic visual recognition (2014)
5. Girshick, R., Donahue, J., Darrell, T., Malik, J.: Rich feature hierarchies for accurate object detection and semantic segmentation. In: CVPR (2014)
6. Mobahi, H., Collobert, R., Weston, J.: Deep learning from temporal coherence in video. In: ICML (2009)
7. Wang, X., Gupta, A.: Unsupervised learning of visual representations using videos. In: ICCV (2015)
8. Ramanathan, V., Tang, K., Mori, G., Fei-Fei, L.: Learning temporal embeddings for complex video analysis. In: ICCV (2015)
9. Goroshin, R., Bruna, J., Tompson, J., Eigen, D., LeCun, Y.: Unsupervised feature learning from temporal data. In: ICLR (2015)
10. Jayaraman, D., Grauman, K.: Slow and steady feature analysis: higher order temporal coherence in video. In: CVPR (2016)
11. Doersch, C., Gupta, A., Efros, A.A.: Unsupervised visual representation learning by context prediction. In: ICCV (2015)
12. Pathak, D., Krahenbuhl, P., Donahue, J., Darrell, T., Efros, A.A.: Context encoders: feature learning by inpainting. In: CVPR (2016)
13. Agrawal, P., Carreira, J., Malik, J.: Learning to see by moving. In: ICCV (2015)
14. Jayaraman, D., Grauman, K.: Learning image representations equivariant to egomotion. In: ICCV (2015)
15. Le, Q.V.: Building high-level features using large scale unsupervised learning. In: ICML (2012)
16. Bengio, Y., Lamblin, P., Popovici, D., Larochelle, H., et al.: Greedy layer-wise training of deep networks. In: NIPS (2007)
17. Bengio, Y.: Learning deep architectures for AI. Found. Trends Mach. Learn. **2**, 1–127 (2009)

18. Hinton, G.E., Salakhutdinov, R.R.: Reducing the dimensionality of data with neural networks. Science **313**, 504–507 (2006)
19. Wiskott, L., Sejnowski, T.J.: Slow feature analysis: unsupervised learning of invariances. Neural Comput. **14**, 715–770 (2002)
20. Bengio, Y., Bergstra, J.S.: Slow, decorrelated features for pretraining complex cell-like networks. In: NIPS (2009)
21. Zou, W., Zhu, S., Yu, K., Ng, A.Y.: Deep learning of invariant features via simulated fixations in video. In: NIPS (2012)
22. Zou, W.Y., Ng, A.Y., Yu, K.: Unsupervised learning of visual invariance with temporal coherence. In: NIPS (2011)
23. Uijlings, J.R., van de Sande, K.E., Gevers, T., Smeulders, A.W.: Selective search for object recognition. IJCV **104**, 154–171 (2013)
24. Olshausen, B.A., Field, D.J.: Sparse coding with an overcomplete basis set: a strategy employed by V1? Vision. Res. **37**, 3311–3325 (1997)
25. Hurri, J., Hyvärinen, A.: Simple-cell-like receptive fields maximize temporal coherence in natural video. Neural Comput. **15**, 663–691 (2003)
26. Srivastava, N., Mansimov, E., Salakhutdinov, R.: Unsupervised learning of video representations using LSTMs. In: ICML (2015)
27. Wang, J., Song, Y., Leung, T., Rosenberg, C., Wang, J., Philbin, J., Chen, B., Wu, Y.: Learning fine-grained image similarity with deep ranking. In: CVPR (2014)
28. Schroff, F., Kalenichenko, D., Philbin, J.: FaceNet: a unified embedding for face recognition and clustering. In: CVPR (2015)
29. Song, H.O., Xiang, Y., Jegelka, S., Savarese, S.: Deep metric learning via lifted structured feature embedding. In: CVPR (2016)
30. Frome, A., Singer, Y., Sha, F., Malik, J.: Learning globally-consistent local distance functions for shape-based image retrieval and classification. In: ICCV (2007)
31. Liang, X., Liu, S., Wei, Y., Liu, L., Lin, L., Yan, S.: Towards computational baby learning: a weakly-supervised approach for object detection
32. Jia, Y., Shelhamer, E., Donahue, J., Karayev, S., Long, J., Girshick, R., Guadarrama, S., Darrell, T.: Caffe: convolutional architecture for fast feature embedding. In: ACM Multimedia (2014)
33. Bay, H., Tuytelaars, T., Gool, L.: SURF: speeded up robust features. In: Leonardis, A., Bischof, H., Pinz, A. (eds.) ECCV 2006. LNCS, vol. 3951, pp. 404–417. Springer, Heidelberg (2006). doi:10.1007/11744023_32
34. Wang, H., Schmid, C.: Action recognition with improved trajectories. In: ICCV (2013)
35. Henriques, J.F., Caseiro, R., Martins, P., Batista, J.: High-speed tracking with kernelized correlation filters. PAMI **37**, 583–596 (2015)
36. Quattoni, A., Torralba, A.: Recognizing indoor scenes. In: CVPR (2009)

Visual Concept Recognition and Localization via Iterative Introspection

Amir Rosenfeld[(✉)] and Shimon Ullman[(✉)]

Department of Computer Science and Applied Mathematics,
Weizmann Institute of Science, Rehovot, Israel
{amir.rosenfeld,shimon.ullman}@weizmann.ac.il

Abstract. Convolutional neural networks have been shown to develop internal representations, which correspond closely to semantically meaningful objects and parts, although trained solely on class labels. Class Activation Mapping (CAM) is a recent method that makes it possible to easily highlight the image regions contributing to a network's classification decision. We build upon these two developments to enable a network to re-examine informative image regions, which we term *introspection*. We propose a weakly-supervised iterative scheme, which shifts its center of attention to increasingly discriminative regions as it progresses, by alternating stages of classification and introspection. We evaluate our method and show its effectiveness over a range of several datasets, where we obtain competitive or state-of-the-art results: on Stanford-40 Actions, we set a new state-of the art of 81.74%. On FGVC-Aircraft and the Stanford Dogs dataset, we show consistent improvements over baselines, some of which include significantly more supervision.

1 Introduction

With the advent of deep convolutional neural networks as the leading method in computer vision, several attempts have been made to understand their inner workings. Examples of pioneering work in this direction include [1,2]; providing glimpses into representations learned by intermediate layers. Specifically, the recent work of Zhou *et al.* [3] provides an elegant mechanism to highlight the discriminative image regions that served the CNN for a given task. This can be seen as a form of introspection, highlighting the source of the network's conclusions. A useful trait we have observed in experiments is that even if the final classification is incorrect, the highlighted image regions are still be informative with respect to the correct target class. This is probably due to the similar appearance of confused classes. See Figs. 1 and 4 for some examples. Motivated by this observation, we propose an iterative mechanism of internal supervision, termed introspection, which revisits discriminative regions to refine the classification. As the process is repeated, each stage further highlights discriminative

Electronic supplementary material The online version of this chapter (doi:10. 1007/978-3-319-54193-8_17) contains supplementary material, which is available to authorized users.

© Springer International Publishing AG 2017
S.-H. Lai et al. (Eds.): ACCV 2016, Part V, LNCS 10115, pp. 264–279, 2017.
DOI: 10.1007/978-3-319-54193-8_17

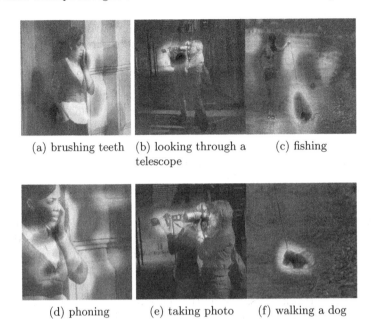

(a) brushing teeth   (b) looking through a   (c) fishing
                         telescope

(d) phoning   (e) taking photo   (f) walking a dog

Fig. 1. (*Top*) Class Activation Maps [3] show the source of a network's classification. The network tends to focus on relevant image regions even if its final prediction is wrong. An SVM trained on features extracted from VGG-GAP [3] misclassified all of these images, while highlighting the discriminative regions. (*a,b,c*) the predicted classes appears in red. (*Bottom, zoomed in version of top*) The proposed method effectively removes many such errors by focusing attention on the highlighted regions. (*d,e,f*) the corrected prediction following the introspection stage appears in green. (Color figure online)

sub-regions. Each stage uses its own classifier, as we found this to be beneficial when compared to using the same classifier for all sub-windows.

We describe strategies for how to leverage the introspection scheme, and demonstrate how these consistently improve results on several benchmark datasets, while progressively refining the localization of discriminative regions. As shown, our method is particularly beneficial for fine-grained tasks such as species [4,5] or model [6] identification and to challenging cases in e.g., action recognition [7], which requires attention to small and localized details.

In the following we will first review some related work. In Sect. 3 we describe our method in detail. Section 4 contains experiments and analysis to evaluate the proposed method, followed by concluding remarks in Sect. 5.

2   Related Work

Supervised methods consistently outperform unsupervised or semi-supervised methods, as they allow for the incorporation of prior knowledge into the learning process. There is a trade-off between more accurate classification results and structured output on the one and, the cost of labor-intensive manual annotations, on the other. Some examples are [8,9], where bounding boxes and part annotations are given at train time. Aside from the resources required for large-scale annotations, such methods elude the question of learning from weakly supervised data (and mostly unsupervised data), as is known to happen in human infants, who can learn from limited examples [10]. Following are a few lines of work related to the proposed method.

2.1 Neural Net Visualization and Inversion

Several methods have been proposed to visualize the output of a neural net or explore its internal activations. Zeiler *et al.* [1] found patterns that activate hidden units via deconvolutional neural networks. They also explore the localization ability of a CNN by observing the change in classification as different image regions are masked out. [2] Solves an optimization problem, aiming to generate an image whose features are similar to a target image, regularized by a natural image prior. Zhou *et al.* [11] aims to explicitly find what image patches activate hidden network units, finding that indeed many of them correspond to semantic concepts and object parts. These visualizations suggest that, despite training solely with image labels, there is much to exploit within the internal representations learned by the network and that the emergent representations can be used for weakly supervised localization and other tasks of fine-grained nature.

2.2 Semi-Supervised Class Localization

Some recent works attempt to obtain object localization through weak labels, i.e., the net is trained on image-level class labels, but it also learns localization. [12] Localizes image regions pertaining to the target class by masking out sub-images and inspecting change in activations of the network. Oquab *et al.* [13] use global max-pooling to obtain points on the target objects. Recently, Zhou *et al.* [3] used global average pooling (GAP) to generate a Class-Activation Mapping (CAM), visualizing discriminative image regions and enabling the localization of detected concepts. Our introspection mechanism utilizes their CAMs to iteratively identify discriminative regions and uses them to improve classification without additional supervision.

2.3 Attention Based Mechanisms

Recently, some attention based mechanisms have been proposed, which allow focusing on relevant image regions, either for the task of better classification [14] or efficient object localization [15]. Such methods benefit from the recent fusion between the fields of deep learning and reinforcement learning [16]. Another method of interest is the spatial-transformer networks in [17]: they designed a network that learns and applies spatial warping to the feature maps, effectively aligning inputs, which results in increased robustness to geometric transformations. This enables fine-grained categorization on the CUB-200-2011 birds [4] dataset by transforming the image so that only discriminative parts are considered (bird's head, body). Additional works appear in [18], who discovers discriminative patches and groups them to generate part detectors, whose detections are combined with the discovered patches for a final classification. In [19], the outputs of two networks are combined via an outer-product, creating a strong feature representation. [20] discovers and uses parts by using co-segmentation on ground-truth bounding boxes followed by alignment.

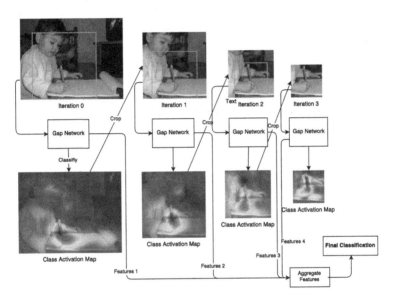

Fig. 2. Overview of proposed method. At each iteration, an image window (*top*) is classified using features from a GAP-network. Top scoring classes are used to generate Class Activation Maps (*bottom*), which are then used to select sub-windows (green rectangles). The process is repeated for a few iterations and features from all visited image windows are aggregated and used in a final classifier. The correct class is "writing on a book". Attention shifts gradually and closes in on the discriminative image region, i.e., the boy's hand holding a pencil. (Color figure online)

3   Approach

Our approach is composed of alternating between two main steps: classification and introspection. In the *classification* step, we apply a trained network to an image region (possibly the entire image). In the *introspection* step, we use the output of a hidden layer in the network, whose values were set during the classification step. This highlights image regions which are fed to the next iteration's classification step. This process is iterated a few times (typically 4, see Sect. 4, Fig. 5), and finally the results of all stages are combined. Training proceeds by learning a specialized classifier for each iteration, as different iterations capture different contexts and levels of detail (but without additional supervision).

Both classification/introspection steps utilize the recent Class-Activation Mapping method of [3]. We briefly review the CAM method and then describe how we build upon it. In [3], a deep neural network is modified so that post-classification, it is possible to visualize the varying contribution of image regions, via a so-called Class Activation Mapping (CAM). A global average pooling was used as a penultimate feature representation. This results in a feature vector which is a spatial averaging of each of the feature maps of the last convolutional layer. Using the notation in [3]: let $f_k(x, y)$ be the k'th output of the last convolutional layer at grid location (x, y). The results of the global-average pooling results in a vector $F = (F_1, F_2, \ldots, F_k)$, defined as:

$$F_k = \sum_{x,y} f_k(x, y) \tag{1}$$

This is followed by a fully connected layer with C outputs (assuming C target classes). Hence the score for class c before the soft-max will be:

$$S_c = \sum_k w_k^c F_k \tag{2}$$

$$= \sum_k w_k^c \sum_{x,y} f_k(x, y) \tag{3}$$

$$= \sum_{x,y} \sum_k w_k^c f_k(x, y) \tag{4}$$

Now, define

$$M_c(x, y) = \sum_k \omega_k^c f_k(x, y) \tag{5}$$

where ω_k^c are class-specific weights. Hence we can express S_c as a summation of terms over (x, y):

$$S_c = \sum_{x,y} M_c(x, y) \tag{6}$$

And the class probability scores are computed via soft-max, e.g., $P_c = \frac{e^{S_c}}{\sum_t e^{S_t}}$. Equation 5 allows us to measure the contribution of each grid cell $M_c(x, y)$ *for each specific class c*. Indeed, [3] has shown this method to highlight informative image regions (with respect to the task at hand), while being on par with the classification performance obtained by the unmodified network (GoogLeNet [21] in their case). See Fig. 1 for some CAMs. Interestingly, we can use the CAM method to highlight informative image regions for classes other than the correct class, providing intuition on the features it has learned to recognize. This is discussed in more detail in Sect. 4.2 and demonstrated in Fig. 4. We name a network whose final convolutional layer is followed by a GAP layer as a GAP-network, and the output of the GAP layer as the GAP features. We next describe how this is used in our proposed method.

3.1 Iterative Classification-Introspection

The proposed method alternates classification and introspection. Here we provide the outline of the method, with specific details such as values of parameters discussed in Sect. 4.6.

For a given image I and window w (initially the entire image), a learned classifier is applied to the GAP features extracted from the window I_w, resulting in C classification scores S_c, $c \in [1 \ldots C]$ and corresponding CAMs $M_c^w(x, y)$. The introspection phase employs a strategy to select a sub-window for the next step by applying a beam-search to a set of putative sub-windows. The sequence of windows visited by the method is a route on an *exploration-tree*, from the root to one of the leaves. Each node represents an image window and the root is the entire image. We next explain how the sub-windows are created, and how the search is applied.

We order the current classification scores S_c by descending order and retain the top k scoring classes. Let $\hat{c}$ be one of these classes and $M_{\hat{c}}^w(x, y)$ the corresponding CAM. We extract a square sub-window w' centered on the maximal value of $M_{\hat{c}}^w(x, y)$. Each such w' (k in total) is added as a child of the current node, which is represented by w. In this way, each iteration of the method expands a selected node in the exploration-tree, corresponding to an image window, until a maximum depth is reached. The tree depth is the number of iterations. We define iteration 0 as the iteration acting on the root. The size of a sub-window w' is of a constant fraction of the size of its parent w. We next describe how the exploration-tree is used for classification.

3.2 Feature Aggregation

Let k be the number of windows generated at iteration $t > 0$. We denote the set of windows by:

$$W_t = (w_i^t)_{i=1}^k \tag{7}$$

And the entire set of windows as:

$$\mathcal{R} = (W_t)_{t=0}^T \tag{8}$$

where W_0 is the entire image. For each window w_i^t we extract features $f_w^t \in \mathbb{R}^K$, e.g., $K = 1024$, the dimension of the GAP features, as well as classification scores $S_{w_i^t} \in \mathbb{R}^C$. The set of windows $\mathcal{R}$ for an image I is arranged as nodes in the exploration-tree. The final prediction is a result of aggregating evidence from selected sub-windows along some path from the root to a tree-leaf. We evaluate variants of both early fusion (combining features from different iterations) or later fusion (combining predictions from different iterations).

3.3 Training

Training proceeds in two main stages. The first is to train a sequence of classifiers that will produce an exploration-tree for each training/testing sample. The second is training on feature aggregations along different routes in the exploration-trees, to produce a final model.

During training, we train a classifier for each iteration (for a predefined number of iterations, 5 total) of the introspection/classification sequence. The automatic training of multiple classifiers at different scales contributes directly to the success of the method, as using the same classifier for all iterations yielded no improvement over the baseline results (Sect. 4.1). For the first iteration, we simply train on entire images with the ground-truth class-labels. For each iteration $t > 1$, we set the training samples to sub-windows of the original images and the targets to the ground-truth labels. The sub-windows selected for training are always those corresponding to the strongest local maximum in $M_c(x, y)$, where $M_{\hat{c}}(x, y)$ is the CAM corresponding to the highest scoring class. Each classifier is an SVM trained on the features the output of the GAP layer of the network (as was done in [3]). We also checked the effect of fine-tuning the network and using additional features. The Results are discussed in the experiments, Sect. 4.

Routes on Exploration Trees. The image is explored by traversing routes on a tree of nested sub-windows. The result of training is a set of classifiers, $\mathcal{E} = (E)_{i=1...T}$. We produce an exploration tree by applying at each iteration

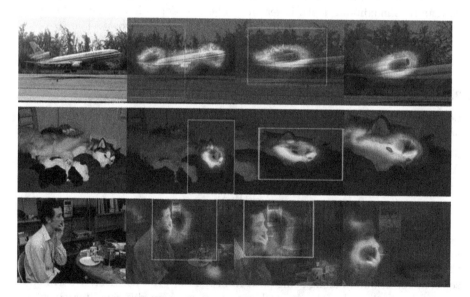

Fig. 3. Exploration routes on images: Each row shows the original image and 3 iterations of the algorithm, including the resulting Class-Activation Maps [3] used to guide the next iteration. The selected sub-window is shown with a green bounding box. Despite being trained and tested without any bounding box annotations, the proposed method closes in on the features relevant to the target class. The first 2 predictions (*columns 2,3*) in each row are mistaken and the last one (*rightmost column*) is correct. (Color figure online)

j the classifier E_j on the features of the windows produced by the previous iteration. The window of iteration 0 is the entire image. A route along the tree will consist of a sequence of windows $w^1, w^2, \ldots w^T$ where T is the number of iterations. We found in experiments that more than 5 iterations (including the first) brings negligible boosts in performance. The image score for a given class is given by either (1) summing the scores of classifiers along a route (late fusion), or (2) learning a classifier for the combined features of all visited windows along the route (early fusion). Features are combined via averaging rather than concatenation. This reduces the training time at no significant change to the final performance; such an effect has also been noted by [22]. See Fig. 2 for an overview of the proposed method. Figure 3 shows some examples of how progressively zooming in on image regions helps correct early classification mistakes.

4   Experiments

4.1   Setup

In all our experiments, we start with a variant of the VGG-16 network [22] which was fined tuned on ILSVRC by [3]. We chose it over GoogLeNet-GAP as it obtained slightly higher classification results on the ILSVRC validation set. In this network, all layers after *conv5-3* have been removed, including the subsequent pooling layer; hence the spatial-resolution of the resultant feature maps/CAM is 14×14 for an input of size 224×224 (leaving the pooling layer would reduce resolution to be 7×7). A convolutional layer of 1024 filters has been added, followed by a fully-connected layer to predict classes. This is our basic GAP-network, called VGG-GAP. Each image window, including the entire original image, is resized so that its smaller dimension is 224 pixels, resulting in a feature map $14 \times n \times 1024$, for which we compute the average along the first two dimensions, to get a feature representation. We resize the images using bilinear interpolation. We train a separate classifier for each iteration of the classification/introspection process; treating all visited image windows with the same classifier yielded a negligible improvement (0.3% in precision) over the baseline. All classifiers are trained with a linear SVM [26] on ℓ_2 normalized feature vectors. If the features are a concatenation of two feature vectors, they are ℓ_2 normalized before concatenation. Our experiments were carried out using the MatConvNet framework [27] and as well as [28,29]. We evaluated our approach on several datasets, including Stanford-40 Actions [7], the Caltech-USCD Birds-200-2011 [4] (a.k.a CUB-200-2011), the Stanford-Dogs dataset, [5] and the FGVC-Aircraft dataset [6]. See Table 1 for a summary of our results compared to recent work. In the following, we shall first show some analysis on the validity of using the CAMs to guide the next step. We shall then describe interesting properties of our method, as well as the effects of different parameterizations of the method.

Table 1. Classification accuracy of our method vs. a baseline for several datasets. VGG-GAP* is our improved baseline using the VGG-GAP network [3]. Ours-late-fusion: we aggregate the scores of image windows along the visited path. Ours-early-fusion: aggregate scores of classifiers trained on feature combination of windows along the visited path. Ours-ft: same as ours-early fusion but fine-tuned. +D: concatenated with fc6 features from VGG-16 at each stage. *Stanford Dogs is a subset of ILSVRC dataset. For this dataset, we compare to work which also used a network pre-trained on ILSVRC. [†]means fine tuning the network on all iterations.

Method	40-Actions	Dogs*	Birds	Aircraft
GoogLeNet-GAP [3]	72.03	-	63.00	-
VGG-16-fc6	73.83	83.76	63.46	60.07
VGG-GAP*	75.31	81.83	65.72	62.95
ours-late-fusion	76.88	82.63	73.52	66.34
ours-early-fusion	77.08	83.55	71.64	68.26
ours-ft	80.37	82.62	78.74	79.15
ours-early-fusion+D	81.04	**86.25**	78.91	77.74
ours-ft+D	**81.74**	84.18	79.55	78.04
Previous work	72.03 [3], 81 [23]	79.92 [24]	77.9 [18], 81.01 [25], 82 [20], **84.1** [17]	**84.1** [19]

4.2   Correlation of Class and Localization

In this section, we show some examples to verify our observation that CAMs tend to highlight informative image locations w.r.t to the target class despite the fact that the image may have been misclassified at the first iteration (i.e., before zooming in on sub-windows).

To do so, we have applied to the test-set of the Stanford-40 actions dataset a classifier learned on the GAP features of VGG-GAP. For each category in turn, we ranked all images in the test set according to the classifier's score for that category. We then picked the top 5 true positive images and top 5 false positive images. See Fig. 4 for some representative images. We can see that the CAMs for a target class tend to be consistent in both positive images and high-ranking non-class images. Intuitively, this is because the classifier gives more weight to patterns which are similar in appearance. For the "writing on a book" category (top-left of Fig. 4) we can see how in positive images the books and especially the hands are highlighted, as they are for non-class images, such as "reading", or "cutting vegetables". For "texting message" (top-right) the hand region is highlighted in all images, regardless of class. For "shooting an arrow" class (bottom-right), elongated structures such as fishing rods are highlighted. A nice confusion appears between an archer's bow and a violinist's bow (bottom-right block, last row, first image), which are also referred to by the same word in some human languages.

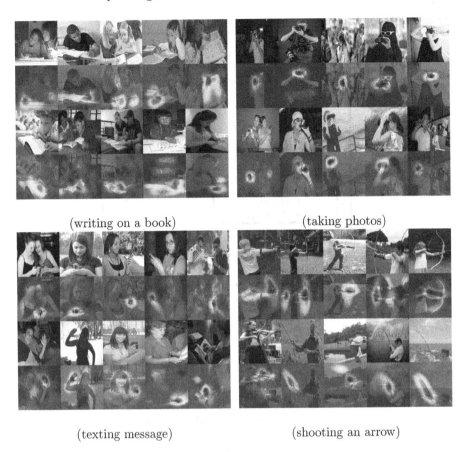

(writing on a book)   (taking photos)

(texting message)   (shooting an arrow)

Fig. 4. Top ranking images (both true positives and false positives) for various action categories along with Class-Activation Maps. Misclassified images still carry information on where additional attention will disambiguate the classification. Each block of images shows, from first to fourth row: high-ranking true-positives and their respective CAMs, high ranking false-positive and their respective CAMs. The target class appears below each block. We recommend viewing this figure online to zoom in on the details.

To check our claim quantitatively, we computed the extent of two square sub-windows for each image in the test-set: one using the CAM of the true class and one using the CAM of the highest scoring non-true class. For each pair we computed the overlap (intersection over union) score. The mean score of all images was 0.638; This is complementary evidence to [3], who shows that the CAMs have good localization capabilities for the correct class.

4.3   Early and Late Fusion

In all our experiments, we found that using the features extracted from a window at some iteration can bring worse results on its own compared to those extracted

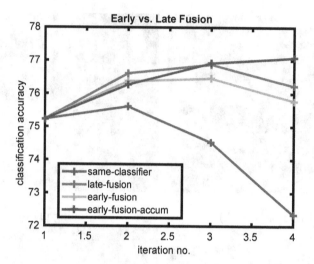

Fig. 5. Effect of learning at different iterations: using the same classifier trained on entire images for all iterations tends to cause overall precision to drop (*same-classifier*). Accumulating the scores of per-iteration learned classifiers along the explored path improves this (*late-fusion*). Using all features along the observed exploration path improves classification as the path length increases (*early-fusion*) and summing all scores along the early-fusion path brings the best performance (*early-fusion-accum*). Performance is shown on the Stanford-40 [7] dataset

from earlier iterations (which include this window). However, the performance tends to improve as we combine results from several iterations, in a late-fusion manner. Training on the combined (averaged) features of windows from multiple iterations further improves the results (early-fusion). Summing the scores of early-fused features for different route lengths further improves accuracy: if S_i is the score of the classifier trained on a route of length i. Then creating a final score from $S_1 + S_t + \ldots S_T$ tends to improve as T grows, typically stabilizing at $T = 5$. See Fig. 5 for an illustration of this effect. Importantly, we tried using the classifier from the first iteration (i.e., trained on entire images) for all iterations. This performed worse than learning a classifier per-iteration, especially in later iterations.

4.4   Fine-Grained vs. General Categories

The Stanford-40 Action dataset [7] is a benchmark dataset made of 9532 images of 40 different action classes, with 4000 images for training and the rest for testing. It contains a diverse set of action classes including transitive ones with small objects (smoking, drinking) and large objects (horses), as well as intransitive actions (running, jumping). As a baseline, we used the GAP-network of [3] as a feature extractor and trained a multi-class SVM [26] using the resulting features. It is particularly interesting to examine the classes for which our

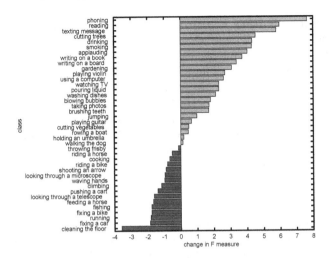

Fig. 6. Our approach improves mainly fine-grained tasks and categories where classification depends on small and specific image windows. The figure shows absolute difference in terms of F-measure over the baseline approach on all categories of the Stanford-40 Actions [7] dataset. It is recommended to view this figure online.

method is most beneficial. We have calculated the F-measure for each class using the classification scores from the fourth and first iteration and compared them. Figure 6 shows this; the largest absolute improvements are on relatively challenging classes such as texting a message (7.64%), drinking (4.32%), smoking (4.3%), etc. For all of these, the discriminative objects are small objects and are relatively hard to detect compared to most other classes. In some cases, performance is harmed by zooming in on too-local parts of an image: for "riding a bike" (−0.8%), a small part of the bicycle will not allow disambiguating the image from e.g., "fixing a bike". Another pair of categories exhibiting similar behavior is "riding a horse" vs. "feeding a horse".

4.5 Top-Down vs. Bottom-Up Attention

To further verify that our introspection mechanism highlights regions whose exploration is worthwhile, we evaluated an alternative to the introspection stage by using a generic saliency measure [30]. On the Stanford-40 dataset, instead of using the CAM after the first classification step, we picked the most salient image point as the center of the next sub-window. Then we proceeded with training and testing as usual. This produced a sharp drop in results: on the first iteration performance dropped from 74.47% when using the CAM to 62.31% when using the saliency map. Corresponding drops in performance were measured in the late-fusion and early fusion steps, which improve results in the proposed scheme but made them worse when using saliency as a guide.

Usage of Complementary Feature Representations. The network used for drawing attention to discriminative image regions need not necessarily be the one used for feature representation. We used the VGG-16 [22] network to extract fc6 features along the GAP features for all considered windows. On the Stanford-40 Actions dataset, when used to classify categories using features extracted from entire images, these features we slightly weaker than the GAP features (73% vs. 75%). However, training on a concatenated feature representation boosted results significantly, reaching a precision of 80%. We observed a similar effect on all datasets, showing that the two representations are complementary in nature. Combined with our iterative method, we were able to achieve 81.74%, compared to the previous best 81% of [23].

Effect of Aspect-Ratio Distortion. Interestingly, our baseline implementation (using only the VGG-GAP network as a feature extractor for the entire image) got a precision score of 75.23% compared to 72.03% of [3]. We suspect that it may be because in their implementation, they modified the aspect ratio of the images to be square regardless of the original aspect ratio, whereas we did not. Doing so indeed got a score more similar to theirs, which is interesting from a practical viewpoint.

4.6 Various Parameters and Fine-Tuning

Our method includes several parameters, including the number of iterations, the width of the beam-search used to explore routes of windows on the image and the ratio between the size of the current window and the next. For the number iterations, we have consistently observed that performance saturates, and even deteriorates a bit, around iteration 4. An example can be seen in Fig. 5 showing the performance vs. iteration number on the Stanford-40 dataset. We observed a similar behavior in every dataset on which we've evaluated the method. This is probably due to the increasingly small image regions considered at each iteration. As for the number of windows to consider at each stage, we tried choosing between 1 and 3 of the windows relating to the highest ranking classes on a validation set. At best, this performed as well as the greedy strategy, which chose only the highest scoring window at each iteration. The size of the sub-window with respect to the current image window was set as $\sqrt{2}m$ where m is the geometric mean of the current window's height and width (in effect, all windows are square, except the entire image). We have experimented with smaller and larger values on a validation set and found this parameter to give a good trade-off between not zooming in too much (risking "missing" relevant features) and too little (gaining too little information with respect to the previous iteration).

We have also evaluated our results when fine-tuning the VGG-GAP network before the first iteration. This improves the results for some of the datasets, i.e., Stanford-40 [7], CUB-200-2011 [4], but did not improve results significantly for others (Dogs [5], Aircraft [6]). Finally, we evaluated the effect of fine-tuning the network for *all* iterations on the CUB-200-2011. This resulted in a competitive results of 79.95%. Some of the best results to date added a massive amount of external data mined from the web [31] (91.9%) and/or strong supervision [32] (84.6%).

5 Conclusions

We have presented a method, which by repeatedly examining the source of the current prediction, decides on informative image regions to consider for further examination. The method is based on the observation that a trained CNN can be used to highlight relevant image areas even when its final classification is incorrect. This is a result of training on multiple visual categories using a shared feature representation. We have built upon Class Activation Maps [3] due to their simplicity and elegance, though other methods for identifying the source of the classification decision (e.g., [1]) could probably be employed as well. The proposed method integrates multiple features extracted at different locations and scales. It makes consistent improvement over baselines on fine-grained classification tasks and on tasks where classification depends on fine localized details. It obtains competitive results on CUB-200-2011 [4], among methods which avoid strong supervision such as bounding boxes or keypoint annotations. The improvements are shown despite the method being trained using only class labels, avoiding the need for supervision in the form of part annotations or even bounding boxes. In future work, it would be interesting to examine the use of recurrent nets (RNN, LSTM [33]) to automatically learn sequential processes, which incrementally improve classification results, extending the approach described in the current work.

Acknowledgement. The work supported by ERC Advanced Grant 269627 Digital Baby and in part by the Center for Brains, Minds and Machines, funded by NSF Science and Technology Centers Award CCF-1231216.

References

1. Zeiler, M.D., Fergus, R.: Visualizing and understanding convolutional networks. In: Fleet, D., Pajdla, T., Schiele, B., Tuytelaars, T. (eds.) ECCV 2014. LNCS, vol. 8689, pp. 818–833. Springer, Heidelberg (2014). doi:10.1007/978-3-319-10590-1_53
2. Mahendran, A., Vedaldi, A.: Understanding deep image representations by inverting them. In: 2015 IEEE Conference on Computer Vision and Pattern Recognition (CVPR), pp. 5188–5196. IEEE (2015)
3. Zhou, B., Khosla, A., Lapedriza, A., Oliva, A., Torralba, A.: Learning deep features for discriminative localization. arXiv preprint arXiv:1512.04150 (2015)
4. Wah, C., Branson, S., Welinder, P., Perona, P., Belongie, S.: The Caltech-UCSD birds-200-2011 dataset. California Institute of Technology (2011)
5. Khosla, A., Jayadevaprakash, N., Yao, B., Fei-Fei, L.: Novel dataset for fine-grained image categorization. In: First Workshop on Fine-Grained Visual Categorization, IEEE Conference on Computer Vision and Pattern Recognition, Colorado Springs, CO (2011)

6. Maji, S., Rahtu, E., Kannala, J., Blaschko, M., Vedaldi, A.: Fine-grained visual classification of aircraft. arXiv preprint arXiv:1306.5151 (2013)

7. Yao, B., Jiang, X., Khosla, A., Lin, A.L., Guibas, L., Fei-Fei, L.: Human action recognition by learning bases of action attributes and parts. In: 2011 IEEE International Conference on Computer Vision (ICCV), pp. 1331–1338. IEEE (2011)

8. Zhang, N., Donahue, J., Girshick, R., Darrell, T.: Part-based R-CNNs for fine-grained category detection. In: Fleet, D., Pajdla, T., Schiele, B., Tuytelaars, T. (eds.) ECCV 2014. LNCS, vol. 8689, pp. 834–849. Springer, Heidelberg (2014). doi:10.1007/978-3-319-10590-1_54

9. Zhang, N., Shelhamer, E., Gao, Y., Darrell, T.: Fine-grained pose prediction, normalization, and recognition. arXiv preprint arXiv:1511.07063 (2015)

10. Lake, B.M., Salakhutdinov, R., Tenenbaum, J.B.: Human-level concept learning through probabilistic program induction. Science **350**, 1332–1338 (2015)

11. Zhou, B., Khosla, A., Lapedriza, A., Oliva, A., Torralba, A.: Object detectors emerge in deep scene CNNs. arXiv preprint arXiv:1412.6856 (2014)

12. Bergamo, A., Bazzani, L., Anguelov, D., Torresani, L.: Self-taught object localization with deep networks. arXiv preprint arXiv:1409.3964 (2014)

13. Oquab, M., Bottou, L., Laptev, I., Sivic, J.: Is object localization for free?-weakly-supervised learning with convolutional neural networks. In: Proceedings of the IEEE Conference on Computer Vision and Pattern Recognition, pp. 685–694 (2015)

14. Mnih, V., Heess, N., Graves, A., et al.: Recurrent models of visual attention. In: Advances in Neural Information Processing Systems, pp. 2204–2212 (2014)

15. Caicedo, J.C., Lazebnik, S.: Active object localization with deep reinforcement learning. In: Proceedings of the IEEE International Conference on Computer Vision, pp. 2488–2496 (2015)

16. Mnih, V., Kavukcuoglu, K., Silver, D., Graves, A., Antonoglou, I., Wierstra, D., Riedmiller, M.: Playing atari with deep reinforcement learning. arXiv preprint arXiv:1312.5602 (2013)

17. Jaderberg, M., Simonyan, K., Zisserman, A., et al.: Spatial transformer networks. In: Advances in Neural Information Processing Systems, pp. 2008–2016 (2015)

18. Xiao, T., Xu, Y., Yang, K., Zhang, J., Peng, Y., Zhang, Z.: The application of two-level attention models in deep convolutional neural network for fine-grained image classification. In: Proceedings of the IEEE Conference on Computer Vision and Pattern Recognition, pp. 842–850 (2015)

19. Lin, T.Y., RoyChowdhury, A., Maji, S.: Bilinear CNN models for fine-grained visual recognition. In: Proceedings of the IEEE International Conference on Computer Vision, pp. 1449–1457 (2015)

20. Krause, J., Jin, H., Yang, J., Fei-Fei, L.: Fine-grained recognition without part annotations. In: Proceedings of the IEEE Conference on Computer Vision and Pattern Recognition, pp. 5546–5555 (2015)

21. Szegedy, C., Liu, W., Jia, Y., Sermanet, P., Reed, S., Anguelov, D., Erhan, D., Vanhoucke, V., Rabinovich, A.: Going deeper with convolutions. In: Proceedings of the IEEE Conference on Computer Vision and Pattern Recognition, pp. 1–9 (2015)

22. Simonyan, K., Zisserman, A.: Very deep convolutional networks for large-scale image recognition. arXiv preprint arXiv:1409.1556 (2014)

23. Gao, B.B., Wei, X.S., Wu, J., Lin, W.: Deep spatial pyramid: the devil is once again in the details. arXiv preprint arXiv:1504.05277 (2015)

24. Zhang, Y., Wei, X., Wu, J., Cai, J., Lu, J., Nguyen, V.A., Do, M.N.: Weakly supervised fine-grained image categorization. arXiv preprint arXiv:1504.04943 (2015)

25. Simon, M., Rodner, E.: Neural activation constellations: unsupervised part model discovery with convolutional networks. In: Proceedings of the IEEE International Conference on Computer Vision, pp. 1143–1151 (2015)

26. Fan, R.E., Chang, K.W., Hsieh, C.J., Wang, X.R., Lin, C.J.: LIBLINEAR: a library for large linear classification. J. Mach. Learn. Res. **9**, 1871–1874 (2008)

27. Vedaldi, A., Lenc, K.: MatConvNet: convolutional neural networks for matlab. In: Proceedings of the 23rd Annual ACM Conference on Multimedia Conference, pp. 689–692. ACM (2015)

28. Dollár, P.: Piotr's Computer Vision Matlab Toolbox (PMT). http://vision.ucsd.edu/~pdollar/toolbox/doc/index.html

29. Vedaldi, A., Fulkerson, B.: VLFeat: an open and portable library of computer vision algorithms. In: Bimbo, A.D., Chang, S.F., Smeulders, A.W.M. (eds.) ACM Multimedia, pp. 1469–1472. ACM, New York (2010)

30. Zhu, W., Liang, S., Wei, Y., Sun, J.: Saliency optimization from robust background detection. In: Proceedings of the IEEE Conference on Computer Vision and Pattern Recognition, pp. 2814–2821 (2014)

31. Krause, J., Sapp, B., Howard, A., Zhou, H., Toshev, A., Duerig, T., Philbin, J., Fei-Fei, L.: The unreasonable effectiveness of noisy data for fine-grained recognition. arXiv preprint arXiv:1511.06789 (2015)

32. Xu, Z., Huang, S., Zhang, Y., Tao, D.: Augmenting strong supervision using web data for fine-grained categorization. In: Proceedings of the IEEE International Conference on Computer Vision, pp. 2524–2532 (2015)

33. Hochreiter, S., Schmidhuber, J.: Long short term memory. Neural Comput. **9**, 1735–1780 (1997)

Aggregating Local Context for Accurate Scene Text Detection

Dafang He[1(✉)], Xiao Yang[2], Wenyi Huang[1], Zihan Zhou[1],
Daniel Kifer[2], and C. Lee Giles[1]

[1] Information Science and Technology, Penn State University, State College, USA
duh188@psu.edu
[2] Department of Computer Science and Engineering,
Penn State University, State College, USA

Abstract. Scene text reading continues to be of interest for many reasons including applications for the visually impaired and automatic image indexing systems. Here we propose a novel end-to-end scene text detection algorithm. First, for identifying text regions we design a novel Convolutional Neural Network (CNN) architecture that aggregates local surrounding information for cascaded, fast and accurate detection. The local information serves as context and provides rich cues to distinguish text from background noises. In addition, we designed a novel grouping algorithm on top of detected character graph as well as a text line refinement step. Text line refinement consists of a text line extension module, together with a text line filtering and regression module. Jointly they produce accurate oriented text line bounding box. Experiments show that our method achieved state-of-the-art performance in several benchmark data sets: ICDAR 2003 (IC03), ICDAR 2013 (IC13) and Street View Text (SVT).

1 Introduction

Scene text provides rich semantic cues about an image. Many useful applications, such as image indexing system and autonomous driving system could be built on top of a robust scene text reading algorithm. Thus detecting and recognizing scene text has recently attracted great attention from both research community and industry.

Even with the increasing attention in reading text in the wild, scene text reading is still a challenging problem. Unusual font, distortion, reflection and low contrast make the problem still unsolved. Algorithms for scene text detection could be classified into three categories [1]: (1) sliding window based methods, (2) region proposal based methods, and (3) hybrid methods. Sliding window based approaches [2,3] try to determine whether a fixed-sized small image patch is a text area or not. Such methods need to examine every locations in different scales in an exhaustive manner, thus are very time consuming. Region-based methods [4–9] propose a set of connected components as candidate text regions with predefined rules. These approaches greatly reduce the number of image

© Springer International Publishing AG 2017
S.-H. Lai et al. (Eds.): ACCV 2016, Part V, LNCS 10115, pp. 280–296, 2017.
DOI: 10.1007/978-3-319-54193-8_18

patches needed to be examined, however, they may miss some text areas or generate regions with too many characters connected. Hybrid methods [10,11] integrate the region based methods with the sliding window methods to combine the advantages of both categories. One of the common components in these methods is a classifier that determines whether a given image patch or a region is text or not. Given the success of the convolutional neural networks (CNN) in object detection and recognition [12], they have also been used as the text/non-text classifier in scene text detection tasks [8,13,14]. The strong power of CNN makes scene text detection more robust to outlier noises. A typical pipeline is to first generate a set of region proposals and then apply a binary classifier trained on text/non-text data. Although this method is efficient but even CNN trained on millions of samples are not stable for robust scene text reading on complex scenes. This is because context is often necessary for disambiguation. It is needed to determine whether a vertical line is an "I", a "1", or part of background noise like the space between bricks or the edge of a window. In Fig. 4(a), we show some examples of text-like background noise which are confusing, even to people, when appearing without their contexts.

Recently, Zhu et al. [15] proposed to use highly-semantic context, which tried to explore the assumption that text are typically on specific background, such as sign board, but seldom on the others, for e.g., sky in natural image. However, modeling text which potentially exists in a wide variety of places is impossible with an exhaustive manner, and whether a new class of semantic object which does not appear in the training set will hurt results is in doubt. In addition, in a lot images that are not purely natural, text could be placed in unusual places, e.g. sky. In those cases, the model might hurt the performance. In this paper, we define a region's local context to be its horizontal surroundings. A text localization algorithm is proposed which efficiently aggregates local context information in detecting candidate text regions. The basic idea is that the surrounding information of a candidate region usually contains strong cues about whether the region is text or background noise similar to text, and thus helping our model localize text more accurately. Some examples of these context images are shown in Fig. 4(b). In addition, we also propose a grouping algorithm to form lines from verified regions and a line refinement step to extend text lines by searching for missing character components, as well as to regress text lines to obtain accurate oriented bounding boxes.

To be more specific, our major contributions are the following aspects:

1. A method that efficiently aggregates local information for cascaded and accurate classification of proposed regions. This step could be part of any other region based framework.
2. An effective grouping algorithm as well as a novel text line refinement step. Text line refinement includes a Gaussian Mixture Model (GMM) based text line extension module to find new character components, and jointly, a sliding window based oriented bounding box regression and filtering module. They are efficient and robust for post processing, and give an accurate oriented bounding box, instead of a mere horizontal bounding box.

3. A cascaded end-to-end detection pipeline for accurate scene text detection in unconstrained images. State-of-the-art performance is achieved in IC03, IC13 and SVT data sets.

In the following sections, we first describe related works on scene text detection in Sect. 2. Our method will be described in detail in Sect. 3 and experimental results are shown in Sect. 4. Several text detection results from images using the proposed algorithm are shown in Fig. 1.

Fig. 1. Scene text that have been successfully detected by our proposed systems. Images are from IC13 and SVT dataset

2   Related Work

Scene text detection is much more challenging than Optical Character Recognition (OCR). Early work in scene text detection focused on sliding window based approaches. Chen and Yuille [3] proposed an end-to-end scene text reading system which used Adaboost algorithm for aggregating weak classifiers trained on several carefully designed, hand crafted features into a strong classifier. They used sliding window to find candidate text regions. However, the scales of text lines in scene images vary a lot so sliding window based methods are typically very inefficient. Region based methods have recently received more attention. Most works focused on two region based methods: (1) Stroke width transform (SWT) [6] and its variants [7,16]; (2) Extreme Region (ER) detector [5,8] and Maximally Extreme Region (MSER) detector [9]. SWT explicitly explores the assumption that text consists of strokes with nearly constant width. It first detects edges in the image and tries to group pixels into regions based on the orientation of the edge. However, its performance is severely decreased when the text are of low contrast or in unusual font style. ER based methods [5,8,9,17] are now popular since they are computationally efficient and achieve high recall as well. Neumann and Matas [5] proposed an ER based text detection algorithm which utilized several carefully designed region features, such as hole area ratio, Euler number, etc. However, the designed region features are not representative enough to remove background noise that is similar to text. Several other works

[17,18] follows a similar patterns in the classification step. These region features are fast to compute, but they typically lack the ability of robust classification.

More recently, CNN based methods have been used with significant success [8,13]. The ability to synthesize data for training, which was specifically explored in [19], has greatly accelerated the research in scene text detection due to the unlimited amount of training data. Several works [9,14], trained on synthetic images, have achieved the state-of-the-art performance in real image as well.

However, even with the introduction of CNN models and synthetic image generation for training, it is still hard for the CNN models to achieve good performance in some complex scenarios. We observe that the main reason of this incorrect classification lies on the fact that some characters have simple shapes that are also contained in non-text objects and CNNs are often fooled into classifying such objects as text. In this work, we explore ways to overcome these challenges and propose a system for efficient and accurate scene text detection.

3   Methodology

Our proposed detection pipeline is as follows. First, an ER detector is conducted on 5 channels of an input image. For each detected region, we first classify it to get a coarse prediction and filter out most non-text regions. Then the local context is aggregated to classify the remaining regions in order to obtain a final prediction. Text lines will be formed on top of a character component graph by grouping the verified regions with similar properties together. A text line refinement step is also designed to further filter out false positive and obtain accurate oriented bounding boxes. Several successive image examples of the proposed method can be seen in Fig. 2.

Fig. 2. Images from the scene text detection pipeline. From left to right: (1) Region proposals; (2) Coarse predictions; (3) Predictions after aggregating local context; (4) Final detected text lines.

3.1   Cascaded Classification and Local Context Aggregation

Context information is critical for object detection and recognition [20, 21]. Previous region based methods often focus on classifying each region independently and the image patch is cropped tightly from a generated region [8]. Here, we consider the local context of a given region as its local surroundings and argue that surrounding information should be incorporated in determining whether a given region should be classified as text or not. We observe that characters, which are often represented as simple shapes such as "I" or "l", cannot be well distinguished from background noises that are similar to them. However, text in an image is often represented as lines of characters, and for a given region, its local surroundings give rich information about whether it is text or not. There have been works that explored relation between text regions before, such as [17, 18]. They proposed to use graphcut on top of MSER region graph to refine the results. Instead, we try to aggregate more higher level information in classification step of each region by the proposed network, and we show that this aggregation provides rich information. Some background regions, which are difficult to be distinguished from text when cropped from a tight bounding box, can be accurately predicted by our model after we aggregate this context information.

Design Rationale: The architecture of the proposed framework is shown in Fig. 3. We call this network a text local feature aggregation network (TLFAN). This is a two-column CNN with joint feature prediction on the top. It is designed for cascaded classification that will be explained in the next part. One CNN branch with fully connected layers is for coarse prediction, and we refer to this branch as standard CNN. The other branch takes an input with aggregated local information to produce a context vector, and we refer it as context CNN. The first column of the architecture is for learning features from the tight image patch which we are focusing at, and the other is for learning features from its surroundings. This CNN structure is specifically designed for scene text reading

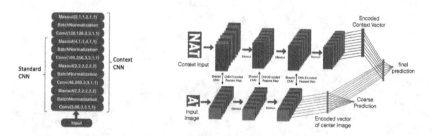

Fig. 3. The proposed TLFAN architecture for scene text detection. Left: CNN structure of the proposed network. The bottom 3 layers of convolution and pooling are shared, and for context branch, another CNN layer and pooling layer is added to produce deeper representation. Right: The whole architecture of the network. One column is for the given patch that we are trying to classify. The other is for extracting context information for this patch, and the generated feature vector will be further used to give an accurate prediction of the region.

and several design rationales are here: (1) Local context typically provides rich information about whether a specific region is text or not. Some background noise can not be well distinguished with text from a mere tight bounding box. In Fig. 4, there are some example image patches cropped from IC13 where traditional text/non-text binary classification will easily fail. But our model, by aggregating local context, can robustly distinguish it from text. (2) Even though here we consider text lines with arbitrary orientation, horizontal neighborhood already provides rich clues of whether the given region is text or not. (3) Since the input to the proposed TLFAN and its context have the same scale, we can use shared CNN parameters for the two columns and thus reduce the number of parameters that need to be learned. We also tried to use central surround network [22] which will consider a larger surrounding region. However, by doing so, we will have to learn more parameters either in CNN part or in fully connected part. In addition, We found that this will not improve the performance as much as the proposed manner, and it is likely to cause overfitting, since it considers information that is mostly unrelated.

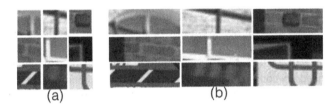

Fig. 4. Cropped image patches which demonstrate that local context helps in distinguishing between background noise and text. (a) The original image patches cropped tightly from the generated regions. (b) The horizontal context images which correspond to the region on the left, respectively. All the examples here are background noises which easily cause false positive if we only consider a tight bounding box for classification. Instead, our model can efficiently aggregate local context, so as to give accurate prediction.

Region Proposal and Cascaded Classification: We use an ER detector as our region proposal because of its efficiency and robustness. We extract ERs from RGB, Gray scale and gradient of intensity channels. In order to achieve high recall, approximately thousands of regions will be generated from the 5 channels for each image. We preprocess each region as described in [8] and resize them into 32×32. We then run the standard CNN branch to remove false positives in a similar manner as [8]. For regions with aspect ratios larger than 3.0 or smaller than 1/3, we will do sliding window on top of it since each region might contain blurred text with several characters connected, and in that case, we should not simply resize and classify them. In our experiment, **91.5%** of the regions will be removed, and it achieves **92%** recall on the IC13 testing set.

After this step, the retained regions will be passed into the context branch to generate context vectors. To be more specific, we calculate the width and height of the region, and extend the patch in the horizontal direction so that the resulting context input patch is with 3 times the width of the original image patch. Mean value is padded when there is not enough space in the image for the context input. Because of the strong ability of CNNs in extracting high level information, this context vector provides rich cues in helping with the classification of the given region. This step can further remove some false positives that are similar to text and only regions with high confidence of text will be retained. In IC13 testing set, **94.5%** regions will be removed and it achieves **91%** recall.

In the final prediction, we still consider the two column structure instead of only the context branch for tow reasons: (1) The generated feature vector of standard CNN already contains rich information and it actually produces the feature of the region we are looking at. (2) It will be much easier to train since the standard CNN already produce a really meaningful result. For the context column, they only need to 'figure out' that in some certain cases, the input is not a text even though the feature produced by standard CNN "says" that it is close to text. Such cases include, but not limited to, repetitive patterns, corner of objects and so on.

Training: The proposed model is more difficult to train than a simple text/non-text classifier. We follow the same manner as described in [14,19] by synthesizing image patches for training which provides unlimited number of training data. Since our proposed architecture needs context information, our synthetic positive images need to cover different situations that will happen in real natural images, such as characters with one or two near neighboring characters. Randomly cropped images with their context from several image sources will be considered as negative samples. Several example images for training are shown in Fig. 5. In order to train a better classifier, a two-step training scheme is introduced. First we train a character recognizer with negative samples. This is a 46 classes classification problem, and the positive 45 classes contain all 10 digits and letters with both capitalized or lower cases. Here we merged several similar shaped characters into one class. For example, 'O' and 'o', '0' will be merged into one class. We train with negative log likelihood as the loss function:

$$\text{NLL}(\theta, D) = -\sum_i log(Y = y^i | x^i, \theta) \tag{1}$$

and the 46 classes training makes the learned filters better than binary text/non-text training.

Fig. 5. Several training samples. Left: positive training samples and their context input patches. Right: negative training samples and their context input patches.

The parameters of the trained convolutional layers are then used to initialize the proposed TLFAN architecture and only the fully connected layers will be tuned. After the loss become stable, we train the two parts jointly with smaller learning rate for finetuning. In addition to this, it is necessary to collect harder examples, and we will explain in Sect. 3.2.

Figure 6 shows several images demonstrating the effectiveness of the proposed network. It can effectively use local context to determine whether a given region is text or not, and thus make the prediction more robust. The generated saliency image is the raw output by sliding the classifier on the whole image. We could see that even a well-trained text/non-text classifier [13] has problem when background is noisy. However, by aggregating local context, our model gives much more robust performance.

Fig. 6. The comparison of performance between a state-of-the-art text/non-text classifier proposed in [13], and our method in two challenging image in IC13 test set. From left to right: (1) Original image; (2) The saliency image generated by [13]; (3) The saliency image generated by our method; (4) Final detected text lines by our method.

3.2   Hard Example Mining

In this section, we are going to describe how we collect hard training samples. This could also been seen as a way of bootstrapping. Mining hard examples is a critical step for training accurate object detector, and here we focus on hard negative examples. One of the reasons is that most training examples cropped from negative images have few geometric patterns. Training on these negative examples will make the model less robust to noisy backgrounds. So here we collect more hard negative training data from two sources: (1) **ImageNet**: We specifically collect images from several challenging topics, such as buildings, fences, and trees, and windows. These are objects that typically cause troubles in text detection, since their shapes are close to text. (2) **Synthetic hard negative samples**: we also synthesized a large bunch of negative samples. These samples are not texts but with similar structures as texts, such as stripes and cluttered rectangles. We follow a typical, iterative manner by training until the loss becomes stable and testing on these data to collect hard examples.

We found that this step improved the robustness of the text detector, especially on some hard testing images. These hard examples are also used in the later part for line refinement training.

3.3 Character Component Graph

Grouping text lines from regions is conducted after classification. This step differentiates text detection and object detection. Previous methods [6] typically used relatively heuristic rules to connect text components. This will cause troubles when there are false positives regions. Here we treat it as an assignment problem on top of a character component graph, and the best assignment without conflicts will be chosen based on scores calculated by several text line features as well as several empirical and useful selection standards. In this section, we first describe the ways of how we build the graph. Then we describe how we optimize on the assignment of each character component. The proposed algorithm is illustrated in Fig. 7.

Character Graph: We first build a connected graph on top of the extracted regions, and each node represents a verified region, and each edge represents the neighborhood relation of the text components. We define a function Sim which calculates the similarity between two regions with several low level properties:

(1) Perceptual divergence p, which is calculated as the KL divergence between the color histogram of two regions: $\sum_i x_p(i) * log(\frac{x_p(i)}{y_p(i)})$ where x and y represent two regions and $x_p(i)$ represent the ith entry in its color histogram.
(2) Relative Aspect ratio: $a = \frac{x_{AspectRatio}}{y_{AspectRatio}}$.
(3) Height ratio $h = \frac{x_{Height}}{y_{Height}}$.
(4) Stroke width ratio $s = \frac{x_{StrokeWidth}}{y_{StrokeWidth}}$, which is calculated by distance transform on the original region.

We trained a logistic regression on these four region features extracted from IC13 training set to determine whether two given regions are similar. For each region, we further define its neighbor as: $y \in N(x)$ if $Dis(x, y) < \theta_1$ and $Sim(x, y) > \theta_2, \forall x, y \in R$, where R represents all the regions that have been verified by previous process, and $N(x)$ represents neighbor of region x. Dis means the distance between the center of the two regions. In our experiment, we set threshold θ_1 as $3 * Max(x_{Height}, x_{Width})$ where x_{Height}, x_{Width} means the height and width of the region. We set θ_2 as 0.5 to filter out regions with less probability as being its neighbor.

Stable Pair: We first define stable pair as pair of regions x and y where they belong to neighbor of each other, and they has a similarity score $Sim(x, y) > 0.8$. In addition, their distance should be no more than twice the shortest distance from all other neighbors to the region. This definition aims at obtaining more "probable pairs", since only pairs which "prefer" each other as their neighbors will be considered as "stable". After going through all the regions in $O(n)$ time complexity, we will obtain a set of stable pairs. Outliers are typically not able to form a stable pair with real characters, since real character will not prefer an outlier as its neighbor. In the first row of image in Fig. 7, the defined stable pair criterion successfully prevent generating vertical lines as possible candidate lines. Note that 0.8 is selected empirically from SVT training set. The overall performance is not too sensitive to it.

Optimization: In order to optimize on the assignment of each region to one of lines, for each stable pair, we estimate its orientation based on their center points, and then conduct a greedy search algorithm to find components that align with the current line. Note that it is possible to find conflicting lines because two lines might share the same region components. In order to resolve the conflict, a score is calculated for each line based on the following properties: the average, standard deviation, max value, min value of the pairwise angle, perceptual divergence, size ratio and distance of neighbor components along the line. We will calculate these 16 features in total and a Random Forest [23] is trained to give each line an alignment score. For conflict alignment, we will choose the the best assignment. Here, several empirical but useful standards are also applied. For example, assignment which creates more long lines will be preferred than assignment which creates more short lines as shown in Fig. 7. This step aims at resolving the different possible alignments and find true text lines.

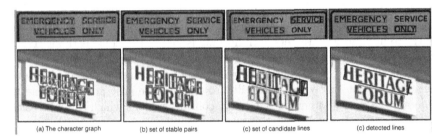

(a) The character graph   (b) set of stable pairs   (c) set of candidate lines   (c) detected lines

Fig. 7. The proposed algorithm which could effectively resolve conflicted candidate lines and find best line assignment of text regions. The final detected lines could be in any orientation. From left to right: (a) The constructed character component graph. (b) A set of generated stable pairs which will be used to create candidate lines. (c) Candidate lines represented as different colored bounding boxes. (d) Detected text lines. (Color figure online)

3.4   Line Extension and Refinement

After we generate lines of regions, a line refinement step is taken to finalize the results. This step aims at two targets: (1) extend lines so as to find missing components. (2) filtering out false positive and predict a tight oriented bounding box. They aim at finding a better bounding box that cover the whole word and it's important for an end-to-end system which incorporates text recognition since the performance of recognition highly relies on accurate bounding boxes.

Gaussian Mixture Model Based Line Extension: Even with carefully designed classifier and grouping algorithm, there still might be some letters in a line that are not found in the previous steps. Here we propose a simple model to recover the proposed lines.

The model is based on the assumption that lines of characters typically have different color distribution with its direct background, and characters in one line typically follow the same color distribution. The algorithm pipeline is shown in Fig. 8. For each line that has been found by previous approaches with more than 2 regions, we crop the patch from the lines and estimate a GMM on the color of the patch. The Gaussian component associated with foreground (text region) is estimated by a voting mechanism: (1) we calculate the skeleton from the region patch; (2) for all pixels in the skeleton patch, we obtain which Gaussian component it belongs to by a simple voting mechanism conducted among these pixels. The reason for only using skeleton pixels lies on the fact that pixels that are close to the boundary are not accurate enough for color distribution estimation. For each line, we consider an square image patch whose side length is twice the height of the line, with its location on the two end of the lines. We estimate the color distribution in the region and a MSER detector is conducted on top of predicted color probability image. We filter out MSER regions whose size is too large or too small when compared to the height of the line. We then classify the retained region as being a text or not using the standard CNN branch in TFLAN. If its probability is larger than 0.4, then we will group it into the lines. If we find one additional character, the GMM is updated and will try to find more characters until nothing is found. Previous methods [17,18] used a graphcut algorithm on top of the extracted regions which serves as similar purpose. However, if the graphcut is directly conducted on all the verified regions, it is still in doubt that whether it will also create more false positives. Here we use a more conservative method and only consider regions which could be easily attached to the current line that we already found.

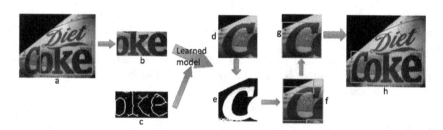

Fig. 8. Our proposed line extension pipeline: (a) The original image with the detected lines; (b) Cropped line image patch we used for estimation of GMM; (c) The skeleton of all the region components; (d) Cropped image patch whose color distribution needs to be estimated; (e) Predicted color probability image. Each pixel of it is predicted from the estimated GMM model; (f) MSER result on the estimation patch. Because of the large contrast in the predicted image patch, there are only few bounding box that we need to verify; (g) After running the standard CNN branch and non-maximum suppression; (h) The final detected line. (Color figure online)

Line Filtering and Sliding Window Regression: In this section, we propose a novel joint line filtering and regression model. Our model is based on making prediction in a sliding window manner on all the text lines that have been verified in the previous steps. Existing methods [14] typically have two steps: (1) word-level verification. (2) bounding box regression. A CNN model is used for filtering and regression. Since fully connected layer only takes fixed sized input, the image patch needs to be resized in order to fit into the network. However, the length of text lines could vary according to the font, and number of characters. It is not desirable to resize an image patch with a text line of 2 to 3 characters to the same size as a text line with 10 characters. Instead, we consider joint regression and filtering in a sliding window manner.

The proposed architecture is shown in Fig. 9(a). The CNN is taken directly from the previous detection architecture. The proposed CNN model takes an input of 48×64 color image patch, and gives 7 prediction. One prediction is a simple part-of-word/non-word prediction which predicts whether the input patch containing part of word, or several characters. It is trained with negative log likelihood loss. The rest of 6 values all represent vertical coordinates because we are doing in a sliding window manner along the text line. Two of them are the minimal and maximal vertical values of the text in the current patch, and they are the same as the vertical coordinates that are predicted in traditional bounding box regression. The other four values represents the minimal and maximal vertical values in left and right side of the patch which are used to predict an oriented bounding box. Some training examples are shown in Fig. 9(b). We train the regression model with standard mean square error loss: $\mathrm{MSE}(x, y) = -\sum_i (x_i - y_i)^2$. Where x, y represent the predicted coordinates, and the ground truth coordinates, respectively. By predicting these four values, an oriented bounding box could be estimated. Since there are only a few lines in an

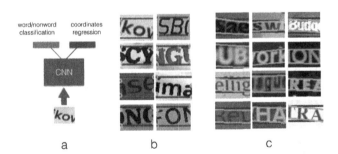

Fig. 9. Our proposed line refinement illustration: (a) The proposed architecture which used the CNN from detection part. Training is only on the fully connected layers for classification and regression. (b) Several examples of training images. The red lines are drawn from minimal and maximal vertical coordinates. The green dots are the vertical coordinates for oriented bounding box. (c) Several testing result images. The oriented bounding box is drawn with green lines instead of dots for better visualization of the orientation. (Color figure online)

image, even sliding window will not need much computation. Several predicted results have been shown in Fig. 9(c) on images cropped from IC13 and SVT data sets.

In order to refine the text lines based on the proposed architecture, we first crop the text line patch from the image, and resize the height of the text line to 48. We slide a window of height 48 and width 64 on the cropped patch. Background noise lines will be filtered out by part-of-word/non-word classification. If the patch is predicted as part-of-word by the classifier, we will perform the oriented bounding box regression on it. A step by step example is shown in Fig. 10. Here we only show lines with text on them.

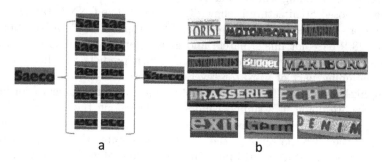

Fig. 10. Our proposed line refinement pipeline. (a) For each cropped lines, we do sliding window prediction and merge the results. (b) Several line regression results based on the proposed framework. The red lines correspond to standard regression, and the green lines represent oriented regression. (Color figure online)

4   Experiments and Evaluation

In this section, we present an evaluation of the proposed method on several benchmark datasets. We report the precision, recall and F-measure scores on our detection results.

Implementation Details: We implemented our algorithms in python and torch 7 [24] on a work station with 64 GB RAM and Nvidia GPU tesla k40 and 16 processors (3000 MHz). All the generation of region proposals and post processing with different channels are parrallized.

ICDAR Robust Reading: We tested our algorithm on IC13 and IC03 testing sets. IC13 and IC03 testing sets contain 233 and 251 testing images, respectively. For IC13, it provides an online evaluation system where we evaluated our proposed method. For IC03, we evaluate our result according to the metric in [25]. The results are shown in Table 1. Evaluation shows that our algorithm gives good performance in both data sets.

Street View Text: The SVT data set contains 249 testing images used for evaluation. It was first introduced by Wang and Belongie [31]. One of the problems om the data set is that it is not fully annotated: some text in the image

Table 1. Localization performances on: left: IC13 (%), right: IC03 (%) data sets. Bold number outperforms previous methods. 'Our model no post' represents final results without line extension and refinement steps [5,6,9,16–18,26–30].

method	precision	recall	F-measure
Neumann[5]	73	65	69
Shi[18]	83	63	72
Bai et[26]	79	68	73
Zamberletti[9]	86	70	77
Tian[27]	85	**76**	80
Zhang[28]	88	74	80
Our model no post	89	73	80
Our model	**90**	75	**81**

method	precision	recall	F-measure
Li[17]	79	64	71
Yao[16]	69	66	67
Kim[29]	78	65	71
Yi[30]	73	67	66
Epshtein[6]	73	60	66
Zamberletti[9]	71	**74**	70
Our model	**84**	70	**76**

are not included in the annotation. This problem has been mentioned in [14], and we call this annotation as *partial annotated*. Our proposed algorithm could efficiently detect most of the text in images and thus the unlabeled text will decrease the precision of detection result and makes it hard to compare with other methods. So we manually labeled all the text in the images following simple rules: (1) text is not too blurry to read by human; (2) it contains more than 2 characters. We call this version *fully annotated* dataset and we tested our algorithm on both versions of the dataset for evaluation. The performance is shown in Table 2. Figure 11 illustrates several examples of *partial annotated dataset*, *fully annotated dataset* as well as our detection results. Experiments on the *fully annotated* dataset shows that our detection algorithm have good performance in SVT dataset as well.

Table 2. Text detection performance on SVT. The bold results outperforms the previous state-of-the-art results. (1) *Partial annotated*: detection recall measured with the partial annotation. The accuracy here does not makes sense, so we only tested its recall. (2) *Fully annotated*: detection precision, recall, F-measure with full annotation.

	Partial annotated	Fully annotated		
	Recall	Precision	Recall	F-measure
Jaderberg et al. [14]	0.71	-	-	-
Our model	**0.75**	**0.87**	**0.73**	**0.79**

Limitation: Our proposed method achieved fairly good results in terms of precision, recall, and F-measure on standard datasets. However, it can still fail on several extremely challenging cases: (1) Text lines that are too blurry will cause problem in accurate region proposal generation as well as classification. (2) Strong reflection, too low contrast will still cause troubles in the detection. (3) Curved text lines might cause incomplete detection. Figure 12 shows several failure cases that our algorithm cannot get good results. They are all challenging images in terms of text reading, and some of them are even hard for human to read.

Fig. 11. For each pair of images, left: the original incomplete annotation, right: detection result of our model as well as the *fully annotated* ground truth. The *fully annotated* dataset provides oriented bounding box annotation. Green boxes represent our detected result which matches the ground truth. Yellow boxes represent the ground truth. (Color figure online)

Fig. 12. Example images where we failed to detect all the lines or detected the wrong lines. The green boxes are the text lines that are correctly detected. The blue boxes are text lines that we fail to detect, and the red boxes are false positives, or incomplete detection. (Color figure online)

5   Conclusions and Future Work

Here we proposed a novel scene text detection algorithm which efficiently aggregates local context information into detection as well as a novel two step text line refinement. Experiments show that our pipeline works well on several challenging images and achieved state-of-the-art performance on three benchmark data sets. Our future work will focus on extending our work in order to combine scene text recognition into an end-to-end scene text reading system. An image indexing system using text information retrieval will be implemented to help the visually impaired with shopping. Other applications for complex text such as sign reading will also be explored.

Acknowledgement. This work was supported by NSF grant CCF 1317560 and a hardware grant from NVIDIA.

References

1. Zhu, Y., Yao, C., Bai, X.: Scene text detection and recognition: recent advances and future trends. Front. Comput. Sci. **10**, 19–36 (2016)
2. Wang, T., Wu, D.J., Coates, A., Ng, A.Y.: End-to-end text recognition with convolutional neural networks. In: 2012 21st International Conference on Pattern Recognition (ICPR), pp. 3304–3308. IEEE (2012)
3. Chen, X., Yuille, A.L.: Detecting and reading text in natural scenes. In: Proceedings of the 2004 IEEE Computer Society Conference on Computer Vision and Pattern Recognition, CVPR 2004, vol. 2, p. II-366. IEEE (2004)
4. Donoser, M., Bischof, H.: Efficient maximally stable extremal region (MSER) tracking. In: 2006 IEEE Computer Society Conference on Computer Vision and Pattern Recognition, vol. 1, pp. 553–560. IEEE (2006)
5. Neumann, L., Matas, J.: Real-time scene text localization and recognition. In: 2012 IEEE Conference on Computer Vision and Pattern Recognition (CVPR), pp. 3538–3545. IEEE (2012)
6. Epshtein, B., Ofek, E., Wexler, Y.: Detecting text in natural scenes with stroke width transform. In: 2010 IEEE Conference on Computer Vision and Pattern Recognition (CVPR), pp. 2963–2970. IEEE (2010)
7. Huang, W., Lin, Z., Yang, J., Wang, J.: Text localization in natural images using stroke feature transform and text covariance descriptors. In: Proceedings of the IEEE International Conference on Computer Vision, pp. 1241–1248 (2013)
8. Huang, W., Qiao, Y., Tang, X.: Robust scene text detection with convolution neural network induced MSER trees. In: Fleet, D., Pajdla, T., Schiele, B., Tuytelaars, T. (eds.) ECCV 2014. LNCS, vol. 8692, pp. 497–511. Springer, Heidelberg (2014). doi:10.1007/978-3-319-10593-2_33
9. Zamberletti, A., Noce, L., Gallo, I.: Text localization based on fast feature pyramids and multi-resolution maximally stable extremal regions. In: Jawahar, C.V., Shan, S. (eds.) ACCV 2014. LNCS, vol. 9009, pp. 91–105. Springer, Heidelberg (2015). doi:10.1007/978-3-319-16631-5_7
10. Neumann, L., Matas, J.: Scene text localization and recognition with oriented stroke detection. In: Proceedings of the IEEE International Conference on Computer Vision, pp. 97–104 (2013)
11. Tonouchi, Y., Suzuki, K., Osada, K.: A hybrid approach to detect texts in natural scenes by integration of a connected-component method and a sliding-window method. In: Jawahar, C.V., Shan, S. (eds.) ACCV 2014. LNCS, vol. 9009, pp. 106–118. Springer, Heidelberg (2015). doi:10.1007/978-3-319-16631-5_8
12. Krizhevsky, A., Sutskever, I., Hinton, G.E.: Imagenet classification with deep convolutional neural networks. In: Advances in Neural Information Processing Systems, pp. 1097–1105 (2012)
13. Jaderberg, M., Vedaldi, A., Zisserman, A.: Deep features for text spotting. In: Fleet, D., Pajdla, T., Schiele, B., Tuytelaars, T. (eds.) ECCV 2014. LNCS, vol. 8692, pp. 512–528. Springer, Heidelberg (2014). doi:10.1007/978-3-319-10593-2_34
14. Jaderberg, M., Simonyan, K., Vedaldi, A., Zisserman, A.: Reading text in the wild with convolutional neural networks. Int. J. Comput. Vis. **116**, 1–20 (2016)
15. Zhu, A., Gao, R., Uchida, S.: Could scene context be beneficial for scene text detection? Pattern Recogn. **58**, 204–215 (2016)
16. Yao, C., Bai, X., Liu, W., Ma, Y., Tu, Z.: Detecting texts of arbitrary orientations in natural images. In: 2012 IEEE Conference on Computer Vision and Pattern Recognition (CVPR), pp. 1083–1090. IEEE (2012)

17. Li, Y., Jia, W., Shen, C., van den Hengel, A.: Characterness: an indicator of text in the wild. IEEE Trans. Image Process. **23**, 1666–1677 (2014)
18. Shi, C., Wang, C., Xiao, B., Zhang, Y., Gao, S.: Scene text detection using graph model built upon maximally stable extremal regions. Pattern Recogn. Lett. **34**, 107–116 (2013)
19. Jaderberg, M., Simonyan, K., Vedaldi, A., Zisserman, A.: Synthetic data and artificial neural networks for natural scene text recognition. arXiv preprint arXiv:1406.2227 (2014)
20. Divvala, S.K., Hoiem, D., Hays, J.H., Efros, A.A., Hebert, M.: An empirical study of context in object detection. In: IEEE Conference on Computer Vision and Pattern Recognition, CVPR 2009, pp. 1271–1278. IEEE (2009)
21. Oliva, A., Torralba, A.: The role of context in object recognition. Trends Cogn. Sci. **11**, 520–527 (2007)
22. Zagoruyko, S., Komodakis, N.: Learning to compare image patches via convolutional neural networks. In: Proceedings of the IEEE Conference on Computer Vision and Pattern Recognition, pp. 4353–4361 (2015)
23. Liaw, A., Wiener, M.: Classification and regression by randomforest. R News **2**, 18–22 (2002)
24. Collobert, R., Kavukcuoglu, K., Farabet, C.: Torch7: a matlab-like environment for machine learning. In: BigLearn, NIPS Workshop, Number EPFL-CONF-192376 (2011)
25. Lucas, S.M., Panaretos, A., Sosa, L., Tang, A., Wong, S., Young, R.: ICDAR 2003 robust reading competitions. In: Null, p. 682. IEEE (2003)
26. Bai, B., Yin, F., Liu, C.L.: Scene text localization using gradient local correlation. In: Proceedings of the ICDAR 2013, pp. 1380–1384. IEEE (2013)
27. Tian, S., Pan, Y., Huang, C., Lu, S., Yu, K., Lim Tan, C.: Text flow: a unified text detection system in natural scene images. In: Proceedings of ICCV 2015, pp. 4651–4659 (2015)
28. Zhang, Z., Shen, W., Yao, C., Bai, X.: Symmetry-based text line detection in natural scenes. In: Proceedings of the CVPR 2015 (2015)
29. Koo, H.I., Kim, D.H.: Scene text detection via connected component clustering and nontext filtering. IEEE Trans. Image Process. **22**, 2296–2305 (2013)
30. Yi, C., Tian, Y.: Localizing text in scene images by boundary clustering, stroke segmentation, and string fragment classification. IEEE Trans. Image Process. **21**, 4256–4268 (2012)
31. Wang, K., Belongie, S.: Word spotting in the wild. In: Daniilidis, K., Maragos, P., Paragios, N. (eds.) ECCV 2010. LNCS, vol. 6311, pp. 591–604. Springer, Heidelberg (2010). doi:10.1007/978-3-642-15549-9_43

Bilinear Discriminant Analysis Hashing: A Supervised Hashing Approach for High-Dimensional Data

Yanzhen Liu[1], Xiao Bai[1], Cheng Yan[1(✉)], and Jun Zhou[2]

[1] School of Computer Science and Engineering, Beihang University, Beijing, China
{lyzeva,baixiao,beihangyc}@buaa.edu.cn
[2] School of Information and Communication Technology,
Griffith University, Nathan, Australia
jun.zhou@griffith.edu.au

Abstract. High-dimensional descriptors have been widely used in object recognition and image classification. How to quickly index high-dimensional data into binary codes is a challenging task which has attracted the attention of many researchers. Most existing hashing solutions for high-dimensional dataests are based on unsupervised schemes. On the other hand, existing supervised hashing methods cannot work well on high-dimensional datasets, as they consume too much time and memory to index high-dimensional data. In this paper, we propose a supervised hashing method *Bilinear Discriminant Analysis Hashing* (BDAH) to solve this problem. BDAH leverages supervised information according to the idea of Linear Discriminant Analysis (LDA), but adopts bilinear projection method. Bilinear projection needs two small matrices rather than one big matrix to project data so that the coding time and memory consumption are drastically reduced. We validate the proposed method on three datasets, and compare it to several state-of-the-art hashing schemes. The results show that our method can achieve comparable accuracy to the state-of-the-art supervised hashing schemes, while, however, cost much less time and memory. What's more, our method outperforms unsupervised hashing methods in accuracy while achieving comparable time and memory consumption.

1 Introduction

Nearest Neighbor (NN) search is a basic and important step in image retrieval. For a large scale dataset of size n, the complexity of NN search $O(n)$ is too high for big data processing. To solve this problem, approximate nearest neighbor (ANN) methods have been proposed, in which hashing is a class of well-behaved methods. The basic idea of hashing is to use binary code to represent high-dimensional data, with similar data pairs having smaller Hamming distance of binary codes. Then the task of finding the nearest neighbors of a data point can be transformed into finding the most similar hash codes. With hashing, we only need to apply XOR operations between binary hash codes, rather than

© Springer International Publishing AG 2017
S.-H. Lai et al. (Eds.): ACCV 2016, Part V, LNCS 10115, pp. 297–310, 2017.
DOI: 10.1007/978-3-319-54193-8_19

calculating the Euclidean distances between data vectors which involves frequent addition and multiplication operations.

To calculate proper hash codes, we should construct a mapping from high-dimension Euclidean space to Hamming space and preserve the original Euclidean distance of data in the Hamming space. There are already many works on hash code learning. According to leveraging supervised information or not, hashing methods can be classified into supervised hashing methods [1–7] or unsupervised hashing methods [8–15]. Hashing methods can also be divided into data-dependent [2,4–6,9,11,16,17] and data-independent [8,12,13] depending on whether to use training data to generate hash codes or not.

A large number of hashing methods are based on linear projection and 0–1 quantization. A simple and basic data-independent method is Locality Sensitive Hashing (LSH) [8]. LSH is based on generating random linear projection vectors in the Euclidean space. After projection on one vector, data are quantized into 0 or 1 by its sign, consequently generating one bit of hash code. This method can also be comprehended as hyperplane partitioning the Euclidean space. A hyperplane correspond to one projection vector and data points on one side of the hyperplane are coded 0, while at the other side are coded 1. LSH is easy to implement and successfully reduces the NN search to sublinear query time with acceptable accuracy. However, because the hyperplanes are randomly set and not depend on the dataset, we need longer codes to achieve accuracy. To overcome this shortcoming, many data-dependent methods have been proposed, such as SSH [2] and ITQ [11], which try to learn projection from the distribution of the dataset. They achieved higher accuracy than LSH.

In projection based hashing schemes, dimension of projection vector is the same as the dimension of the original data. Nowadays, features of thousands of dimensions appear with high retrieval and classification accuracy, such as Fisher Vectors (FV) [18], Vectors of Locally Aggregated Descriptors (VLAD) [19] and Deep Neural Networks Features (DNN) [20]. Projection vectors for these data are also high-dimensional. What's more, hash codes of these high-dimensional datasets are also longer to ensure accuracy. As a consequence, the projection matrix is huge for high-dimensional dataset, and more time is needed to calculate hash codes as well as train the projection matrix. Several works are concentrating on learning hash codes for high-dimensional data [21–23]. However, they are all unsupervised methods without utilizing supervised information, and supervised hashing scheme for high-dimensional datasets has seldom be studied. Existing supervised hashing schemes like Kernel-Based Supervised Hashing (KSH) [5] and Supervised Discrete Hashing (SDH) [6] will cost too much time and memory to calculate hash codes. They are unfit for high-dimensional datasets.

In this papaer, we propose a supervised hashing scheme specific for high-dimensional datasets, and we name it Bilinear Discriminant Analysis Hashing (BDAH). In the dimension reduction part, we utilize a bilinear projection model from 2DLDA [24] to minimize within-class distance and maximize between-class distance of projected data. According to 2DLDA, we resize long 1-D vectors into 2-D matrices and use two projection matrices to project the 2-D matrices. In this way, although the number of matrices increases, total elements in the projection matrices are drastically reduced, and meanwhile the time and memory consumption for hash code calculation is reduced. Then we optimize the two projection matrices iteratively. After projecting the data, we quantize the projected vectors into binary codes. To minimize quantization loss, Iterative Quantization [11] is adopted to calculate an optimal rotation matrix.

There are three contributions in this paper. Firstly, BDAH projects data in 2-D form, so it can better protect the inner structure of 2-D form descriptors such as LLC [25] and VLAD [19]. Secondly, bilinear projection drastically reduces time and memory cost for hash code generation, so it is suitable for hashing on high-dimensional dataset. Finally, our BDAH is a supervised hashing scheme, and the utilization of label information help to get better retrieval accuracy of the hash codes than unsupervised hashing schemes. We validate the effectiveness of our BDAH on 3 datasets: AwA [26], MNIST [27] and ILSVRC2010 [28], comparing to CCA-ITQ [11], KSH [5], BPBC [21], CBE-opt [22] and PCA-ITQ [11]. The experiments show that our method reduces time and memory consumption while keeping comparable accuracy to other state-of-the-art hashing schemes.

The rest of this paper is organized as follows. Section 2 introduces related work on high-dimensional data hashing. Section 3 describes the technical process of BDAH. Section 4 introduces our experiments and the performances of our proposed methods. Finally, conclusions are drawn in Sect. 5.

2   Related Work

2.1   Hashing Schemes for High-Dimensional Dataset

Recently, several works have been reported which aim at accelerating high-dimensional projection process in different ways [21–23]. The essence of these method is to reduce the actual number of variables in the projection matrix. However they are all unsupervised hashing schemes and cannot leverage supervised information, so they can't achieve particularly high performance.

Bilinear Projection based Binary Codes (BPBC) [21] reshapes the data vectors into matrices, and learns the projection matrix (orthogonal projection matrix) with the 2-D data form. Then BPBC can use two small projection matrices instead of one big projection matrix. In this way, the size of rotation matrix decreases from n^2 to $n_1^2 + n_2^2 (n_1 + n_2 = n)$.

Circulant Binary Embedding (CBE) [22] assumes that the rotation matrix has circulant inner structure and transforms the linear projection into form of circulant convolution. The elements in the projection matrix of CBE is constructed by one of the row vectors in the matrix, so the storage of projection matrix reduces to the size of a vector. Then it uses FFT to speed up the coding process. For code training, CBE adds a regularizer in the quantization loss function to make the rotation matrix approximate orthogonal matrix. CBE is quicker than BPBC in hash coding.

Sparse Projections approach (SP) [23] adds a sparse regularizer to the projection matrix in the quantization loss function. The sparse projection matrix reduces the redundant information in the high-dimensional projection matrix, so SP can achieve comparable accuracy to ITQ but has much less encoding time.

2.2   Linear Discriminant Analysis

For a d-dimensional training set $X = \{x_1, x_2, \ldots, x_n\}$ belonging to c different classes $\Pi_1, \Pi_2, \ldots, \Pi_c$, LDA [29] tries to project the data points into a b-dimensional space with a linear projection matrix $W \in \mathbb{R}^{d \times b}$, in which data points in the same class become nearer while data points in different classes becomes farther. Suppose there are a_i different data points in class Π_i, we get our objective formulation

$$J_L(W) = \max(D_w)^{-1} D_b, \tag{1}$$

where

$$D_w = \sum_{i=1}^{c} \sum_{j \in \Pi_i} \|W^T x_j - W^T m_i\|_F^2 \tag{2}$$

is the within-class distance, and

$$D_b = \sum_{i=1}^{c} a_i \|W^T (m_i - m_0)\|_F^2 \tag{3}$$

is the between-class distance.

Equation (1) can be further derived into

$$J_L(W) = \max trace((W^T S_w W)^{-1} W^T S_b W). \tag{4}$$

This optimization problem is equivalent to the generalized eigenvalue problem $S_b w_i = \lambda_i S_w w_i$, where λ_i denotes the ith largest eigenvalue and w_i is its corresponding eigenvector as well as the ith column of the projection matrix W. We want the projected space to be b-dimensional, so we pick b eigenvectors.

3   Bilinear Discriminant Analysis Hashing

For a d-dimensional training set $\{x_1, x_2, \ldots, x_n\}$ having c classes $\Pi_1, \Pi_2, \ldots, \Pi_c$, we want to map the data vectors into b-bit hash codes $\{b_1, b_2, \ldots, b_n\}$, where data in the same class have smaller hamming distances while data in different classes having larger hamming distances. It is too hard for us to compute the projection directly. So we relax the problem into two steps. Firstly, we project the data vectors into a lower b-dimensional vectors $\{y_1, y_2, \ldots, y_n\}$ to make within-class distance smaller and between-class distance higher (Sect. 3.1). Secondly, we quantize the b-dimensional data into binary codes (Sect. 3.2).

3.1   Learning 2DLDA Projection Matrices

When there is a need to project a data vector $x \in \mathbb{R}^d$ to a lower b-dimensional space, people often use a matrix $W \in \mathbb{R}^{d \times b}$ to multiply the data vector x, and get the low-dimensional projected vector $y \in \mathbb{R}^b$:

$$y = W^T x. \tag{5}$$

If the dimension of the data d is very high, the projection matrix W will be very large as well.

In our method, we adopt another projection form [24]. Firstly, we need to express data vector $\boldsymbol{x}$ in a matrix form. Then the dataset changes from a vector set into a matrix set $\{A_1, A_2, \ldots, A_n\} \in \mathbb{R}^{d_1 \times d_2}$. The elements in matrix A_i are the same as the elements in vector $\boldsymbol{x}_i$, so $d = d_1 \times d_2$. And $\boldsymbol{x} = vec(A)$ where $vec(A)$ denotes converting the matrix A into a vector by concatenating the columns of A.

For the matrix set $\{A_1, A_2, \ldots, A_n\}$, the projection can change into the following form

$$\boldsymbol{y} = vec(L^T A R). \tag{6}$$

In Eq. (6), $L \in \mathbb{R}^{d_1 \times b_1}$ and $R \in \mathbb{R}^{d_2 \times b_2}$ are the projection matrices, where $b = b_1 \times b_2$. We call it bilinear projection.

It is easy to prove that

$$vec(L^T A R) = (R^T \otimes L^T) vec(A) = \hat{W}^T \boldsymbol{x},$$

where $\otimes$ denotes the Kronecker product. So the bilinear projection with two projection matrices is equivalent to common projection with one matrix.

In common projection method, the projection matrix W have $d * b = (d_1 d_2) * (b_1 b_2) = d_1 b_1 * d_2 b_2$ elements, the complexity of projection is $O(d_1 d_2)$. But in bilinear projection method, two projection matrices L and R have $d_1 b_1 + d_2 b_2$ elements in total, and the bilinear projection procedure has $O(d_1^2 d_2 + d_2^2)$ time complexity. We can see that both the time and memory complexity of the projection procedure are drastically reduced.

With two bilinear projection matrices, the between-class distance D_b and within-class distance D_w in LDA turn into the following form

$$D_b = \sum_{i=1}^{c} a_i \| L^T (M_i - M_0) R \|_F^2, \tag{7}$$

$$D_w = \sum_{i=1}^{c} \sum_{A \in \Pi_i} \| L^T (A - M_i) R \|_F^2, \tag{8}$$

where

$$M_i = \frac{\sum_{j \in \Pi_i} A_j}{n} (1 \leqslant i \leqslant c) \tag{9}$$

and

$$M_0 = frac \sum_{i=1}^{n} A_i n. \tag{10}$$

D_b and D_w can be further derived into the following form

$$D_b = trace(\sum_{i=1}^{c} a_i L^T (M_i - M_0) R R^T (M_i - M_0)^T L), \tag{11}$$

$$D_w = trace(\sum_{i=1}^{c} \sum_{A \in \Pi_i} L^T(A - M_i)RR^T(A - M_i)^T L).\tag{12}$$

We want to maximize the between-class distance D_b while minimize the within-class distance D_w. That is to maximize J(L,R) given below.

$$J(L, R) = \max trace((D_w)^{-1}D_b)\tag{13}$$

It is difficult to optimize L and R simultaneously. So we optimize L and R iteratively.

Fix R and update L.

With fixed R, we can derive our object function $J(L)$

$$J_L(L) = \max trace((L^T S_w^R L)^{-1}(L^T S_b^R L)) = \max trace(S_w^R)^{-1} S_b^R),\tag{14}$$

where

$$S_b^R = \sum_{i=1}^{c} a_i(M_i - M_0)RR^T(M_i - M_0)^T,\tag{15}$$

$$S_w^R = \sum_{i=1}^{c} \sum_{A \in \Pi_i} (A - M_i)RR^T(A - M_i)^T.\tag{16}$$

Then we get the optimized L by doing eigen decomposition on the matrix $(S_w^R)^{-1}S_b^R$.

Fix L and update R.

Analogously, with fixed L, we can derive our object function $J(R)$

$$J(R) = \max trace((R^T S_w^L R)^{-1}(R^T S_b^L R)),\tag{17}$$

where

$$S_b^L = \sum_{i=1}^{c} a_i(M_i - M_0)^T LL^T(M_i - M_0),\tag{18}$$

$$S_w^L = \sum_{i=1}^{c} \sum_{A \in \Pi_i} (A - M_i)^T LL^T(A - M_i).\tag{19}$$

The iteration stops after convergence and we got the locally optimized L and R.

3.2 Quantization into Binary Codes

After the 2DLDA projection L and R are recovered, we can calculate the projected vectors $\boldsymbol{y}_i = vec(L^T A_i R)$. The low-dimensional dataset is denoted by

$$Y = [\boldsymbol{y}_1, \boldsymbol{y}_2, \ldots, \boldsymbol{y}_n]^T.\tag{20}$$

Next, we try to quantize Y into binary codes $B \in \{1, -1\}^{n \times b}$. The most direct method is just using the sign function. That is

$$B = \text{sign}(Y).\tag{21}$$

Direct quantization may cause great loss as explained in [11]. So we try to calculate an optimized rotation matrix $U \in \mathbb{R}^{b \times b}$ to minimize the quantization loss Q.

$$Q = \|B - YU\|_F \tag{22}$$
$$UU^T = I$$

Algorithm 1. Training process of Bilinear Linear Discriminant Hashing

---

Data: training data X, code length b, resized matrix size d_1,d_2, column of $L - b_1$, column of $R - b_2$, number of iteration $iter$

Result: projection matrices L and R, rotation matrix U, hash codes of training set B

Resize the dataset X into matrix set $\{A_1, A_2, ..., A_n\}$;

Calculate the mean of the dataset M_0 and the mean of each class M_i;

$R = (I_{b_2}, \mathbf{0})^T$ (I_{b_2} is a $b_2 \times b_2$ identity matrix);

for $j = 1, ..., iter$ **do**

    Calculate S_b^R, S_w^R by Eqs. (15), (16);

    Do eigen decomposition on $(S_w^R)^{-1}S_b^R$ and pick b_1 eigenvectors $\{l_1, l_2, ..., l_{b_1}\}$ with the largest b_1 eigenvalues ;

    $L = [l_1, l_2, ..., l_{b_1}]$;

    Calculate S_b^L, S_w^L by Eqs. (18), (19);

    Perform eigen decomposition on $(S_w^L)^{-1}S_b^L$ and pick b_2 eigenvectors $\{r_1, r_2, ..., r_{b_1}\}$ with the largest b_2 eigenvalues ;

    $R = [r_1, r_2, ..., r_{b_1}]$;

end

for $j = 1, ..., n$ **do**

    $y_j = vec(L^T A_i R)$;

end

$Y = [y_1, y_2, ..., y_n]^T$;

Initialize U with a random orthogonal matrix;

for $j=1,...,50$ **do**

    Update B by Eq. (24);

    Update R by Eq. (25);

end

$b_i = R^T vec(L^T A_i R)$ (i from 1 to n);

$B = [b_1, b_2, ...b_n]^T$;

And the solution is like the iterative process in [11]. We do the iterations between B and U for several times and get a locally optimal solution.

Fix U and update B.

For a fixed U, we can further derive Eq. (22) into

$$Q = \|B\|_F^2 + \|Y\|_F^2 - 2tr(BU^TY^T)$$
$$= -2tr(BU^TY^T) + const. \tag{23}$$

Because B and Y is fixed, Eq. (23) is equivalent to

$$\max tr(BU^T Y^T).$$

And we can derive the optimal B from

$$B = sign(YU). \tag{24}$$

Fix B and update U.

For a fixed B, Eq. (22) becomes the classical Orthogonal Procrustes Problem [30]. Compute SVD of the matrix $B^T Y = S\Omega\hat{S}^T$ and we get the optimal U by

$$U = \hat{S}S^T \tag{25}$$

After the Q converges, we get the locally optimal solution of Eq. (22). Here 50 iterations of optimization are adopted, which is enough to achieve convergence [11].

The above is the training process of our BLDH. Algorithm 1 is the pseudo-code on training of BLDH. Through training, we get the bilinear projection matrice L, R and rotation matrix U. For a new query q, we can get its hash code by first resize it to a $d_1 \times d_2$ matrix Q_q and then calculate by the equation

$$b = sign(U^T \ vec(L^T Q_q R)). \tag{26}$$

3.3   Addition LDA Projection for Further Accuracy

To achieve further accuracy of our hash code, we can project the data once more with LDA after the bilinear projection of 2DLDA [24]. For a d-dimensional dataset $\{x_i\}(i = 1\ldots n)$, we want to get b-bit hash code. We can first project the data into $b1 * b2$-dimensional vectors $\{y_i\}(i = 1\ldots n)$ by bilinear projection of 2DLDA (Sect. 2.2), where $b1 * b2 = b$. Then we project $\{y_i\}(i = 1\ldots n)$ into b-dimensional vector $\{z_i\}(i = 1\ldots n)$ by LDA (Sect. 2.2). Finally, we quantize $\{z_i\}(i = 1\ldots n)$ into binary code (Sect. 3.2). In this paper, we denote this hashing scheme with 2-step projection BDAH^0. Experiments show that, combined with LDA, BDAH^0 can achieve much better retrieval accuracy.

4   Experiment

To test the effectiveness of our method, we conducted experiments on 3 datasets: **AwA** [26], **CIFAR-10** [31] and **ILSVRC2010** [28]. Our BDAH and BDAH^0 were compared against several state-of-the-art methods including two supervised method CCA-ITQ [11] and KSH [5], and three unsupervised methods PCA-ITQ [11], BPBC [21] and CBE-opt [22]. We used the publicly available codes of these methods. All our experiments were run on a PC with a 3.5 GHz Intel Core CPU and 32 GB RAM.

4.1   MNIST: Retrieval with Raw Image

The **MNIST** [27] dataset has 60000 28*28 small greyscale images of handwritten digits from '0' to '9'. Each small image is represented by a 784-dimensional vector and a single digit label. Each 784-dimensional vector is stretched from a corresponding 28*28 greyscale image. We used 54000 random images as the training set, and 6000 random images as the query samples.

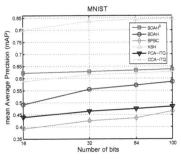
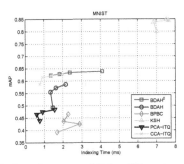

(a) mAP for different code bits.   (b) Coding time vs. mAP for different code bits.

Fig. 1. Results on the MNIST dataset.

We set the hash code length to 16 bits, 36 bits, 64 bits and 100 bits in this dataset. Figure 1 reports the results on this dataset. Table 2 shows the training time and coding time of each method. In supervised methods, our $BDAH^0$ exceeds CCA-ITQ in mAP, but needs a little more time on binary coding. $BDAH^0$ has lower mAP than KSH, but consumes much less time than KSH. For unsupervised methods, $BDAH^0$ and BDAH achieve comparable time consumption to them while having much higher mAP.

4.2   AwA: Retrieval with DNN Features

Animals with Attributes (**AwA**) [26] is a dataset with different kinds of animals. It has 30475 animal images belonging to 50 animal classes. In the experiments, we used a subset containing 9460 images. A seven-layer CaffeNet [20] was adopted to extract 4096-dimensional feature from AwA. The 4096-dimensional output of the fully-connected layer was used as our feature to calculate hash codes. In our $BDAH^0$ and BDAH methods and BPBC in [21], we resized every DNN-4096 feature into a 64*64 matrix and then used the matrices to generate hash codes in our retrieval experiment.

We tested all methods on AwA with 16 bits, 36 bits, 64 bits, 100 bits and 256 bits codes. The memory consumption of indexing is reported on Table 1. For $BDAH^0$, BDAH and BPBC, memory consumption is bilinear projection matrices, and for PCA-ITQ and CCA-ITQ, is the full projection matrix. KSH needs to

Table 1. Memory consumption (KB) to hash new data into binary codes on AwA datset.

Methods	16 bits	36 bits	64 bits	100 bits	256 bits
BDAH^0	16	29	46	67	152
BDAH	6	16	40	88	528
BPBC	4	6	8	10	16
KSH	9640	9687	9752	9837	10202
CBE-opt	64	64	64	64	64
PCA-ITQ	512	1152	2048	3200	8192
CCA-ITQ	512	1152	2048	3200	8192

Table 2. Time consumption (milliseconds) of different methods on the MNIST dataset.

Methods		16 bits	25 bits	64 bits	100 bits
BDAH^0	Training	8254.40	9373.46	9930.19	11624.55
	Test	4.18	4.95	5.03	6.31
BDAH^0	Training	8541.15	9172.20	9463.68	10156.50
	Test	3.93	5.95	5.28	6.70
BDAH	Training	8026.10	8865.57	9219.09	10324.81
	Test	3.42	3.88	4.47	5.34
KSH	Training	38055.16	72090.40	128932.07	199796.28
	Test	16.70	17.67	16.76	17.41
CCA-ITQ	Training	625.52	1126.64	1901.62	3211.55
	Test	1.74	2.10	3.86	4.98
PCA-ITQ	Training	441.86	669.27	1077.90	1815.74
	Test	1.02	1.26	2.08	3.34

store several anchor points and a kernel projection matrix, thus acquiring more memory. CBE-opt has the least memory consumption because it only needs to store a vector, rather than matrices in other methods. We can see clearly that, our BDAH^0 and BDAH require less memory than PCA-ITQ and CCA-ITQ. The longer codes we needs, the more memory consumption BDAH^0 and BDAH save. Figure 2 shows the P-R curves and time vs. mAP for the above-mentioned methods on AwA and Table 3 shows the training time and coding time of each method. BDAH^0 still shows good balance in accuracy and time consumption. KSH is still best in mAP, but worst in indexing time than all the other methods. CCA-ITQ has the minimum coding time while PCA-ITQ trains faster, but both PCA-ITQ and CCA-ITQ cannot achieve as good accuracy as BDAH^0.

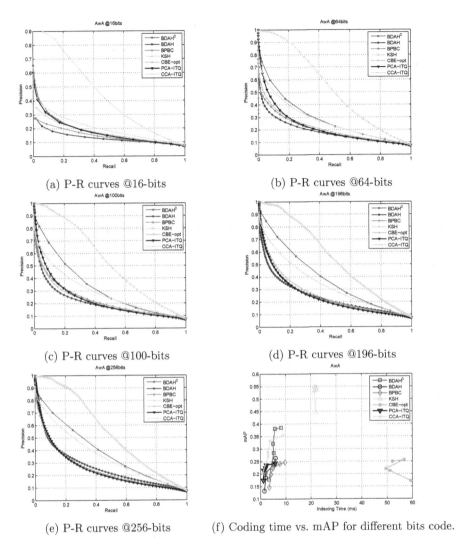

(a) P-R curves @16-bits

(b) P-R curves @64-bits

(c) P-R curves @100-bits

(d) P-R curves @196-bits

(e) P-R curves @256-bits

(f) Coding time vs. mAP for different bits code.

Fig. 2. P-R curves and mAP vs time for different methods with experiments on the AwA dataset.

4.3   ILSVRC2010: Retrieval with VLAD Features

ILSVRC2010 is a subset of ImageNet [28]. ImageNet is a hierarchical image dataset organized following the hierarchy of WordNet [32]. In our experiments, we randomly chose 50 classes, and picked 100 images per class as the training set, and 10 image per class as the test set. On the chosen data in ILSVRC2010, we utilized the publicly available SIFT features, clustered them into 500 centers and calculated 64000-dimensional VLAD features of each image. In our experiments, we used 64000-dimensional VLAD feature vectors to calculate hash codes.

Table 3. Total training time (milliseconds) of different methods on the AWA dataset.

Methods		16 bits	25 bits	64 bits	100 bits	256 bits
BDAH^0	Training	46043.92	68347.63	69742.61	70820.15	79177.46
	Coding	3.30	5.19	4.95	5.72	8.12
BDAH	Training	43357.59	45544.10	46823.71	69739.86	75129.19
	Coding	1.48	2.05	1.83	5.42	5.91
BPBC	Training	19041.92	22239.26	23627.65	59662.54	65745.51
	Coding	3.40	4.09	4.85	7.65	9.64
KSH	Training	33409.12	73810.13	132546.38	20771.29	524378.30
	Coding	21.91	21.48	22.25	21.43	22.39
CBE-opt	Training	255049.87	255402.27	254791.00	258713.2595	253481.55
	Coding	59.35	51.08	49.52	56.63	52.14
PCA-ITQ	Training	3650.91	4479.65	5570.59	6892.80	16345.93
	Coding	1.10	1.25	1.96	2.30	5.75
CCA-ITQ	Training	20432.65	20912.83	21528.06	23202.85	32025.13
	Coding	0.82	3.01	2.92	3.78	9.06

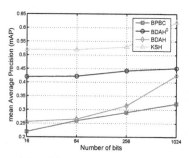
(a) mAP for different bits code.

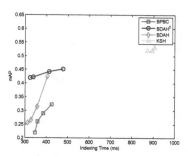
(b) Coding time vs. mAP for different bits code.

Fig. 3. Results on the ILSVRC2010 dataset.

We set the hash code length to 16 bits, 64 bits and 256 bits and 1024 bits on this dataset. CCA-ITQ and PCA-ITQ need to calculate the covariance matrix of the dataset. The 64000 dimension of this dataset is too high that our storage cannot contain the covariance matrix. So we did not compare with these two methods in this experiment. The results of retrieval on ILSVRC2010 are reported in Fig. 3. Compared to supervised methods, our BDAH^0 indexes more quickly than KSH. Compared to unsupervised methods, BDAH^0 performs better than BPBC in accuracy but consumes comparable time and memory as BPBC. Table 4 shows the specific training time and coding time of each method.

Table 4. Time consumption (seconds) of different methods on the ILSVRC2010 dataset.

Methods	BDAH^0		BDAH		BPBC		KSH	
bit	Training	Coding	Training	Coding	Training	Coding	Training	Coding
16 bits	488.74	0.32	482.98	0.31	151.18	0.35	39.65	0.87
64 bits	507.35	0.34	494.52	0.33	163.51	0.36	138.60	0.89
256 bits	543.44	0.41	514.64	0.36	182.84	0.39	553.51	0.90
1024 bits	611.91	0.48	579.24	0.40	237.09	0.43	2193.67	0.93

5 Conclusions

In this paper, we have introduced a supervised hashing scheme with bilinear projection. With bilinear projection, our method can better deal with high-dimensional dataset with less time and memory consumption than traditional supervised hashing schemes. Furthermore, our method processes input data in 2D form, thus it is very convenient to index 2-D structure raw-images and features. The experiments show that our method exceeds other unsupervised hashing schemes for high-dimensional data in accuracy. Moreover, our hashing method achieves comparable accuracy than other supervised methods while occupying much less memory and time.

Acknowledgement. This work was supported by NSFC project No. 61370123 and BNSF project No. 4162037.

References

1. Kulis, B., Darrell, T.: Learning to hash with binary reconstructive embeddings. In: NIPS, pp. 1042–1050 (2009)
2. Wang, J., Kumar, S., Chang, S.F.: Semi-supervised hashing for scalable image retrieval. In: CVPR, pp. 3424–3431. IEEE (2010)
3. Norouzi, M., Blei, D.M.: Minimal loss hashing for compact binary codes. In: ICML, pp. 353–360 (2011)
4. Strecha, C., Bronstein, A.M., Bronstein, M.M., Fua, P.: LDAHash: improved matching with smaller descriptors. TPAMI **34**(1), 66–78 (2012)
5. Liu, W., Wang, J., Ji, R., Jiang, Y.G., Chang, S.F.: Supervised hashing with kernels. In: CVPR, pp. 2074–2081 (2012)
6. Shen, F., Shen, C., Liu, W., Tao Shen, H.: Supervised discrete hashing. In: CVPR, pp. 37–45 (2015)
7. Yang, H., Bai, X., Liu, Y., Wang, Y., Bai, L., Zhou, J., Tang, W.: Maximum margin hashing with supervised information. Multimedia Tools Appl. **75**(7), 3955–3971 (2016)
8. Andoni, A., Indyk, P.: Near-optimal hashing algorithms for approximate nearest neighbor in high dimensions. In: 47th Annual IEEE Symposium on Foundations of Computer Science, FOCS 2006, pp. 459–468. IEEE (2006)

310   Y. Liu et al.

9. Weiss, Y., Torralba, A., Fergus, R.: Spectral hashing. In: NIPS, pp. 1753–1760 (2008)
10. Liu, W., Wang, J., Kumar, S., Chang, S.F.: Hashing with graphs. In: ICML, pp. 1–8 (2011)
11. Gong, Y., Lazebnik, S., Gordo, A., Perronnin, F.: Iterative quantization: a procrustean approach to learning binary codes for large-scale image retrieval. TPAMI **35**(12), 2916–2929 (2013)
12. Kulis, B., Grauman, K.: Kernelized locality-sensitive hashing. TPAMI **34**(6), 1092–1104 (2012)
13. Heo, J.P., Lee, Y., He, J., Chang, S.F., Yoon, S.E.: Spherical hashing. In: CVPR, pp. 2957–2964 (2012)
14. Yang, H., Bai, X., Zhou, J., Ren, P., Zhang, Z., Cheng, J.: Adaptive object retrieval with kernel reconstructive hashing. In: CVPR, pp. 1955–1962 (2014)
15. Bai, X., Yang, H., Zhou, J., Ren, P., Cheng, J.: Data-dependent hashing based on p-stable distribution. TIP **23**(12), 5033–5046 (2014)
16. Xiao, B., Hancock, E.R., Wilson, R.C.: Graph characteristics from the heat kernel trace. Pattern Recogn. **42**(11), 2589–2606 (2010)
17. Zhang, H., Bai, X., Zhou, J., Cheng, J., Zhao, H.: Object detection via structural feature selection and shape model. TIP **22**(12), 4984–4995 (2013)
18. Perronnin, F., Dance, C.: Fisher kernels on visual vocabularies for image categorization. In: CVPR, pp. 1–8 (2007)
19. Jgou, H., Douze, M., Schmid, C., Prez, P.: Aggregating local descriptors into a compact image representation. In: CVPR, pp. 3304–3311 (2010)
20. Jia, Y., Shelhamer, E., Donahue, J., Karayev, S., Long, J., Girshick, R., Guadarrama, S., Darrell, T.: Caffe: convolutional architecture for fast feature embedding. arXiv preprint arXiv:1408.5093 (2014)
21. Gong, Y., Kumar, S., Rowley, H.A., Lazebnik, S.: Learning binary codes for high-dimensional data using bilinear projections. In: CVPR, pp. 484–491 (2013)
22. Yu, F.X., Kumar, S., Gong, Y., Chang, S.F.: Circulant binary embedding. In: ICML, pp. 946–954 (2014)
23. Yan, X., Kaiming, H., Pushmeet, K., Jian, S.: Sparse projections for high-dimensional binary codes. In: CVPR, pp. 3332–3339 (2015)
24. Ye, J., Janardan, R., Li, Q.: Two-dimensional linear discriminant analysis. In: NIPS, pp. 1569–1576 (2004)
25. Jinjun, W., Jianchao, Y., Kai, Y., Fengjun, L., Thomas, H., Yihong, G.: Locality-constrained linear coding for image classification. In: CVPR, pp. 3360–3367 (2010)
26. Lampert, C.H., Nickisch, H., Harmeling, S.: Learning to detect unseen object classes by between-class attribute transfer. In: CVPR, pp. 951–958. IEEE (2009)
27. Lécun, Y., Bottou, L., Bengio, Y., Haffner, P.: Gradient-based learning applied to document recognition. Proc. IEEE **86**(11), 2278–2324 (1998)
28. Deng, J., Dong, W., Socher, R., Li, L.J., Li, K., Li, F.F.: Imagenet: a large-scale hierarchical image database. In: CVPR, pp. 248–255 (2009)
29. Fisher, R.A., et al.: The use of multiple measurements in taxonomic problems. Ann. Eugen. **7**, 179–188 (1936)
30. Schönemann, P.H.: A generalized solution of the orthogonal procrustes problem. Psychometrika **31**(1), 1–10 (1966)
31. Krizhevsky, A., Hinton, G.: Learning multiple layers of features from tiny images. Master's thesis, Department of Computer Science, University of Toronto (2009)
32. Leacock, C., Chodorow, M.: WordNet: an electronic lexical database (1998)

Signature of Geometric Centroids for 3D Local Shape Description and Partial Shape Matching

Keke Tang, Peng Song$^{(\boxtimes)}$, and Xiaoping Chen

University of Science and Technology of China, Hefei, China
kktang@mail.ustc.edu.cn, {songpeng,xpchen}@ustc.edu.cn

Abstract. Depth scans acquired from different views may contain nuisances such as noise, occlusion, and varying point density. We propose a novel *Signature of Geometric Centroids* descriptor, supporting direct shape matching on the scans, without requiring any preprocessing such as scan denoising or converting into a mesh. First, we construct the descriptor by voxelizing the local shape within a uniquely defined local reference frame and concatenating geometric centroid and point density features extracted from each voxel. Second, we compare two descriptors by employing only corresponding voxels that are both non-empty, thus supporting matching incomplete local shape such as those close to scan boundary. Third, we propose a descriptor saliency measure and compute it from a descriptor-graph to improve shape matching performance. We demonstrate the descriptor's robustness and effectiveness for shape matching by comparing it with three state-of-the-art descriptors, and applying it to object/scene reconstruction and 3D object recognition.

1 Introduction

The recent development in depth sensing devices offers a convenient and flexible way to acquire depth scans of an object or a scene that represent their partial shapes. In practice, we need to register these scans into a common coordinate system to better understand the object's or scene's geometry [1] or compare known object models with these scans for 3D object recognition [2]. All these applications require solving the partial shape matching problem [3,4].

Depth scans (i.e., 3D point clouds) lack topology information of the shape and usually contain noise, holes, and/or varying point density. To facilitate partial shape matching, one common way is to convert the point cloud into a mesh to remove the noise and fill the holes, and then perform shape matching on the mesh instead [5–8]. Although this conversion simplifies the matching process, it brings several drawbacks. First, original partial shape could be modified and/or downsampled by the conversion, e.g., when smoothing the depth scan for denoising. Second, the mesh topology generated by the conversion could be different from the real one such as incorrectly filled holes, misleading the shape matching.

Therefore, other researchers seek to perform shape matching directly on the point cloud data. This is generally achieved by representing and matching the scans using local shape descriptors. Although existing descriptors [9–12] work

© Springer International Publishing AG 2017
S.-H. Lai et al. (Eds.): ACCV 2016, Part V, LNCS 10115, pp. 311–326, 2017.
DOI: 10.1007/978-3-319-54193-8_20

well on clean depth scans, they have difficulties dealing with original scans acquired under various conditions such as occlusion, clutter, and varying lighting. This is because these descriptors are sensitive to noise and/or varying point density due to their encoded shape features such as point density [9,10] and surface normals [11], or are sensitive to scan boundary and holes due to their descriptor comparison scheme that is based on the vector distance [11,12].

To address above limitations, we propose a *Signature of Geometric Centroids* (SGC) descriptor for partial shape matching with three novel components:

- *A Robust Descriptor.* We construct the SGC descriptor by voxelizing the local shape within a uniquely defined local reference frame (LRF) and concatenating the geometric centroid and point density features extracted from each non-empty voxel. Thanks to the extracted shape features, our descriptor is robust against noise and varying point density.
- *A Descriptor Comparison Scheme.* Rather than simply computing the Euclidean distance between two descriptors, we compute a similarity score between two descriptors based on comparing the extracted features from corresponding voxels that are both non-empty. By this, the comparison scheme supports shape matching between local shape that are incomplete.
- *Descriptor Saliency for Shape Matching.* Different from keypoint detection [13] that identifies distinct points locally on a single scan/model, we propose descriptor saliency to measure distinctiveness of SGC descriptors across all input scans and compute it from a descriptor-graph. Guided by the descriptor saliency, we improve shape matching performance by intentionally selecting distinct descriptors to find corresponding feature points.

We evaluate the robustness of SGC against various nuisances including scan noise, varying point density, distance to scan boundary, occlusion, and the effectiveness of using SGC and descriptor saliency for partial shape matching. Experimental results show that SGC outperforms three start-of-the-art descriptors (i.e., spin image [9], 3D shape context [10], and signature of histograms of orientations (SHOT) [11]) on publicly available datasets. We further apply SGC to two typical applications of partial shape matching, i.e., object/scene reconstruction and 3D object recognition, to demonstrate its usefulness in practice.

2   Related Work

Shape Matching. Shape matching aims at finding correspondences between complete or partial models by comparing their geometries. Many shape matching approaches apply global shape descriptors to characterize the whole shape, for example, using Reeb graphs [14] or skeleton graphs [15] for articulated objects and shape distributions [16] for rigid objects. However, depth scans acquired from each single view usually have significant missing data. Matching these partial shapes is a difficult task because, before computing the correspondences of the shapes, we first need to find the common portions among them [1]. This requires a careful design of local shape descriptors [17] that are less sensitive to occlusion.

Local Shape Descriptors. Local shape descriptors can be classified as low- and high-dimensional, according to the richness of encoded local shape information. Low-dimensional descriptors such as surface curvature [18] and surface hashes [19], are easy to compute, store, and compare, yet have limited descriptive ability. Compared with them, high-dimensional descriptors provide a fairly detailed description of the local shape around a surface point. We classify high-dimensional descriptors into three classes according to their attached LRF [20].

Descriptors without an LRF. Early local shape descriptors are generated by directly accumulating some geometric attributes into a histogram, without building an LRF. Hetzel et al. [21] represented local shape patches by encoding three local shape features (i.e., pixel depth, surface normals, and curvatures) into a multi-dimensional histogram. Yamany et al. [22] described local shape around a feature point by generating a signature image that captures surface curvatures seen from that point. Kokkinos et al. [23] generated an intrinsic shape context descriptor by shooting geodesic outwards from a keypoint to chart the local surface and creating a 2D histogram of features defined on the chart.

Due to the missing of an LRF, the correspondence built by matching the descriptors is limited to the point spatial position only. Thus, to match two scans by estimating a rigid transform, at least three pairs of corresponding points need to be found, making the space of searching corresponding points large.

Descriptors with a non-unique LRF. Researchers later attached an LRF for local shape descriptors to enrich the correspondence with spatial orientation. By this, two scans can be matched by finding a single pair of corresponding points using the descriptors and estimating the transform based on aligning associated LRFs. However, since the attached LRF is not unique, a further disambiguation process is required for the generated transform.

Johnson and Hebert. [24] proposed a spin image descriptor by spinning a 2D image about the normal of a feature point and summing up the number of points that fall into the bins of that image. Frome et al. [10] proposed a 3D shape context (3DSC) descriptor by generating a 3D histogram of accumulated points within a partitioned spherical volume centered at a feature point and aligned with the feature normal. Mian et al. [5] proposed a 3D tensor descriptor by constructing an LRF from a pair of oriented points and encoding the intersected surface area into a multidimensional table. Zhong [25] proposed intrinsic shape signatures by improving [10] based on a different partitioning of the 3D spherical volume and a new definition of LRF with ambiguity.

Descriptors with a unique LRF. Recently, researchers constructed a unique LRF from the local shape around a feature point and further describe the local shape relative to the LRF. Thanks to the unique LRF, the transform to match two scans can be uniquely defined based on aligning corresponding LRFs.

Tombari et al. [11] proposed a SHOT descriptor by concatenating local histograms of surface normals defined on each bin of a partitioned spherical volume aligned with a unique LRF. Guo et al. [7] constructed a RoPS descriptor by rotationally projecting the neighboring points of a feature point onto 2D planes and calculating a set of statistics within a unique LRF. Guo et al. [12] later generated three signatures representing the point distribution in three cylindrical coordinate systems and concatenated and compressed these signatures into a Tri-Spin-Image descriptor. Song and Chen [8] developed a local voxelizer descriptor by voxelizing local shape within a unique LRF and concatenating an intersected surface area feature in each voxel, and applied it to surface registration [26].

SGC is also constructed within a unique LRF. Compared with above descriptors, the geometric centroid feature that we extract for constructing the descriptor is more robust against noise and varying point density. Moreover, our descriptor comparison scheme supports matching local shape that is close to the scan boundary. By this, SGC is more robust for shape matching on point cloud data than state-of-the-art descriptors [9–11], see Sect. 5 for the comparisons.

3   Signature of Geometric Centroids Descriptor

This section presents the method to construct an SGC descriptor for the local shape (i.e., support) around a feature point p, a scheme to compare a pair of SGC descriptors, and the parameters tuned for generating SGC descriptors.

3.1   LRF Construction

Given a feature point p on a scan and a radius r, a local support is defined by intersecting the scan with a sphere centered at p with radius r. Taking this support as input, we construct a unique LRF based on principal component analysis (PCA) on the support by using the approach in [11], see Fig. 1(a). When the normal of p is available, we further improve the disambiguation of LRF axes by enforcing the principal axis associated with the smallest eigenvalue (i.e., the blue axis in Fig. 1(a)) to be consistent with the normal [8].

3.2   SGC Construction

Given the unique LRF, a general way to construct a descriptor is to partition a support into bins, extract shape features from each bin, and concatenate the values representing the shape features into a descriptor vector (or a histogram).

Partition the Support. Given a support S_p around a feature point p, there are three typical approaches to partition S_p into small local patches. The first one is to partition the bounding spherical volume of S_p into girds evenly [11] or logarithmically [10] along azimuth, elevation and radial dimensions. The second one is to partition the angular space of the spherical volume into relatively homogeneously distributed bins [25]. However, the bins generated by these two approaches have varying sizes, which need to be compensated when constructing a descriptor. In addition, the irregular shape of these bins complicates the segmentation of local shape within each bin for extracting local shape features.

The third approach is to construct a bounding cubical volume of S_p that is aligned with the LRF and partition the cubical volume into regular bins (i.e., voxels) [8]. These regular bins simplify the extraction of local shape features and thus the descriptor construction. Therefore, we employ the third approach to

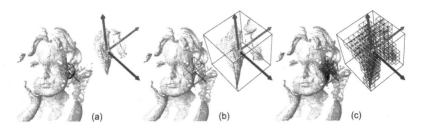

Fig. 1. Constructing an SGC descriptor. (a) Construct a unique LRF from a spherical support centered at a feature point (in pink); (b) segment a cubical support centered at the feature point and aligned with the LRF; (c) voxelize the support and extract centroid features from non-empty voxels; the centroid color indicates point density in the voxel, where small and large densities are colored in blue and red respectively. (Color figure online)

partition S_p for constructing the SGC descriptor, see Fig. 1(b and c). Note that the edges of the cubical volume have a length of $2R$, where $R \geq r$.

Extract Bin Features. Due to the missing of topology information, point clouds have limited types of shape features that can be extracted, e.g., surface normal feature in SHOT [11] and point density feature in 3DSC [10]. This paper proposes extracting a geometric centroid feature from each non-empty voxel for constructing SGC due to following reasons. First, centroid is an integral feature [27], thus can be more robust against noise and varying point density. Second, centroid can be computed simply by averaging the positions of all points staying within a voxel. Note that we do not realize any existing work that employs centroid features for constructing a usable descriptor.

Construct the Descriptor. We divide the cubical volume evenly into $K \times K \times K$ bins (i.e., voxels) with the same size, see Fig. 1(c). For each voxel V_i, we identify all N_i points staying within the voxel and then calculate the centroid (X_i, Y_i, Z_i) for the points. Note that, the position of the centroid is relative to the minimum corner of V_i in the LRF. We save the extracted feature as (X_i, Y_i, Z_i, N_i) for non-empty voxels, and $(0,0,0,0)$ for empty ones. An SGC descriptor is generated by concatenating all these values assigned for each voxel. The dimension of an SGC descriptor saved in this way is $4 \times K \times K \times K$.

Thanks to the unique LRF, the three positional values of V_i's centroid (X_i, Y_i, Z_i) can be compressed into a single value using $C_i = (Z_i \times L + Y_i) \times L + X_i$, where $L = 2R$ denotes the edge length of V_i. By this, we compress the dimension of the descriptor to $2 \times K \times K \times K$, saving 50% storage space.

3.3 Comparing SGC Descriptors

Ideally, SGC descriptors generated for two corresponding points in different scans should be exactly the same. However, due to variance of sampling, noise and occlusion, the two descriptors usually have a certain amount of difference. Unlike existing approaches that compare descriptors by computing their Euclidean distance [7,8,11], we develop a new scheme for comparing two SGC descriptors.

When constructing an SGC descriptor, most of the voxels are likely to be empty (see again Fig. 1(c)). We classify each pair of corresponding voxels into three cases: (1) empty voxel vs empty voxel; (2) non-empty voxel vs empty voxel; and (3) non-empty voxel vs non-empty voxel. In all three cases, only case 3 should contribute to computing a similarity score between two descriptors. Thus, to compare two SGC descriptors quantitatively, we propose to accumulate a similarity score for every pair of corresponding voxels that are both non-empty.

In detail, we denote two SGC descriptors as D_m and D_n. The similarity between the i-th voxel of D_m, V_m^i, and the i-th voxel of D_n, V_n^i, is defined as:

$$s(V_m^i, V_n^i) = \begin{cases} \ln \frac{N_m^i N_n^i}{\|C_m^i - C_n^i\|^2 + \epsilon}, & \text{for } N_m^i > 0 \text{ and } N_n^i > 0 \\ 0 & \text{for } N_m^i = 0 \text{ or } N_n^i = 0 \end{cases} \quad (1)$$

where N_m^i and N_n^i represent the number of points in V_m^i and V_n^i respectively, while C_m^i and C_n^i represent the centroid of V_m^i and V_n^i respectively. Here we directly employ the number of points in each voxel to represent its point density as all voxels have the same size. The formula can be explained as follows. Whenever V_m^i and/or V_n^i are empty (i.e., $N_m^i = 0$ or $N_n^i = 0$), $s(V_m^i, V_n^i) = 0$. Otherwise, when two corresponding voxels contain similar local shape, their centroids should be close to each other, making $s(V_m^i, V_n^i)$ large. When N_m^i and/or N_n^i are large, $s(V_m^i, V_n^i)$ is large also as the estimated centroid(s) are more accurate. By this, the formula encourages to find matches based on denser parts of input scans when the scans are irregularly sampled.

The overall similarity score between D_m and D_n can be obtained by accumulating the similarity value for every pair of corresponding voxels:

$$S(D_m, D_n) = \sum_{i=1}^{K \times K \times K} s(V_m^i, V_n^i) \quad (2)$$

3.4   SGC Generation Parameters

The SGC descriptor has two generation parameters: (i) the support radius R; and (ii) the voxel grid resolution K. According to our experiments, we choose $R = 20\ pr$ as a tradeoff between the descriptiveness and sensitivity to occlusion, where pr denotes the point cloud resolution (i.e., average shortest distance among neighboring points in the scan). And we choose $K = 8$ as a tradeoff between the descriptiveness and efficiency since a larger K increases the descriptiveness and computational cost simultaneously. Note that in these experiments, we let the LRF and the descriptor have the same support radius, i.e., $r = R$.

4   Partial Shape Matching Using SGC

In this section, we describe the general pipeline to match two scans using SGC descriptors and propose a descriptor saliency measure for improving shape matching performance. We also highlight the advantage of using SGC descriptors for matching supports that are close to scan boundary.

4.1   General Shape Matching Pipeline

Given a data scan S_d and a reference scan S_r, the goal of shape matching between S_d and S_r is to find a rigid transform on S_d to align it with S_r. By employing the SGC descriptors, we can find such a transform with following steps:

(1) Represent Scans with SGC Descriptors. We first conduct a uniform sampling on each of S_d and S_r to generate M feature points that cover the whole scan surface. Next, for each feature point p, we construct the LRF and SGC descriptor for the support around p. By this, we represent each of S_d and S_r with M descriptor vectors and the corresponding LRFs, see Fig. 2(a and b).

(2) Generate Transform Candidates. When a point on S_d corresponds to another point on S_r, their associated SGC descriptors should be similar to each other. Hence, we compare each feature descriptor of S_d with each feature descriptor of S_r by calculating a similarity score using Eq. 2. A feature point on S_d and its closest feature point on S_r are considered as a match if the similarity score is higher than a threshold. Each match generates a rigid transform candidate (i.e., a 4×4 transformation matrix) by aligning the associated LRFs.

(3) Select the Optimal Transform. By matching the descriptors of S_d and S_r, we obtain a number of candidate transforms. We sort these transforms based on the descriptor similarity score and then pick the top five candidates with the highest scores. We apply each of the five selected transforms on S_d to align it with S_r. We evaluate the transform by computing a scan overlap ratio. We first find all point-to-point correspondences by checking if the distance between a point on transformed S_d and a point on S_r is sufficiently small, and further compute the overlap ratio as the number of corresponding points divided by the total number of points in S_d or S_r (smaller one). We select the transform that ensures the largest overlap ratio as the optimal one, see Fig. 2(c and d).

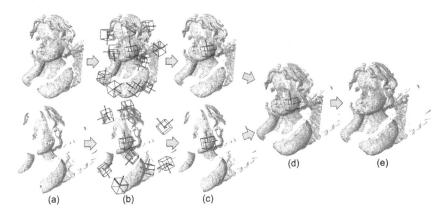

Fig. 2. Matching two scans using SGC descriptors: (a) sampled feature points (in purple) on two input scans (only part of samples are shown for clarity); (b) calculated LRFs and descriptors; (c) a pair of matched descriptors; (d) match the two scans based on aligning the associated LRFs; and (e) refine the scan alignment using ICP. (Color figure online)

(4) Refine the Scan Alignment. Optionally, we can apply iterative closest point (ICP) to refine the alignment generated by the selected optimal transform, see Fig. 2(e). By comparing Fig. 2(d and e), we can see that the transform calculated by aligning LRFs is very close to the one refined using ICP.

4.2   Improve Shape Matching Using Descriptor Saliency

To ensure corresponding points to be found on different scans, we need to sample a large number of feature points on each scan, e.g., $M = 1000$ in our experiments. However, among the M descriptors on a single scan, there could exist some descriptors close to one another since their corresponding supports are similar, see Fig. 3(a). Moreover, among descriptors from all input scans, there could exist a larger number of descriptors with high similarities, see Fig. 3(a–c).

Our observation is that when there exist a large number of descriptors with high similarities, it means their corresponding supports are less distinctive (e.g., flat or spherical shape), see the zooming views in Fig. 3(a). Thus, it has a lower chance to match the scans correctly by using such supports and their descriptors. On the other hand, when a descriptor is quite different from others, it means its support is distinctive (see the top zooming views in Fig. 3(b and c)).

Inspired by this observation, we propose a measure of *descriptor saliency* to improve the shape matching performance and compute it based on a descriptor-graph. The key idea is to find descriptors (and the corresponding supports) that are distinctive by measuring their saliency and apply these descriptors to find corresponding feature points. We first describe our approach to build a descriptor-graph, present our definition on the descriptor saliency, and then show how we apply the descriptor saliency to enhance shape matching.

Build a Descriptor-Graph. For a given reference scan S_r, we build a descriptor-graph for all the descriptors sampled from S_r based on their similarities computed using Eq. 2. Formally, let $G = (V, E)$ be a descriptor-graph, each node $u \in V$ represents an SGC descriptor on S_r. while each directed edge $(u, v) \in E$ represents that v is one of k-nearest neighbours (k-NN) of u in the descriptor similarity space. Note that we do not require u also to be one of k-NN of v, which means there may not exist a directed edge (v, u) in G.

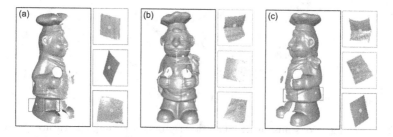

Fig. 3. Supports on three different scans of a CHEF model, where feature points are rendered in pink. The correspondence between a scan support on the left and its zooming view on the right is indicated by the same 2D box color. (Color figure online)

To build such a graph, a straightforward way is to exhaustive search all descriptors on S_r to retrieve k-NN for each descriptor in G. However, this approach is time-consuming, especially when G is large. We speed up the creation of the graph following [28], and the basic idea is to initially fill the nearest neighbors by randomly sampling descriptors in G, and iteratively optimize the nearest neighbors locally via similarity propagation and random search until convergence.

Define Descriptor Saliency. We define descriptor saliency as the distinctiveness among a set of given descriptors. The larger difference between a descriptor and others, the higher its saliency. Thus, we measure saliency of a descriptor D_i in a descriptor-graph G using $sali(D_i) = \frac{1}{1+e^{(I_i - \bar{I})}}$, where I_i denotes the number of nodes in G that considers D_i as a k-NN and $\bar{I}$ is the mean value of all I_i that is larger than zero. Note that although D_i has k nearest neighbors in G, these neighbors could be very different from D_i. By fixing k, the value I_i can reveal how many descriptors are close to D_i (i.e., D_i's distinctiveness). Figure 4 shows descriptor saliency in a simple descriptor-graph with $k = 3$.

Shape Matching with Descriptor Saliency. For a given reference scan S_r, we first create a descriptor-graph G_r for it and compute a saliency value for every descriptor D_r^i in G_r using $sali(D_i)$. For a given descriptor on the data scan S_d, say D_d^j, we enhance the similarity score between D_d^j and D_r^i by using $sali(D_r^i)$, i.e., $\bar{S}(D_d^j, D_r^i) = sali(D_r^i)^\alpha \, S(D_d^j, D_r^i)$, where α is a weight to control the impact of saliency on the descriptor similarity. We set $\alpha = 0.2$ in our experiments.

Intuitively, we can find the descriptor on S_r corresponding to D_d^j on S_d by simply comparing every D_r^i on S_r with D_d^j and selecting the one with the largest $\bar{S}(D_d^j, D_r^i)$. We speed up the search of the corresponding descriptor by taking advantage of G_r with the idea of leveraging existing matches to find better ones. This is achieved by randomly selecting a set of nodes in G_r and updating the nodes by a few iterations of similarity propagation and random search [29], guided by the similarity score (using Eq. 2) between D_d^j and the nodes. After obtaining a small set of descriptors on S_r that are similar to D_d^j, we conduct re-ranking using $\bar{S}(D_d^j, D_r^i)$ to select the final correspondence.

We have illustrated applying descriptor saliency for shape matching between a pair of scans. Descriptor saliency is more suitable for shape matching among a number of scans, with following changes. First, we build a large descriptor-graph

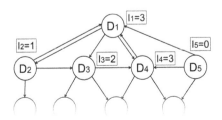

Fig. 4. An example descriptor-graph (outdegree = 3 for every node).

G for descriptors from all the scans. Second, we compare a descriptor on scan S_m with nodes in G that are not from S_m. By this, the larger the number of scans, the higher shape matching performance can be improved by descriptor saliency.

4.3   Matching Supports Close to Scan Boundary

Depth scans captured from a certain view are mostly incomplete due to a limited viewing angle, sensor noise, and occlusion. This results in a surface boundary for a scan. Matching supports close to the boundary is a challenging task. First, the support is likely to be incomplete, see examples in Fig. 2(b). This affects an LRF's repeatability since support is the only input to construct the LRF. Further, deviation of the LRF affects the construction of the descriptor since support partitioning is performed within the LRF. Second, the incomplete support directly affects the construction of the descriptor since voxels locating at the missing part(s) become empty, where no shape feature can be extracted.

Due to the above challenges, many existing descriptors are sensitive to the boundary points according to the evaluation in [17]. Therefore, boundary points are usually ignored when applying existing descriptors to partial shape matching [7,30], assuming that there is sufficient non-boundary scan surface for the matching. On the other hand, matching boundary points will improve the chance to correctly align different scans, especially when the scan overlap is small.

Our SGC descriptor is especially suitable for handling boundary points for shape matching. First, the centroid feature that SGC employs is robust against noise and varying point density, which usually happen at scan boundary. Second, our descriptor comparison scheme allows matching descriptors computed from either a complete or an incomplete support, see Fig. 5. Third, we allow using two different radii for constructing the LRF and the descriptor, i.e., $r \leq R$, see supports with varying sizes in Fig. 5(left). By this, a smaller yet complete support can be employed for constructing a repeatable LRF while a larger support allows encoding more (complete or incomplete) local shape for constructing the descriptor. Based on our experiments, we find that $r = 0.5R$ achieves the best performance for matching boundary points when setting $R = 20\ pr$.

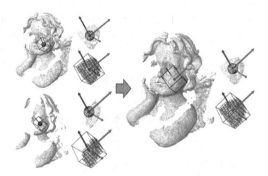

Fig. 5. (left) Match a support containing holes (in gray scan) with a support close to boundary (in cyan scan) using SGC descriptors; (right) aligned scans and supports. (Color figure online)

5 Performance of the SGC Descriptor

This section evaluates the robustness of SGC with respect to various nuisances, including noise, varying point density, distance to scan boundary, and occlusion. We compare SGC with three state-of-the-art descriptors that work on point cloud data: spin image (SI) [24], 3DSC [10] and SHOT [11]. Table 1 presents a detailed description of the parameter settings.

Table 1. Parameter settings of the four descriptors.

	Radius	Bin feature	Dimensionality	Length
SGC	20pr	Geometric centroid	$8 \times 8 \times 8 \times 2$	1024
SHOT	20pr	Histogram of normals	$8 \times 2 \times 2 \times 10$	320
3DSC	20pr	Point density	$15 \times 11 \times 12$	1980
SI	20pr	Point density	15×15	225

We perform the experiments on three publicly available datasets: the Bologna dataset [31], UWA dataset [30], and Queen's dataset [32]. Unlike the Bologna dataset that synthesizes complete object models to generate scenes, the scenes in the UWA and Queen's dataset contain partial shape of object models. We employ the Bologna dataset to evaluate the descriptors' performance with respect to noise and varying point density (Subsects. 5.1 and 5.2), the UWA dataset to evaluate the descriptors' performance with respect to distance to scan boundary and occlusion (Subsects. 5.3 and 5.4), and the Queen's dataset to evaluate improved performance by using descriptor saliency (Subsect. 5.5).

We compare the descriptors' performance using RP curves [33]. In detail, we randomly select 1000 feature points in each model and find their corresponding points in the scenes via the physical nearest neighbouring search. By matching the scene features against the model features using each of the four descriptors, an RP curve of the descriptor is generated.

5.1 Robustness to Noise

To evaluate robustness of the descriptors against noise, we add four different levels of Gaussian noise with standard deviations of 0.1, 0.3, 0.5, and 1.0 pr to each scene. The RP curves of the four descriptors are presented in Fig. 6(a–d). Thanks to the robust centroid feature, the RP curves show that SGC performs the best under all levels of noise, followed by SHOT and 3DSC.

5.2 Robustness to Varying Point Density

To evaluate robustness of the descriptors with respect to varying point density, we downsample the noise free scenes to 1/2, 1/4 and 1/8 of their original point density (pd). The RP curves in Fig. 6(e–g) show that SGC outperforms all other descriptors under all levels of downsampling. Figure 6(h) shows that SGC performs the best when the input scans are downsampled and contain noise.

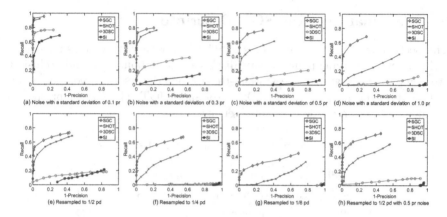

Fig. 6. RP curves of the four descriptors in the presence of (a–d) noise, (e–g) point cloud downsampling, and (h) their combination.

5.3   Robustness to Distance to Scan Boundary

We perform experiments for feature points within different ranges of distance to the boundary, i.e., $(0, 0.25R]$, $(0.25R, 0.5R]$, $(0.5R, 0.75R]$, and $(0.75R, R]$. Note that we set tuned $r = 0.5R$ for SGC and $r = R$ for all the other descriptors. Thanks to the varying support radius and descriptor comparison scheme, Fig. 7 shows that SGC achieves the best performance for all the four cases.

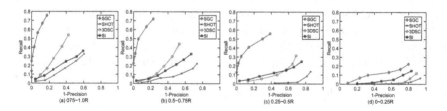

Fig. 7. RP curves of feature points in different ranges of distance to the scan boundary.

5.4   Robustness to Occlusion

To evaluate performance of the descriptors under occlusion, we group sampled feature points into two categories following [17], i.e., (60%, 70%] and (70%, 80%] occlusions. Figure 8(a and b) shows that SGC outperforms all the other descriptors with a large margin since SGC allows handling feature points at scan boundary.

5.5   Effectiveness of Descriptor Saliency

To demonstrate effectiveness of descriptor saliency, we compare our shape matching approach with an exhaustive search to find corresponding feature points.

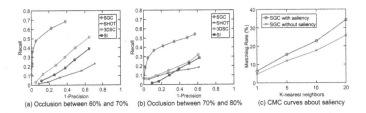

(a) Occlusion between 60% and 70%   (b) Occlusion between 70% and 80%   (c) CMC curves about saliency

Fig. 8. (a and b) RP curves about occlusion. (c) CMC curves about descriptor saliency.

First, we build a descriptor-graph for descriptors sampled from all the five models in the Queen's dataset [32] with $k = 16$. Next, we randomly select 1000 feature points on a scene and calculate their SGC descriptors. For each scene descriptor, we retrieve its neighbours by searching the descriptor-graph with saliency or exhaustive searching all the model descriptors. Here, we concern how many neighbours we need to retrieve to ensure the corresponding descriptor is included. Figure 8(c) shows standard Cumulated Matching Characteristics (CMC) curves [34] by using the two approaches. The curves show that descriptor saliency brings a certain amount of improvement in shape matching. In addition, descriptor-graph speeds up the search of corresponding descriptors, where each query process takes 0.5 ms, much faster than the exhaustive search (62 ms).

6   Applications

3D Object/Scene Reconstruction. To reconstruct a more complete model from a set of scans, we build a descriptor-graph for all the scans. As the graph has encoded k-NN for each descriptor (and the feature point), we search the corresponding feature point (and its associated scan ID) locally within the k-NN, and align the two scans based on the correspondence and merge them into a larger point cloud. We keep aligning each of the remaining scans with the point cloud and merging them until all scans are registered. Figure 9 shows two objects and one scene reconstructed by our approach on different datasets [11,35].

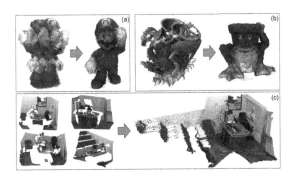

Fig. 9. Our reconstruction results. (a) SUPER MARIO; (b) FROG; and (c) STAGE SCENE.

324   K. Tang et al.

3D Object Recognition. We conduct this experiment on the challenging Queen's dataset [32]. To represent the model library well with SGC, we remove the noise in each model point cloud and build a descriptor-graph for descriptors sampled from all the models. For a give scene scan, we also sample a number of SGC descriptors. By searching a corresponding descriptor in the graph for a given scene descriptor, we know the correspondence between a model in the library and a partial scene, thus recognizing the object in the scene scan. Note that we recognize a single object at a time and segment the object once recognized.

Figure 10(a and b) show the recognition result on an example scene. Figure 10(c) shows that SGC based algorithm outperforms most existing methods including VD-LSD [32], 3DSC [10] and spin image [24] based algorithms. RoPS based algorithm is the current best 3D object recognition approach and it achieves slighter better performance than SGC with additional mesh information of the scene scans. In particular, the performance of our algorithm without using descriptor saliency decreases about 10%, indicating the usefulness of the saliency.

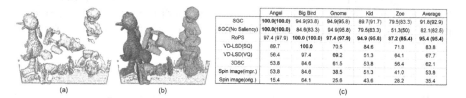

	Angel	Big Bird	Gnome	Kid	Zoe	Average
SGC	100.0(100.0)	94.9(93.8)	94.9(95.8)	89.7(91.7)	79.5(83.3)	91.8(92.9)
SGC(No Saliency)	100.0(100.0)	84.6(83.3)	94.9(95.8)	79.5(83.3)	51.3(50)	82.1(82.5)
RoPS	97.4 (97.9)	100.0 (100.0)	97.4 (97.9)	94.9 (95.8)	87.2 (85.4)	95.4 (95.4)
VD-LSD(SQ)	89.7	100.0	70.5	84.6	71.8	83.8
VD-LSD(VQ)	56.4	97.4	89.2	51.3	64.1	67.7
3DSC	53.8	84.6	61.5	53.8	56.4	62.1
Spin image(impr.)	53.8	84.6	38.5	51.3	41.0	53.8
Spin image(orig.)	15.4	84.1	25.6	43.6	28.2	35.4

(a)　　　　(b)　　　　　　　(c)

Fig. 10. Recognition results on the Queen's dataset. (a) An example scene; (b) our recognition results; and (c) recognition rates of the five models (values in brackets are the results on the whole dataset while others are the results on the subset as in [32]).

7 Conclusion

We have presented a novel SGC descriptor for matching partial shapes represented by 3D point clouds. SGC integrates three novel components: (1) a local shape description that encodes robust geometric centroid features; (2) a descriptor comparison scheme that allows comparing supports with missing parts; and (3) a descriptor saliency measure that can identify distinct descriptors. By this, SGC is robust against various nuisances in point cloud data when performing partial shape matching. We have demonstrated SGC's performance by comparisons with state-of-the-art descriptors and two partial matching applications.

Acknowledgments. This work is supported in part by the Fundamental Research Funds for the Central Universities (WK0110000044), Anhui Provincial Natural Science Foundation (1508085QF122), National Natural Science Foundation of China (61403357, 61175057), and Microsoft Research Asia Collaborative Research Program.

References

1. Aiger, D., Mitra, N.J., Cohen-or, D.: 4-points congruent sets for robust pairwise surface registration. ACM Trans. Graph. **27** Article no. 85 (2008). (Proceedings of SIGGRAPH)
2. Bariya, P., Nishino, K.: Scale-hierarchical 3D object recognition in cluttered scenes. In: CVPR, pp. 1657–1664(2010)
3. Donoser, M., Riemenschneider, H., Bischof, H.: Efficient partial shape matching of outer contours. In: Zha, H., Taniguchi, R., Maybank, S. (eds.) ACCV 2009. LNCS, vol. 5994, pp. 281–292. Springer, Heidelberg (2010). doi:10.1007/978-3-642-12307-8_26
4. Rodolà, E., Cosmo, L., Bronstein, M.M., Torsello, A., Cremers, D.: Partial functional correspondence. In: Computer Graphics Forum (2016)
5. Mian, A.S., Bennamoun, M., Owens, R.A.: A novel representation and feature matching algorithm for automatic pairwise registration of range images. Int. J. Comput. Vis. **66**, 19–40 (2006)
6. Wu, H.Y., Zha, H., Luo, T., Wang, X.L., Ma, S.: Global and local isometry-invariant descriptor for 3D shape comparison and partial matching. In: CVPR, pp. 438–445 (2010)
7. Guo, Y., Sohel, F., Bennamoun, M., Lu, M., Wan, J.: Rotational projection statistics for 3D local surface description and object recognition. Int. J. Comput. Vis. **105**, 63–86 (2013)
8. Song, P., Chen, X.: Pairwise surface registration using local voxelizer. In: Pacific Graphics, pp. 1–6 (2015)
9. Johnson, A.E.: Spin-images: a representation for 3-D surface matching. Ph.D. thesis, Robotics Institute, Carnegie Mellon University, Pittsburgh, PA (1997)
10. Frome, A., Huber, D., Kolluri, R., Bülow, T., Malik, J.: Recognizing objects in range data using regional point descriptors. In: Pajdla, T., Matas, J. (eds.) ECCV 2004. LNCS, vol. 3023, pp. 224–237. Springer, Heidelberg (2004). doi:10.1007/978-3-540-24672-5_18
11. Tombari, F., Salti, S., Stefano, L.: Unique signatures of histograms for local surface description. In: Daniilidis, K., Maragos, P., Paragios, N. (eds.) ECCV 2010. LNCS, vol. 6313, pp. 356–369. Springer, Heidelberg (2010). doi:10.1007/978-3-642-15558-1_26
12. Guo, Y., Sohel, F., Bennamoun, M., Wan, J., Lu, M.: A novel local surface feature for 3D object recognition under clutter and occlusion. Inf. Sci. **293**, 196–213 (2015)
13. Guo, Y., Bennamoun, M., Sohel, F., Lu, M., Wan, J.: 3D object recognition in cluttered scenes with local surface features: a survey. IEEE Trans. Pattern Anal. Mach. Intell. **36**, 2270–2287 (2014)
14. Hilaga, M., Shinagawa, Y., Kohmura, T., Kunii, T.L.: Topology matching for fully automatic similarity estimation of 3D shapes. In: SIGGRAPH, pp. 203–212 (2001)
15. Chao, M.W., Lin, C.H., Chang, C.C., Lee, T.Y.: A graph-based shape matching scheme for 3D articulated objects. Comput. Anim. Virtual Worlds **22**, 295–305 (2011)
16. Osada, R., Funkhouser, T., Chazelle, B., Dobkin, D.: Shape distributions. ACM Trans. Graph. **21**, 807–832 (2002)
17. Guo, Y., Bennamoun, M., Sohel, F., Lu, M., Wan, J., Kwok, N.M.: A comprehensive performance evaluation of 3D local feature descriptors. Int. J. Comput. Vis. **116**, 66–89 (2016)

18. Gal, R., Cohen-Or, D.: Salient geometric features for partial shape matching and similarity. ACM Trans. Graph. **25**, 130–150 (2006)

19. Albarelli, A., Rodolà, E., Torsello, A.: Loosely distinctive features for robust surface alignment. In: Daniilidis, K., Maragos, P., Paragios, N. (eds.) ECCV 2010. LNCS, vol. 6315, pp. 519–532. Springer, Heidelberg (2010). doi:10.1007/978-3-642-15555-0_38

20. Petrelli, A., Stefano, L.D.: On the repeatability of the local reference frame for partial shape matching. In: ICCV, pp. 2244–2251 (2011)

21. Hetzel, G., Leibe, B., Levi, P., Schiele, B.: 3D object recognition from range images using local feature histograms. CVPR **2**, 394–399 (2001)

22. Yamany, S.M., Farag, A.A.: Surface signatures: an orientation independent free-form surface representation scheme for the purpose of objects registration and matching. IEEE Trans. Pattern Anal. Mach. Intell. **24**, 1105–1120 (2002)

23. Kokkinos, I., Bronstein, M.M., Litman, R., Bronstein, A.M.: Intrinsic shape context descriptors for deformable shapes. In: CVPR, pp. 159–166(2012)

24. Johnson, A.E., Hebert, M.: Using spin images for efficient object recognition in cluttered 3D scenes. IEEE Trans. Pattern Anal. Mach. Intell. **21**, 433–449 (1999)

25. Zhong, Y.: Intrinsic shape signatures: a shape descriptor for 3D object recognition. In: 12th International Conference on Computer Vision Workshops, pp. 689–696 (2009)

26. Song, P.: Local voxelizer: a shape descriptor for surface registration. Comput. Vis. Media **1**, 279–289 (2015)

27. Pottmann, H., Wallner, J., Huang, Q.X., Yang, Y.L.: Integral invariants for robust geometry processing. Comput. Aided Geom. Des. **26**, 37–60 (2009)

28. Barnes, C., Shechtman, E., Goldman, D.B., Finkelstein, A.: The generalized patch-match correspondence algorithm. In: Daniilidis, K., Maragos, P., Paragios, N. (eds.) ECCV 2010. LNCS, vol. 6313, pp. 29–43. Springer, Heidelberg (2010). doi:10.1007/978-3-642-15558-1_3

29. Gould, S., Zhao, J., He, X., Zhang, Y.: Superpixel graph label transfer with learned distance metric. In: Fleet, D., Pajdla, T., Schiele, B., Tuytelaars, T. (eds.) ECCV 2014. LNCS, vol. 8689, pp. 632–647. Springer, Heidelberg (2014). doi:10.1007/978-3-319-10590-1_41

30. Mian, A.S., Bennamoun, M., Owens, R.: Three-dimensional model-based object recognition and segmentation in cluttered scenes. IEEE Trans. Pattern Anal. Mach. Intell. **28**, 1584–1601 (2006)

31. Tombari, F., Salti, S., Stefano, L.D.: Unique shape context for 3D data description. In: ACM Workshop on 3D Object Retrieval, pp. 57–62 (2010)

32. Taati, B., Greenspan, M.: Local shape descriptor selection for object recognition in range data. Comput. Vis. Image Underst. **115**, 681–694 (2011)

33. Mikolajczyk, K., Schmid, C.: A performance evaluation of local descriptors. IEEE Trans. Pattern Anal. Mach. Intell. **27**, 1615–1630 (2005)

34. Wang, X., Doretto, G., Sebastian, T., Rittscher, J., Tu, P.: Shape and appearance context modeling. In: ICCV, pp. 1–8 (2007)

35. Mellado, N., Aiger, D., Mitra, N.J.: Super 4PCS fast global pointcloud registration via smart indexing. Comput. Graph. Forum **33**, 205–215 (2014). (Proceedings of SGP)

Video Understanding

Unsupervised Crowd Counting

Nada Elassal[✉] and James H. Elder

Centre for Vision Research, York University, Toronto, Canada
nada.elassal@gmail.com, jelder@yorku.ca

Abstract. Most crowd counting methods rely on training with labeled data to learn a mapping between image features and the number of people in the scene. However, the nature of this mapping may change as a function of the scene, camera parameters, illumination etc., limiting the ability of such supervised systems to generalize to novel conditions. Here we propose an alternative, unsupervised strategy. The approach is anchored on a 3D simulation that automatically learns how groups of people appear in the image. Central to the simulation is an auto-scaling step that uses the video data to infer the distribution of projected sizes of individual people detected in the scene, allowing the simulation to adapt to the signal processing parameters of the current viewing scenario. Since the simulation need only run periodically, the method is efficient and scalable to large crowds. We evaluate the method on two datasets and show that it performs well relative to supervised methods.

1 Introduction

Systems for estimating the number of people in a visual scene have applications in urban planning, transportation, event management, retail, security, emergency response and disaster management. The problem can be challenging for a number of reasons (Fig. 1): (1) the projected size of people in the scene varies over the image, (2) inevitable errors in signal processing may lead to partial detection, (3) people in the scene will often overlap on projection and thus be detected

(a)	(b)	(c)	(d)

Fig. 1. Example frames and background subtraction results from (a–b) York Indoor Pedestrian dataset. (c–d) PETS 2009 dataset.

Electronic supplementary material The online version of this chapter (doi:10. 1007/978-3-319-54193-8_21) contains supplementary material, which is available to authorized users.

© Springer International Publishing AG 2017
S.-H. Lai et al. (Eds.): ACCV 2016, Part V, LNCS 10115, pp. 329–345, 2017.
DOI: 10.1007/978-3-319-54193-8_21

as a single cluster rather than individuals, (4) due to variations in pose and distance, detailed features of the human body may not be discriminable, making it difficult to accurately parse these clusters, and (5) the computational budget is typically limited by the need to run at frame rate.

Here we propose to meet these challenges by embracing two key principles: (1) Reasoning in 3D and (2) Unsupervised adaptive processing.

2   Previous Work

It is important to be clear on the goal. Some prior work (e.g., [1,2]), aims only to report some relative measure of crowd density. Here, we focus on the more challenging problem of crowd *counting*, i.e., estimating the absolute number of people in each frame of the scene. To place our paper in context, we focus on the two key issues: reasoning in 3D and unsupervised adaptive processing.

2.1   Reasoning in 3D

The importance of accounting for the effects of perspective projection has long been recognized, however mainly this has consisted of weighting pixels by 'perspective maps' that take into account distance scaling over the image [1,3–9].

Unfortunately, perspective scaling does not fully account for the complexities of occlusion: clusters of people, especially when seen higher in the image, tend to occlude each other (Fig. 1), and as a consequence the number of people in an image cluster can vary in a complex way with the size of the image cluster. Failing to account for this effect will lead to systematic bias.

Ryan et al. [10] addressed this problem by explicitly incorporating the angle of the view vector with respect to the ground plane as a predictor of the size of the crowd within a standard supervised regression framework. However, their approach requires annotation of each individual within each frame of the training dataset, and is subject to the limitations of supervised approaches (see below).

There are several studies that attempt to address the occlusion problem through a more complete 3D modeling of the scene. Zhao [11] employed 3D human models consisting of four ellipsoids, matching to detected foreground regions. However, due to the combinatorial complexity of the method it is infeasible for real-time applications for large crowds.

Kilambi et al. [12,13] (also see Fehr et al. [14]) avoided this complexity by modeling image segments as aggregates, back-projecting to head and ground planes (Fig. 2b) to identify the ground plane footprint representing the 3D extent of the group, and thus properly accounting for occlusion. However, the method requires labelled training data and supervision and assumes all people to be of exactly the same height and separated by a gap of exactly one foot when in the same segment. Neither of these assumptions will be correct in practice.

One way to overcome these limitations is through direct simulation. Dong et al. [15] used 3D motion capture data of walking humans and rendered the models to the image to compute an average 2D human shape. They then simulated the image appearance of groups of people in the scene. However, there is no mechanism to account for variations in the appearance of image segments due to partial failures in background subtraction. Also, group size was limited to six individuals: it is unclear whether the method could generalize to larger crowds.

2.2 Unsupervised Adaptive Processing

Most prior methods for crowd estimation are supervised [7–10,12,13,15–29]. As has been noted previously [10] this approach can be problematic, as the system learned from the training dataset may not generalize to different cameras, illumination, viewing geometries, etc.

Acknowledging this problem, a few groups have proposed methods that do not involve explicit training [5,6,30]. The key problem here is clustering and occlusion in densely crowded scenes. Without supervision, some general principle must be identified that allows features in a connected cluster or blob in the image to be mapped to an estimate of the number of people in the cluster.

Celik et al. [6] fit segments in the image with perspective-scaled rectangular models of fixed size. However this approach is likely to lead to bias, as there is no way to adjust the model based on biases in the segmentation (e.g., missing heads or feet), and occlusions and clustering are not handled explicitly.

Rittscher et al. [5] attempted to improve individuation within clusters with an adaptive mixture model over rectangular shapes approximating individuals within a cluster. The height and width of the shapes are governed by a Gaussian prior, allowing some adaptability. However, the parameters of the prior must still be selected in advance, reducing generality. Moreover, while the proposed system was never evaluated on (and likely was not intended for) crowd counting per se, as for other individuation approaches [11,15,20,27,30], it is likely to break down for larger and more dense crowds where occlusions obscure individual features.

2.3 Our Contributions

In this paper, we propose an efficient, fully unsupervised method for crowd counting that handles issues of clustering and occlusion by reasoning in 3D. Following Kilambi et al [12,13], we account for both perspective scaling and occlusions by projecting to the ground plane. However, rather than using training data to learn a fixed model of the appearance of an individual in the image and fixing the spacing between individuals, we use an adaptive method to learn a distribution over the appearance of individuals back-projected to the scene. As in Dong et al [15], we use a 3D simulation approach to learn, in unsupervised fashion, how to relate image observations to numbers of people. However, to factor out the effects of perspective projection, we learn the mapping from detected segments to numbers of people in ground plane coordinates, and use simpler features of these segments that will generalize to dense crowds. Finally, our adaptive unsupervised method for learning the distribution of individual appearance will account for signal processing noise and errors, avoiding the bias that would otherwise result from hardwiring the human model. The method is fast because the simulation need only be done periodically, while inference amounts to detection of connected segments (background subtraction), back-projection of these to form ground plane footprints and linear prediction of the number of people in each group based on simple features of these footprints.

3   Geometry

The viewing geometry is shown in Fig. 2a. We assume a camera with known focal length and principal point, square pixels and zero skew. To simplify notation, we assume that the principal point has already been subtracted off from the image coordinates (x, y). We also assume negligible camera roll, which is reasonable for many installations.

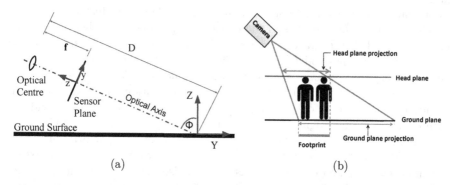

(a)   (b)

Fig. 2. (a) Viewing geometry. The X axis of the world coordinate frame and the x axis of the image coordinate frame point out of the page. (b) Back-projection using head and ground planes.

We assume a planar horizontal ground surface and adopt a right-hand world coordinate system $[X, Y, Z]$ with the Z-axis in the upward normal direction. We assume the camera is tilted down and locate the origin of the world coordinate system at the intersection of the optical axis of the camera with the ground plane. We align the X axis of the world coordinate system with the x-axis of the image coordinate system, so that the y-axis of the image is the projection of the Y-axis of the world frame. Under these conditions, a point $[X, Y]^T$ on the ground plane projects to a point $[x, y]^T$ on the image plane according to $\lambda[x, y, 1]^T = H[X, Y, 1]^T$ where λ is a positive scaling factor and the homography H is given by ([31], p. 328, Eq. 15.16):

$$H = \begin{bmatrix} f & 0 & 0 \\ 0 & f\cos\phi & 0 \\ 0 & \sin\phi & D \end{bmatrix} \tag{1}$$

Here f is the focal length, D is the distance of the camera from the ground plane along the optic axis, and ϕ is the tilt angle of the camera.

Conversely, points in the image can be back-projected to the ground plane using the inverse of this homography, $\lambda[X, Y, 1]^T = H^{-1}[x, y, 1]^T$, where

$$H^{-1} = \begin{bmatrix} 1/f & 0 & 0 \\ 0 & (1/f)\sec\phi & 0 \\ 0 & (1/Df)\tan\phi & (1/D)\cos\phi \end{bmatrix} \qquad (2)$$

We will use this back-projection to support 3D analysis of image segments. First, we use it to roughly localize the person or group in the scene by identifying the lowest point (x_l, y_l) in the image segment and back projecting it to a point (X_l, Y_l) on the ground plane. This allows us to refer the image height h and width w of the segment to scene height $H = (D_l/f)\, h$ and width $W = (D_l/f)\, w$ in a vertical plane intersecting the ground plane at (X_l, Y_l), where D_l is the distance from the optical centre to (X_l, Y_l): $D_l = \sqrt{X_l^2 + Y_l^2 + 2Y_l D \sin\phi + D^2}$.

We use these back-projected segment dimensions (H, W) to identify segment fragments too small to correspond to a whole person, which are then reconnected to form larger groups (Sect. 4.2). We also use these dimensions to identify the subset of segments that appear to correspond to individual people (singletons), and then use this subset to auto-scale our 3D simulation (Sect. 4.3).

We also use Eq. 2 to map image segments to polygonal footprints on the scene ground plane. To perform this mapping, we back-project the top and bottom of each column of pixels in the segment. To account for the fact that people are typically standing vertically on the ground plane, the bottom of each column is mapped to the ground plane, but the top is mapped to a head plane, located at a nominal average human height of 1.7 m, and this head-plane point is then projected vertically down to the ground plane. For short vertical segment columns that back-project to less than human height, we map the centre of the column to the normative mid-body plane (0.85 m), and represent each as a single point specified by the vertical projection of this mid-body point to the ground plane. The sequence of these back-projected points sweep out a closed polygonal ground plane footprint for the segment (Fig. 7(b)).

4   Algorithm

4.1   Overview

Detection is based on background subtraction, as is common in crowd estimation systems. Analysis of the resulting foreground segments then unfolds in two stages: an unsupervised learning stage and an inference stage (Fig. 3). We assume that internal and external camera parameters have been fully identified, as is the norm [32]. In practice, many pan/tilt cameras do not provide motor encoder feedback, however there are auto-calibration methods for estimating tilt angle online (e.g., [33–35]), which we consider in our evaluations (Sect. 7).

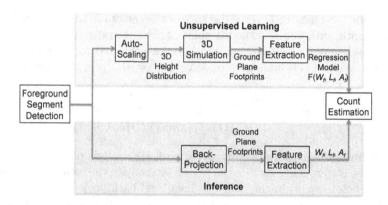

Fig. 3. Overview of the proposed algorithm

4.2 Foreground Segment Detection

We employ the background subtraction algorithm of Elder et al. [36], which is based on a pixel-wise two-component Gaussian mixture model estimated online using an approximation of the EM algorithm. The algorithm operates in the 2D colour subspace spanned by the second and third principal components of the colour distribution, and thus achieves a degree of insensitivity to shadows. Pixels with foreground probability greater than 0.5 are labelled as foreground, and segments are identified as eight-connected foreground regions. Figure 1 shows example output. (As later stages of our pipeline are adaptive, alternative segmentation methods could be substituted.)

Partial inactivity or colour similarities between foreground and background can lead to fragmentation of single individuals into disconnected image segments (Fig. 4(b)). To correct for this problem, we note that at least one part of a fragmented body must be less than half body height, and so identify all small segments that, when back-projected to the scene (Sect. 3), have vertical subtense less than half a normative human height of 1.7 m. These small segments are considered candidate fragments and are thus iteratively dilated until reaching

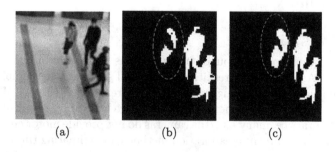

(a)   (b)   (c)

Fig. 4. Selective dilation to correct for fragmentation.

half height, using a standard 3×3 dilation kernel. If through dilation a segment merges with other segments, these segments are assume to project from the same individual/group and the new merged segment is retained (Fig. 4(c)). If no merger occurs, the segment is restored to its original state.

4.3   Unsupervised Learning

The goal of the unsupervised learning stage of the algorithm is to learn, without labelled training data, how to relate foreground segments in the image to the number of people in the scene. In a deployed system running continuously, this stage would only be invoked periodically to recalibrate the system. (We assess the effects of delays between unsupervised learning and inference in Sect. 7.) The unsupervised learning stage consists of three sequenced computations: auto-scaling, 3D simulation and feature extraction.

Auto-Scaling. Assuming that individuals detected in the image can be mapped directly to a single normative human height in the scene is risky for several reasons. First, human height varies broadly, especially when children are considered. Second, even the best background subtraction algorithm will miss some extremal pixels, leading to segments smaller than predicted, and may include false positive pixels projecting from shadows, leading to segments larger than predicted.

In prior work this problem is handled by supervised learning. Here we take an unsupervised, adaptive approach based upon the image segments that have been thus far observed. The strategy is to identify and use segments that are likely to contain only one individual to fine-tune the scaling of the system. We represent the scene scale S of each segment by the square root of the product of back-projected height and width $S = \sqrt{HW}$ (Sect. 3). The resulting distribution of back-projected scales (Fig. 5(a)) will generally be composed of a mixture of components from groups of different sizes, but the left part of the distribution will be dominated by groups of size 1 (singletons). Our objective is to estimate this component of the mixture, in order to scale the whole distribution.

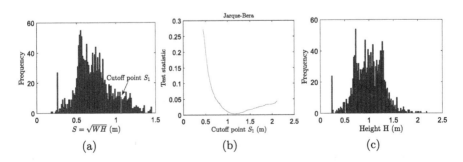

Fig. 5. Example of auto-scaling on the York Indoor Pedestrian Dataset. (a) Histogram of back-projected segment scales S. (b) Normalized Jarque-Bera test statistic. (c) Back-projected segment heights H below the scale cutoff S_1.

To do this, we appeal to the central limit theorem and assume that this component will be close to normal. We assess normality with the Jarque-Bera test statistic [37] for subsets $X(S')$ of the distribution on a series of intervals $[0, S']$ as the maximum scale S' is varied from a lower bound S_0 (we use $S_0 = 0.5\,\mathrm{m}$ here) to the maximum scale observed S_{max}. The Jarque-Bera statistic $J(X(S'))$ is a weighted sum of skewness and kurtosis:

$$J(X(S')) = \frac{n}{6}\left(Skew\,(X(S'))^2 + \frac{1}{4}\,(Kurt\,(X(S')) - 3)^2 \right) \tag{3}$$

where $n = |X(S')|$ is the sample size. $J(X(S'))$ tends to zero as the sample $X(S')$ approaches normality (Fig. 5(b)), and can thus be used to find an appropriate upper cutoff point S_1 for the singleton distribution: $S_1 = arg\,min_{S' \in [S_0 \ldots S_{max}]} J(X(S'))$ Selecting all segments below this cutoff allows us to estimate the distribution $p(H)$ of scene heights H for all singletons (Fig. 5(c)).

This method for identifying singleton segments will not always be correct: there may be some group segments that are only partially detected and fall under the threshold, while some singleton segments may cast long shadows and exceed the threshold. However, here we rely only upon the approximate correctness of the *statistics* of the singleton density, which will serve as a generative distribution from which we sample in our simulation phase.

3D Simulation. Figure 6 illustrates the 3D simulation process. The fine-tuned distribution of heights estimated by auto-scaling (Step 1) reflects the portion of the human body that was successfully detected, which we model as a 3D prolate spheroid (ellipsoid with circular symmetry about the major axis), with a 3:1 ratio of the vertical major axis to horizontal minor axes, reflecting the vertical elongation of human bodies: $(3X/H)^2 + (3Y/H)^2 + (Z/H)^2 = 1$.

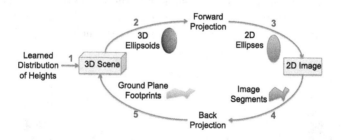

Fig. 6. Simulation.

Sampling H from the estimated singleton height distribution thus yields a distribution of ellipsoids of various sizes. To simulate crowds, we sample fairly from this distribution, placing each ellipsoid randomly and uniformly over the ground plane (Step 2). Sampled ellipsoids that intersect with existing ellipsoids are discarded. Since we do not know which portion of the body was successfully detected, we place all ellipsoids taller than a normative height of 1.7 m on the ground plane, and all ellipsoids less than this height centred at a mid-body height of 0.85 m above the ground plane. For the experiments reported here we simulated crowds from $n = 1 \ldots n_s$ people, repeating the simulation 20 times. We set $n_s = 200$ in the experiments reported here, which generated roughly 400,000 ellipsoids in all. (Results are roughly independent of n_s as long as it substantially exceeds the maximum number of people observed at inference time.)

To map these ellipsoids to the image (Step 3), we project the midpoint (X_m, Y_m, Z_m) of the ellipsoid to the image using the forward homography (Eq. 1) and approximate the projection of the ellipsoid as an ellipse of height $h = (D_m/f)H$ and width $w = (D_m/f)W$, where (H, W) are the height and width of the ellipsoid and D_m is the distance from the camera to the midpoint of the ellipsoid: $D_m = \sqrt{X_m^2 + Y_m^2 + Z_m^2 + D^2 - 2D\left(Y_m \sin\phi + Z_m \cos\phi\right)}$.

This forward projection stage is crucial to modeling the occlusion process: ellipsoids representing distinct individuals in the scene may project to intersecting ellipses in the image, forming larger group segments (Step 4, Fig. 7(a)). Both the shape and size of these image segments will vary with location in the scene. To factor this variation out, in Step 5 we use Eq. 2 to back-project each of these segments to a ground plane footprint, as described in Sect. 3 and illustrated in Fig. 2b. These footprints are expected to be roughly invariant to the location of the group within the scene, and thus variation in the size of the footprint can be largely attributed to the number of individuals within the group.

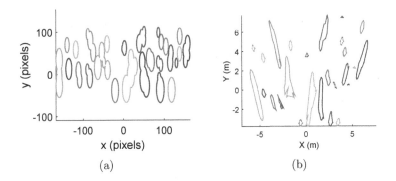

(a)   (b)

Fig. 7. (a) Image segments formed by occlusions of ellipsoidal models of human bodies in the simulated scene. (b) Back-projected ground plane footprints.

Feature Extraction. We use this 3D simulation to learn a simple model relating a ground plane footprint polygon to the number of people generating it. We use three size features: the width W_f of the footprint in the direction normal to the ground plane projection of the view vector, the length L_f of the footprint in the orthogonal direction, and the area A_f of the footprint, calculated using MATLAB function *polyarea* (Fig. 7b). The number of individuals in the segment is then predicted as a linear regression on these three variables:

$$n = b_0 + b_W W_f + b_L L_f + b_A A_f \tag{4}$$

Figure 8 shows the projection of the simulated data on the three variables. The regression model provides an estimate of the error variance, which can be used to provide a confidence interval on the estimated count (Sect. 7).

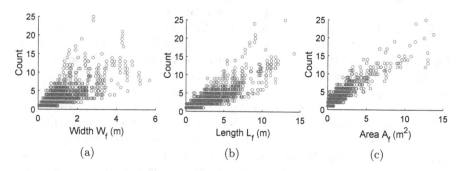

(a)   (b)   (c)

Fig. 8. Example simulation, for the York Indoor Pedestrian Dataset, showing relationship between number of simulated people in a segment and the footprint cues: (a) Width W_f, (b) Length L_f and (c) Area A_f.

4.4   Inference

Once unsupervised learning is complete, inference is relatively straightforward and fast. Detected segments are back-projected to the ground plane as described in Sect. 3 and illustrated in Fig. 2b, and width, length and area features of the ground plane footprints are computed. These features are then entered into the regression model (Eq. 4) to compute the estimated number of people n in the segment. Summing over all segments in the image yields an estimate of the number of people in the frame.

5   Datasets

We used two datasets to evaluate the proposed method (Fig. 1). We recorded the York Indoor Pedestrian Dataset using a Canon EOS Rebel T3i camera with a 40 mm lens at 30 fps. The frames were down-sampled to 320×182 pixels. The camera/lens system was calibrated in the lab using a standard calibration procedure [38]. A tripod level was used to zero the camera roll and a digital inclinometer was used to accurately measure tilt angle at the scene: $\phi = 60.7$ deg. Camera height was measured to be 10.3 m. The number of people per frame in this dataset ranges from $n = 5-16$. This dataset is available at elderlab.yorku.ca.

We also evaluated on the PETS 2009 dataset, commonly used to evaluate crowd estimation algorithms. We evaluated on three different sequences: (1) View 1, Region R0 of the S1.L1.13-57 sequence. (2) View 1, Region R0 of the S1.L1.13-59 sequence. (3) View 1, Region R1 of the challenging S1.L2.14-06 sequence, which involves a very dense crowd with high occlusion levels. We used the available camera calibration data. The number of people in these sequences ranges from $n = 0 - 38$ per frame.

6   Run Time

Run time for inference, including backround subtraction, is $O\left(m + k\right)$ per frame, where m is the number of pixels and k is the number of segments in the frame. The system was implemented in unoptimized MATLAB code, and all experiments were conducted on a 4-core desktop computer (3.40 GHz CPU). We report run time at inference for each dataset tested below. Autoscaling and 3D simulation phases of the unsupervised calibration stage take roughly 14 s and 65 s respectively.

7   Evaluation

Figure 9 shows the estimated number of people in each frame over time, compared to ground truth, for the three datasets. The method performs well: estimates typically remain within a 95% confidence interval of ground truth, with a mean absolute error (MAE) ranging from 1.04–2.47 people per frame (10.8–12.4%), and a relatively low bias (mean signed error) of -1.2–1.9 people per frame (-7.9–9.8%) (Table 1(a)). Average runtime for inference was 0.04 s per frame for the York Indoor Pedestrian Dataset and 0.31 s per frame for the PETS datasets.

Table 1(b) evaluates the influence of key components of the algorithm on performance (MAE). Error increases substantially without the segment dilation stage due to interpretation of multiple fragments projecting from the same person as separate people.

Fixing the height to a normative value of 1.7 m was found to increase the MAE for all three datasets. We also tried using a normal height distribution with a mean of 1.7 m and standard deviation of 0.1 m [41], but this increased the error even further. We believe this is due to segmentation errors, which cause

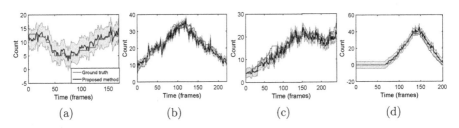

Fig. 9. Performance of the proposed algorithm over time. Blue shading indicates the 95% confidence interval for the estimate. (a) York Indoor Pedestrian Dataset. (b) S1.L1.13-57 sequence. (c) PETS S1.L1.13-59 sequence. (d) PETS S1.L2.14-06 sequence (Color figure online)

Table 1. Evaluation results. (a) Summary performance of the proposed method over all four sequences. $\bar{n}$ denotes the mean ground truth count per frame, over frames containing one or more person. (b) Contribution of algorithm components to performance. Mean absolute error (MAE) per frame reported. See text for details. (c) MAE per frame on York Indoor Pedestrian dataset for different combinations of ground plane footprint features. (d) MAE per frame as a function of delay between unsupervised learning and inference. See text for details. (e) Comparison (MAE per frame) with previous algorithms on three PETS 2009 sequences.

(a)

Dataset	$\bar{n}$	MAE	MAE (%)	Bias	Bias (%)
York Indoor Pedestrian	9.0	1.04	11.5	0.39	4.3
PETS S1.L1.13-57	22.4	2.47	10.8	-0.60	-2.7
PETS S1.L1.13-59	16.0	1.97	12.4	-1.20	-7.9
PETS S1.L2.14-06	17.5	2.13	12.2	1.86	9.8

(b)

Dataset	Proposed	No Dilation	$H = 1.7$m	$H \sim \mathcal{N}(1.7, 0.1)$m
York Indoor Pedestrian	1.04	2.13	1.33	1.16
PETS S1.L1.13-59	1.97	5.89	2.34	2.59
PETS S1.L2.14-06	2.13	–	2.48	2.7

(c)

Width	Length	Area	Width + Length	Area + Width	Area + Length	**All**
1.96	1.83	1.44	1.65	1.40	1.13	**1.04**

(d)

Dataset	MAE (No Delay)	Delay	MAE (Delay)	Change
York Indoor Pedestrian	1.04	5 sec	1.16	12%
PETS S1.L1.13-59	1.97	2 min	2.47	25%
PETS S1.L2.14-06	2.13	25 min	2.29	7%

(e)

Method	S1.L2.13-57	S1.L2.13-59	S2.L1.14-06
Albiol [25]	2.80	3.86	5.14
Fradi [9]	1.78	3.16	2.89
Li [39]	1.91	2.02	2.87
Conte [7]	1.91	2.24	4.66
Subburaman [27]	5.95	2.08	2.40
Jeong [28]	2.10	1.88	–
Riachi [29]	2.28	1.81	–
Rao [40]	2.78	1.62	2.47
Conte [8]	**1.14**	**1.59**	1.99
Proposed method	2.47	1.97	2.13
Proposed method + temporal smoothing	1.93	1.70	**1.90**

the back-projected height distribution to be considerably smaller than predicted by the normative distribution. This underlines the importance of adapting to these segmentation errors.

We also assessed the individual contributions of the three groundplane footprint features on performance (Table 1(c)). The area carries most weight, but the length and width features also contribute, at least for this dataset.

To deploy our system, a schedule for running the periodic unsupervised learning stage must be established. To inform that decision, we assessed the impact of delay between learning and inference on performance (Table 1(d)). For the No Delay condition, the same frames used for learning were used for inference. For the Delay condition, the delay varied from 5 s to 25 min depending upon the dataset. The results suggest that a delay between learning and inference does lower performance somewhat, but in these experiments there was no clear systematic dependence on the length of the delay.

Most crowd counting algorithms, including ours, depend on intrinsic and extrinsic camera parameters. Tilt angle will greatly affect results, and many deployed pan/tilt cameras are not equipped with encoders that can provide accurate online tilt angle readings. To address this, we analyzed the feasibility of using our system in combination with an automatic algorithm for estimating camera tilt [35]. The algorithm yielded a tilt estimate for the York Indoor Pedestrian Dataset of 59.2 deg, representing an error of 1.5 deg. Using this biased tilt estimate increased the MAE from 1.04 to 1.29 people per frame, a fairly graceful degradation. This suggests that the method may be deployed on common pan/tilt systems in combination with such auto-calibration algorithms.

8   Comparison with Prior Algorithms

Table 1(e) compares the proposed method against state-of-the-art methods that have reported accuracy on the PETS datasets. As a number of these methods use temporal smoothing to improve their results, we also compared against our method combined with a median temporal smoothing filter with a window size of 36 frames.

For the first PETS sequence (S1.L2.13-57), the proposed unsupervised method outperforms 3 of the 9 supervised methods without smoothing, and 5 of the 9 with smoothing. For the second PETS sequence (S1.L2.13-59), our method outperforms 4 of the 9 prior supervised methods tested without smoothing, and 6 of the 9 with smoothing. For the third sequence (S1.L2.14-06), it outperforms 6 of the 7 prior supervised methods tested without smoothing, and outperforms **all** prior methods with smoothing. Importantly, given its unsupervised nature, we expect that the proposed method will generalize more readily to a broad range of conditions (different cameras, tilt angles, illumination etc.). This is already suggested by its strong performance on both indoor and outdoor datasets.

9   Implications, Limitations, Future Work

For accurate crowd understanding, the effects of perspective projection must be accurately accounted for. Most prior work handles scaling but not the effects of occlusion. Systems that attempt to model occlusion tend to break down for larger, denser crowds, require extensive supervised training, and make unreasonable assumptions about the people in the scene. Here we have shown that through a periodic 3D recalibrating simulation of the scene, the effects of perspective projection and occlusion can be accurately accounted for. Central to this is the identification of singletons in the image that allow the simulation to be properly scaled. The result is a highly efficient inference method that does not require training, has low bias and scales easily to denser crowds.

In the experiments reported here, auto-scaling was based on between 1–8 singleton individuals, observed over multiple frames. Thus a relatively small number of singleton observations appears to be sufficient to calibrate the system. Since singletons may not be apparent in very dense crowds, in extended surveillance scenarios recalibration should be timed to coincide with sparser crowds. Optimization of the unsupervised learning schedule is a topic for future study.

In the datasets tested here most people are standing or walking. Statistics would clearly change if people were sitting; it is hoped that the unsupervised learning stage would allow the system to adapt to these different statistics, but we have not yet verified this. Note also that the background subtraction algorithm will fail to detect people who remain stationary for long periods.

We see many opportunities to improve the method. In our 3D simulation we centred all smaller ellipsoids at mid-body height. A more accurate regression model might be learned by randomizing the vertical offset of these smaller ellipsoids uniformly between ground plane and normative head plane contact. It would also make sense to sample both height and width from the singleton distribution, rather than assuming a fixed 3:1 ratio.

Our system uses width, length and area features of the ground plane footprints as predictors of the number of people in the segment. We have not systematically explored other features - there may be additional information in the shape of the footprint that would improve performance.

Finally, tracking image segments over time would allow counts to be smoothed independently for each segment and this might yield greater accuracy.

Acknowledgement. This research was supported by an NSERC Discovery research grant and by the NSERC CREATE training program in Vision Science and Applications.

References

1. Paragios, N., Ramesh, V.: A MRF-based approach for real-time subway monitoring. In: Proceedings of the IEEE Computer Society Conference on Computer Vision and Pattern Recognition, vol. 1, pp. 1034–1040 (2001)

2. Rahmalan, H., Nixon, M.S., Carter, J.N.: On crowd density estimation for surveillance. In: The Institution of Engineering and Technology Conference on Crime and Security, pp. 540–545. IET (2006)
3. Ma, R., Li, L., Huang, W., Tian, Q.: On pixel count based crowd density estimation for visual surveillance. In: 2004 IEEE Conference on Cybernetics and Intelligent Systems, vol. 1, pp. 170–173. IEEE (2004)
4. Lin, S., Chen, J., Chao, H.: Estimation of number of people in crowded scenes using perspective transformation. IEEE Trans. Syst. Man Cybern. - Part A: Syst. Hum. **31**, 645–654 (2001)
5. Rittscher, J., Tu, P.H., Krahnstoever, N.: Simultaneous estimation of segmentation and shape. In: IEEE Computer Society Conference on Computer Vision and Pattern Recognition, CVPR 2005, vol. 2, pp. 486–493. IEEE (2005)
6. Celik, H., Hanjalic, A., Hendriks, E.A.: Towards a robust solution to people counting. In: 2006 IEEE International Conference on Image Processing, pp. 2401–2404. IEEE (2006)
7. Conte, D., Foggia, P., Percannella, G., Tufano, F., Vento, M.: A method for counting people in crowded scenes. In: 2010 Seventh IEEE International Conference on Advanced Video and Signal Based Surveillance (AVSS), pp. 225–232. IEEE (2010)
8. Conte, D., Foggia, P., Percannella, G., Vento, M.: A method based on the indirect approach for counting people in crowded scenes. In: 2010 7th IEEE International Conference on Advanced Video and Signal Based Surveillance (2010)
9. Fradi, H., Dugelay, J.L.: Low level crowd analysis using frame-wise normalized feature for people counting. In: 2012 IEEE International Workshop on Information Forensics and Security (WIFS), pp. 246–251. IEEE (2012)
10. Ryan, D., Denman, S., Sridharan, S., Fookes, C.: Scene invariant crowd counting. In: International Conference on Digital Image Computing: Techniques and Applications, pp. 237–242 (2011)
11. Zhao, T., Nevatia, R.: Bayesian human segmentation in crowded situations. In: Proceedings of the 2003 IEEE Computer Society Conference on Computer Vision and Pattern Recognition, vol. 2, p. II-459. IEEE (2003)
12. Kilambi, P., Masoud, O., Papanikolopoulos, N.: Crowd analysis at mass transit sites. In: IEEE Intelligent Transportation Systems Conference, ITSC 2006, pp. 753–758. IEEE (2006)
13. Kilambi, P., Ribnick, E., Joshi, A.J., Masoud, O., Papanikolopoulos, N.: Estimating pedestrian counts in groups. Comput. Vis. Image Underst. **110**, 43–59 (2008)
14. Fehr, D., Sivalingam, R., Morellas, V., Papanikolopoulos, N., Lotfallah, O., Park, Y.: Counting people in groups. In: Sixth IEEE International Conference on Advanced Video and Signal Based Surveillance, AVSS 2009, pp. 152–157. IEEE (2009)
15. Dong, L., Parameswaran, V., Ramesh, V., Zoghlami, I.: Fast crowd segmentation using shape indexing. In: IEEE 11th International Conference on Computer Vision, 2007, ICCV 2007, pp. 1–8. IEEE (2007)
16. Marana, A.N., da Fontoura Costa, L., Lotufo, R., Velastin, S.A.: Estimating crowd density with Minkowski fractal dimension. In: Proceedings of the 1999 IEEE International Conference on Acoustics, Speech, and Signal Processing, vol. 6, pp. 3521–3524. IEEE (1999)
17. Cho, S.Y., Chow, T.W.: A fast neural learning vision system for crowd estimation at underground stations platform. Neural Process. Lett. **10**, 111–120 (1999)
18. Cho, S.Y., Chow, T.W., Leung, C.T.: A neural-based crowd estimation by hybrid global learning algorithm. IEEE Trans. Syst. Man Cybern. Part B: Cybern. **29**, 535–541 (1999)

19. Huang, D., Chow, T.W., Chau, W.: Neural network based system for counting people. In: IEEE 2002 28th Annual Conference of the Industrial Electronics Society, IECON 2002, vol. 3, pp. 2197–2201. IEEE (2002)
20. Leibe, B., Seemann, E., Schiele, B.: Pedestrian detection in crowded scenes. In: IEEE Computer Society Conference on Computer Vision and Pattern Recognition, CVPR 2005, vol. 1, pp. 878–885. IEEE (2005)
21. Kong, D., Gray, D., Tao, H.: A viewpoint invariant approach for crowd counting. In: 18th International Conference on Pattern Recognition, ICPR 2006, vol. 3, pp. 1187–1190. IEEE (2006)
22. Rabaud, V., Belongie, S.: Counting crowded moving objects. In: 2006 IEEE Computer Society Conference on Computer Vision and Pattern Recognition, vol. 1, pp. 705–711. IEEE (2006)
23. Chan, A.B., Liang, Z.S.J., Vasconcelos, N.: Privacy preserving crowd monitoring: counting people without people models or tracking. In: IEEE Conference on Computer Vision and Pattern Recognition, CVPR 2008, pp. 1–7. IEEE (2008)
24. Jones, M.J., Snow, D.: Pedestrian detection using boosted features over many frames. In: 19th International Conference on Pattern Recognition, ICPR 2008, pp. 1–4. IEEE (2008)
25. Albiol, A., Silla, M.J., Albiol, A., Mossi, J.M.: Video analysis using corner motion statistics. In: Proceedings of the IEEE International Workshop on Performance Evaluation of Tracking and Surveillance, pp. 31–38 (2009)
26. Chan, A.B., Morrow, M., Vasconcelos, N.: Analysis of crowded scenes using holistic properties. In: Performance Evaluation of Tracking and Surveillance Workshop at CVPR, pp. 101–108 (2009)
27. Subburaman, V.B., Descamps, A., Carincotte, C.: Counting people in the crowd using a generic head detector. In: 2012 IEEE Ninth International Conference on Advanced Video and Signal-Based Surveillance (AVSS), pp. 470–475. IEEE (2012)
28. Jeong, C.Y., Choi, S., Han, S.W.: A method for counting moving and stationary people by interest point classification. In: 2013 20th IEEE International Conference on Image Processing (ICIP), pp. 4545–4548. IEEE (2013)
29. Riachi, S., Karam, W., Greige, H.: An improved real-time method for counting people in crowded scenes based on a statistical approach. In: 2014 11th International Conference on Informatics in Control, Automation and Robotics (ICINCO), vol. 2, pp. 203–212. IEEE (2014)
30. Masoud, O., Papanikolopoulos, N.P.: A novel method for tracking and counting pedestrians in real-time using a single camera. IEEE Trans. Veh. Technol. **50**, 1267–1278 (2001)
31. Prince, S.: Computer Vision: Models, Learning and Inference. Cambridge University Press, Cambridge (2012)
32. Ryan, D., Denman, S., Sridharan, S., Fookes, C.: An evaluation of crowd counting methods, features and regression models. Comput. Vis. Image Underst. **130**, 1–17 (2015)
33. Wildenauer, H., Hanbury, A.: Robust camera self-calibration from monocular images of Manhattan worlds. In: 2012 IEEE Conference on Computer Vision and Pattern Recognition (CVPR), pp. 2831–2838. IEEE (2012)
34. Tal, R., Elder, J.H.: An accurate method for line detection and manhattan frame estimation. In: Park, J.-I., Kim, J. (eds.) ACCV 2012. LNCS, vol. 7729, pp. 580–593. Springer, Heidelberg (2013). doi:10.1007/978-3-642-37484-5_47

35. Corral-Soto, E.R., Elder, J.H.: Automatic single-view calibration and rectification from parallel planar curves. In: Fleet, D., Pajdla, T., Schiele, B., Tuytelaars, T. (eds.) ECCV 2014. LNCS, vol. 8692, pp. 813–827. Springer, Heidelberg (2014). doi:10.1007/978-3-319-10593-2_53

36. Elder, J.H., Prince, S., Hou, Y., Sizintsev, M., Olevskiy, E.: Pre-attentive and attentive detection of humans in wide-field scenes. Int. J. Comput. Vis. **72**, 47–66 (2007)

37. Jarque, C., Bera, A.: Efficient tests for normality, homoscedasticity and serial independence of regression residuals. Econ. Lett. **6**, 255–259 (1980)

38. Zhang, Z.: Flexible camera calibration by viewing a plane from unknown orientations. In: Proceedings of the Seventh IEEE International Conference on Computer Vision, vol. 1, pp. 666–673 (1999)

39. Li, Y., Zhu, E., Zhu, X., Yin, J., Zhao, J.: Counting pedestrian with mixed features and extreme learning machine. Cogn. Comput. **6**, 462–476 (2014)

40. Rao, A.S., Gubbi, J., Marusic, S., Palaniswami, M.: Estimation of crowd density by clustering motion cues. Vis. Comput. **31**, 1533–1552 (2015)

41. Schilling, M., Watkins, A., Watkins, W.: Is human height bimodal? Am. Stat. **56**, 223–229 (2002)

Long-Term Activity Forecasting Using First-Person Vision

Syed Zahir Bokhari[✉] and Kris M. Kitani

Robotics Institute, Carnegie Mellon University, Pittsburgh, PA, USA
sbokhari@andrew.cmu.edu

Abstract. Long-term activity forecasting deals with the problem of predicting how an agent will complete a full activity, defined as a continuous trajectory and a discrete sequence of sub-actions. While previous data-driven methods only dealt with forecasting 2D trajectories, we present a method that leverages common sense prior knowledge and minimal data. In order to forecast the trajectories, we learn a policy function that maps from states to actions the agent should perform next. Through the use of deep reinforcement learning, our method is able to learn a highly non-linear mapping from agent states to actions. We develop the first forecasting framework that uses ego-centric video input, which is an optimal vantage point for understanding human activities over large spaces. Given an annotated first person video sequence for the activity, we construct a 3D point cloud of the environment and activity paths through 3D space. Based on a limited number of examples, we use reinforcement learning to derive a policy for the entire environment, even for areas that have never been visited during the demonstrated examples. We explore the use of deep reinforcement learning to recover a direct mapping from environmental features to best action. Our approach makes it possible to combine a high dimensional continuous state (namely the local point could density surrounding the agent) with a discrete state portion (action stage of an activity) into a single state for behavior forecasting. The result is a policy that generalizes very well from only a few activity samples. We validate our approach on our First-Person Office Behavior Dataset and show that our method of encoding more prior knowledge leads to an increase in forecasting accuracy. We also demonstrate that the deep reinforcement learning approach is able to achieve higher forecasting accuracy than the traditional alternatives.

1 Introduction

There has been recent interest in computer vision algorithms that have the capability to predict human activities into the future at various time scales. In particular, there has been significant work concentrated on predicting very short term actions, on the order of a few seconds [1–4]. In contrast, we focus on the forecasting human activities over a longer time horizon (several minutes) while retaining the ability to perform highly accurate trajectory forecasting at a fine time resolution.

© Springer International Publishing AG 2017
S.-H. Lai et al. (Eds.): ACCV 2016, Part V, LNCS 10115, pp. 346–360, 2017.
DOI: 10.1007/978-3-319-54193-8_22

Developing the technology needed to forecast human activity over a longer time horizon is a critical feature necessary for advanced intelligent agents. The ability to forecast human activity minutes in advance of execution will allow automated homes and personal robotic systems to take pre-emptive actions to better meet the needs of the user. Not only would such forecasting technology allow robotic agents to veer around potential short-term collisions but could also enable them to turn on the air conditioning or warm up the bath water minutes in advance of our arrival. The ability to see things long before they happen is an essential technology to enable higher levels of human-computer interaction.

How does one obtain such an ability to forecast human activities minutes in advance of execution? One straightforward approach is to address this task completely in a data-driven fashion without any prior knowledge. This is performed by observing a large number of activity sequences in an environment to build a forecasting model over all possible activities in that scene. However, there is no guarantee that the system will be able to observe every possible trajectory or sub-action that can be performed at every location in the scene. Moreover, such a data-driven approach may not make use of prior information about the scene. When prior information about the scene is available, we would like to use it to inform the forecasting model.

We would prefer an approach that can generalize from only a few samples while also making use of common sense prior knowledge about human activities. To this end, we make use of ideas from reinforcement learning to build a model for accurate long-term activity forecasting from limited examples and common sense prior knowledge. In particular, we implement the concept of temporal difference learning in the form of Q-learning, which allows the forecasting model to generalize to new activity trajectories through the use of off-policy exploration. The RL framework also allows us to encode prior information about the scene in terms of the reward function. Common sense knowledge such as the fact that people will avoid walls and obstacles in the scene can be encoded as part of the reward function during the learning process. With the combination of a few observed activities sequences and proper prior knowledge in the form of a reward function, our proposed approach is able to accurately forecast long-term human activities.

This paper proposes a reinforcement learning approach to learn a model for long-term activity forecasting. Our method is able to encode prior knowledge about the scene through the use of the reward function. Moreover, we are able to learn from only a few examples through the use of off-policy reinforcement learning. We validate our approach on real human activity data recorded with a wearable camera using our First-Person Office Behaviour Dataset (see Fig. 1). We show that our approach of encoding more prior knowledge in the problem formulation leads to an increase in forecasting accuracy. Additionally, our proposed approach can forecast plausible sub-action sequences along with their detailed motion trajectories for common office activities such as making coffee or picking up a package from the mail room.

Fig. 1. Sample images from our First-Person Office Behaviour dataset. The dataset contains twelve activity sequences with three main activities: getting coffee, printing a page, and going to mail room. Each activity has a variety of sub actions such as washing cup, using computer, posting letter etc.

2   Related Work

The problem of trajectory forecasting has been tackled several times before. Kitani *et al.* introduced the problem of activity forecasting and proposed a solution based on semantic segmentation of the environment and inverse optimal control [5]. Karasev *et al.* used a Markov Decision Process to predict the motion of pedestrians on a street, in order to help the decision making of driverless cars [6]. They also make use of a slightly higher dimensional state than [5] by incorporating the orientation angle of the pedestrians. A simpler approach was taken by Walker *et al.* where a goal probability is learned for each type of agent in a specific scene [4]. This method's use of standard tracking algorithms makes it completely unsupervised, which permits the use of unlabeled data. Xie *et al.* predicts trajectory as well as functional objects in the environment that draw agents to approach them [7]. The problem is modelled as a physics problem, where attractive objects in the scene emit energy that draws the agent towards them. Interestingly Huang *et al.* approach forecasting from a different angle. Instead of predicting trajectories, they instead attempt to forecast interactions between two people by hallucinating the pose of one person given the pose of the other person. Their approach used inverse optimal control, modified for high dimensional situations [1,8]. Our approach not only predicts trajectory through the environment, but also the actions that happen along the trajectory, with the use of reinforcement learning.

Reinforcement learning in general has been a very well explored field. One of the earliest and most popular examples of practical reinforcement learning was TD-Gammon [9], where a neural network was used as a function approximator in the TD(λ) algorithm. The effects of function approximation on value iteration, such as the propagation of the approximation error, are investigated in [10–12]. It is useful to look at value iteration to learn how function approximation affects standard reinforcement learning because value iteration is deterministic, as opposed to Q-Learning which typically follows an ϵ-greedy policy. In the realm of Q-Learning, Riedmiller *et al.* looked at an efficient way of training Q-value functions approximated by multi-layer perceptrons [13], while Farahmand *et al.* looked at applying L^2 regularization to Q-Learning when using a function approximator, in order to control the complexity of the learned models [14].

3   Preliminaries

Given several egocentric demonstrations of a specific activity from a dataset, our goal is to learn the human activity model for performing the activity in the environment. Reinforcement learning is one natural approach for inferring this policy from a sequence of demonstrated states and actions. In our problem setup, we make use of use of a rich vision-based state representation which encodes local geometric features. As such, the state space is a large dimensional space in which the state transition dynamics are non-trivial to derive. When dealing with large state spaces it is common to use function approximators (in place of traditional tabular functions). Furthermore, for situations in which the state transition dynamics are unknown, it is typical to use Temporal Difference (TD) methods, such as Q-learning, to learn a policy from demonstrations. In the following we review the classic Q-learning algorithm and describe a Q-learning framework using deep neural networks as the value function approximator, which is used to learn egocentric activity policies.

3.1   Q-Learning

In reinforcement learning we have an environment referred to as the state space S, as well as a set of actions A that can be performed in each state. The reward function $r(s, a, s')$ describes the reward an agent receives when transitioning from one state to another upon performing a certain action. Over the course of an activity with N states, the agent receives various reward values for each action it performs. Let R_i indicate the reward for the ith action performed by the agent. The goal of an optimal control algorithm is to find an optimal policy $\pi^*(s) = a$ which describes the best action to perform in each state in order to maximize the expected future reward.

Q-Learning is an off-policy TD method, meaning that the optimal policy can be found and evaluated while exploring the environment based on another policy. It does so by computing the action-value function of the form

$$Q(s_k, a_k) = \mathbf{E} \left[\sum_{i=0}^{N} \gamma^i R_{k+i+1} \Big| s_k, a_k \right] \tag{1}$$

where s_k is the kth state the agent is in, a_k is the action performed in that state, R_{k+i+1} is the reward for all subsequent actions that may be performed, and $0 < \gamma \leq 1$ is the discount factor. Equation 1 defines the value of an action as the discounted future reward the agent is expected to receive. Given the optimal action-value function Q, the optimal greedy policy is simply $\pi^*(s) = \text{argmax}_a Q(s, a)$. The Q-Learning algorithm first initializes $Q(s, a)$ to be some constant value (such as 0). It then computes Q by allowing the agent to explore the environment in multiple steps. With each step, an experience tuple of the form (s, a, R, s') is collected. We can then update Q with

$$Q(s, a) \leftarrow Q(s, a) + \alpha \left[R + \gamma \max_A Q(s', A) - Q(s, a) \right] \tag{2}$$

as described in [15], where α is the learning rate. While this experience tuple can be discarded now, it is common to reuse it through experience replay [16].

3.2 Deep Q-Network

Deep Q-Learning is an extension to classic Q-Learning that models the Q-function using a deep network. This is an attractive approach as classic Q-Learning is not well suited for high dimensional state spaces, whereas deep networks are very good at dealing with high dimensional, low level features. This way we can take full advantage of our 3D point cloud that represents the structure of the environment. A deep network is also needed as our policy is highly non-linear, and deep networks have been shown to be effective means of learning non-linear functions.

In deep Q-Learning, we replace the table used to keep track of the action-values in the Q function with a deep network. This way, large inputs can be dealt with and observed states can generalize to similar unseen states. Following the same formulation used in [17], we define a Q-Network with parameters θ as $Q(s, a; \theta) = v$. Then given an experience tuple (s, a, R, s'), we can compute the target value as

$$v^* = \begin{cases} R & \text{if } s' \text{ is a terminal state} \\ R + \gamma \max_A Q(s', A; \theta_-) & \text{if } s' \text{ is not a terminal state} \end{cases} \tag{3}$$

θ_- are the network parameters for the target network (the current parameters or parameters from a previous iteration for delayed updates). Once the target value v^* is computed, gradient descent can be performed on the parameters with respect to the squared loss $(v^* - Q(s, a; \theta))^2$. Using this to change the update step of the batch Q-learning algorithm, we get the deep Q-learning algorithm.

4   Long-Term Activity Forecasting

In order to use deep Q-Learning to forecast the agent's behaviour, we simulate the exploration dynamics of an agent in the environment. This allows us to gather experience that the deep Q-Learning algorithm will use to learn the policy. We must thus model the agent's sequential decision making process. Many previous approaches model the agent's behaviour as a Markov Decision Process (MDP), and this is well suited for our problem as well. To define the MDP, we need to define the state space S, action set A, and reward function r. It is in the definition of these components that we are able to encode our prior knowledge of the environment, which will be leveraged by the reinforcement learning algorithms.

4.1   State Space and Action Set

In order to forecast the agents decisions in the environment, the state space needs to be able to express all of the locations in the environment, as well as the agent's behaviour state. This means that our state can be represented as a tuple $(\mathbf{x}, w)$ where $w \in W$ is the stage of the activity, and $\mathbf{x} \in X$ is the position portion. X represents the set of all possible position states. It is possible for $\mathbf{x}$ to simply be the position on the map, however we can also take advantage of a more rich representation. This can also include the local structure of the environment around the position, and distances to objects and obstacles. While this is more information for the learning algorithm to take advantage of, this also causes the size of the state space to increase. W represent all possible states the activity alone can be in, regardless of location. For example, the first column of Table 1 shows the possible stages in a simple coffee making activity. With both X and W defined, the total state space can be expressed as the Cartesian product $S = X \times W$.

The action set can also be defined in a similar way, where each action either affects the location portion or the activity portion of the state. The action set M that affect the location portion can be considered movement actions, while the

Table 1. Coffee making activity.

Coffee making stages	Activity sequence frame actions	RL actions
No Cup	Standing	Move North
Has Dirty Cup	Walking	Move East
Has Clean Cup	Pickup Cup	Move South
Has Unstirred Coffee	Wash Cup	Move West
Has Coffee	Put Down Cup	Pickup Cup
Finished	Make Coffee	Wash Cup
	Pickup Staw and Stir	Make Coffee
	Finish	Stir Coffee
		Finish

action set E that affect the activity portion can be though of as environmental interactions. In our coffee making example, M would be represented by the first 4 actions in the third column of Table 1, while E is represented by the rest. The total action set in this case is $A = M \cup E$. Since the action set comprises of actions that either change the position portion or stage portion, it might seem like the transition dynamics are simple. However this depends highly on the representation of $\mathbf{x}$. If $\mathbf{x}$ is simply the 3D position in the environment, then the transition function is trivial. However if we wish to encode the environment structure and obstacle positions, then the transition function becomes non-trivial, and we must make use of a deep Q-Learning.

4.2 Reward Function

The definition of the reward function determines which policies are encouraged and which are not. This makes the reward function a good candidate for encoding our prior knowledge of how humans typically navigate an environment. We know that for most activities with an end goal, agents must move to certain positions in the environment (around obstacles if need be) in order to complete a task. In our coffee making example, the cup may only be washed at a sink, and coffee can only be made at a coffee machine. In order to encourage a policy that mimics human behaviour, we need to encode these three aspects into the reward function: the end goal (controlled by the R_{end} term), locations where actions may be performed (controlled by the R_{act} term), and obstacles in the environment (controlled by the R_{wall} term).

$$
\begin{aligned}
r(s, a, s') = {} & R_{end} \cdot \mathrm{isTerm}(s') \\
& + R_{act} \cdot \sum_{i=0}^{N} \mathbb{1}(a = a_i) \min_{j} \mathrm{dist}(s, x_i^j) \\
& + R_{wall} \cdot \mathrm{wallScore}(s')
\end{aligned}
\tag{4}
$$

In order to encourage a policy that tries to mimic the activity training sequences, we put a large positive reward R_{end} at the end of all example sequences. This gives the Q-Learner a large reward for terminating the activity in a similar state as one of the example sequences. This provides the algorithm motivation for completing the activity.

However only including the goal reward will encourage a policy that transitions through the activity stages at any location (such as making coffee far away from the coffee machine). It will take any actions needed to arrive at a terminal state, regardless of environment. In order to prevent this from happening, we have a punishment (negative reward) R_{act} for performing the stage change actions far away from the demonstrated locations in the training data. This punishment constant is multiplied by the distance from the nearest example location order to encourage being closer to the correct locations for performing actions. Since we are not giving our exploration agent access to the true environment, this penalty is needed to enforce the common sense notion that not all actions (such as **Wash Cup** and **Make Coffee**) can be successfully performed anywhere. This will encourage the agent to learn a policy that only performs the stage change actions near the locations where it is actually possible, and demonstrated in the training data.

Desirable policies must also avoid obstacles in the environment. We do not want policies that try to move through walls and tables. To prevent this, we introduce a penalty term R_{wall} for the movement actions. This way, we can punish the agent for trying to move through a space that seems to contain a wall or other obstacle, since a human in the environment is unlikely to try and move through that same space. The penalty is a constant multiplied by the point density around the location $\mathbf{x}$ in the environment point cloud.

Equation 4 shows the final form of the reward function. In the equation:

- isTerm(s') is an indicator function that has the value 1 if s' is the terminal state of one of the training examples, and 0 otherwise
- wallScore(s') is the point density at the location of the state s'
- $a_i \in A$ is all of the actions that can be performed and a_i^j are all of the locations in the training data that action was observed. The third term penalized actions that are performed at a distance from where they were demonstrated in the training data

The end sequence reward, action penalty, and density penalty are the main reward types needed for this problem. Then encode our intuition of how a person will generally navigate any environment, avoiding obstacles and moving towards their final goal. These, along with the state space and transition function are enough to encourage a simple environment specific policy for a given activity (such as making coffee).

5   Experiments/Results

In this section, we seek to evaluate the effectiveness of encoding intuition and prior knowledge into the reward function. We also seek to validate the use of deep reinforcement learning as an effective and flexible means of activity forecasting. As no other papers have done activity forecasting with 3D point cloud data generated from ego-centric video, there are no pre-exsisting baselines to compare against. Instead we validate our approach on our First-Person Office Behaviour dataset. We demonstrate that encoding more prior knowledge into the reward function results in a lower forecasting error. We also show that the deep Q-Learning approach, with its access to low level state information produces the lowest forecasting error in our tests.

5.1   First-Person Office Behavior Dataset

Since we seek to do trajectory and action forecasting with egocentric video, a dataset with long egocentric video demonstrations of multi-step activities was needed. As no existing dataset meets these needs, we collected a our own First-Person Office Behavior Dataset.

The data consists of two types of first-person videos. The first type of video is a mapping sequence, which simply contains a detailed view of all corners and viewpoints of the environment. The second type of video is the activity sequences. Twelve activity sequence videos were collected for three types of activities: getting a mug and making coffee, printing a page from an office computer then picking up the printout, and picking up a package from the mail room. The mapping sequence is 24 min long and the activity sequences are each around 13 min long. The printing activity contains 8 sub-actions, the coffee making activity contains 8 sub-actions, and the package collecting activity contains 7 sub-actions. All videos were recorded at 60fps. We only report our results on the coffee making activity. All of these activity sequences take place in the same large environment, and are annotated in with the action being performed in each video frame. The first column of Table 1 shows the dataset video frame actions for the coffee making activity.

With the mapping sequence we can use any structure from motion algorithm to build a dense point cloud of the environment. For our experiments, we used VisualSFM [18–20] to construct the dense point cloud and activity paths. Then the activity sequences can be registered against the environment model images, and the 3D trajectory for each of the actions sequences can be reconstructed. We first use VisualSFM on the mapping sequence videos to construct the environment features and point cloud. Then to register the activity sequences, we reuse the environment features and register each activity frame one by one, allowing us to recover the camera positions for each frame. This gives us the full trajectory of the agent through the environment over the course of the activity. Since the video is labeled with actions on each frame, the reconstructed data will indicate where in the environment the agent was when performing each action.

While the mapping sequences is not explicitly needed, they do provide a more detailed and noise free trajectory through the environment than if the activity sequences were used alone in the structure from motion. This provides the basic dataset that will be used in both the classic and deep Q-Learning algorithms.

5.2   Including Prior Knowledge in the Reward Function

In order to validate our approach of including prior knowledge of activity dynamics in the formulation of the reward function, we conduct an ablative analysis across various settings of the reward function. We show that as more common sense prior knowledge is included in the reward function, the forecasting error is reduced. In all experiments, we decide only to deal with 3 activity stages (before washing cup, before making coffee, after making coffee). This allows for simple comparison between the tabular method and deep network method of Q-learning and for simplicity of visualization. The state representation used for the deep Q-Network approach consisted of a 13×13 patch of the point cloud density centered at the location corresponding to the state. This representation is a powerful and low level way to represent the obstacles in the environment near the state. However since this state representation cannot be used with the tabular reinforcement learning methods, we must simply use the 2D position as the state for those methods.

Tables 2, 3, and 4 show generated policies using various settings for the reward function, and the resulting mean Modified Hausdorff Distance (MHD). The number reported is the mean MHD over 5000 random paths sampled from each learned policy. The sampling is done by specifying the start as the start of the test path, and then randomly selecting an action based on the softmax probability of all actions for the current state. Once an entire sequence has been formed, the minimum distance from a point on the sampled path to a point on the true path is computed. This is done for all points on the generated path, and the MHD is computed by taking the sum. Thus a lower MHD indicates a better conformance to the true path. The MHD can also be thought of as the forecasting error, with a larger distance indicating a trajectory very different from the data sample.

Value Iteration Results. In using value iteration, we update all states with respect to the best action that can be performed. To ensure that the evaluation metric is giving sensible results, we also computed the Modified Hausdorff Distance with respect to simpler reward functions that we know are likely to give incorrect policies. Table 2 show that as we introduce more useful prior knowledge in the form of a more complex reward function, our forecasting accuracy increases (the MHD decreases).

Table 2. Value iteration method with different goal parameters

Goal reward	Map penalty	Action distance penalty	MHD
30	0	0	3.2187
30	0	-100	1.7937
30	-1	0	4.3866
30	-1	-100	4.6255
100	**-1**	**-100**	**1.6816**

Q-Learning and Q-Network Results. In using Q-Learning we simulate the exploration behavior of the agent to collect experience. This involves building a state-action trajectory piece by piece. For the tabular Q-Learning method, we are required to use a simple state representation. We used (x, y, w) where x and y were the integer positions in the environment grid. In simulating the exploration, actions that change the position increment and decrement the x and y values. Table 3 shows the results for using tabular Q-Learning to learn the prediction model. As expected, the more complex reward function performed better. We also see that an extreme setting for the goal reward is not good.

The policies computed with deep Q-Learning used the exact same exploration dynamics as the tabular Q-Learning method. In our implementation, we tested

Table 3. Discrete Q-Learning with different goal parameters

Goal reward	Map penalty	Action distance penalty	MHD
30	0	0	6.0381
30	0	−100	3.8177
30	−1	0	5.5639
30	**−1**	**−100**	**2.1859**
100	−1	−100	2.8341

the Q-Network with the local state (p, w) where p is a patch of the voxel density grid centered at the position (x, y) of the agent during the simulation. Since most motion of the agent is in a 2D plane, we found it better to flatten the voxel density grid into a 2D density map, and use a 2D density image centered at the agent position instead of the 3D density volume. p is then represented as 13×13 grayscale image, with lighter pixels indicating higher point density, and thus a likely obstacle for the agent. This gives the deep Q-Network access to more low level data thus making it easier to fit a good model.

Table 4. Deep Q-Learning with different goal parameters

Goal reward	Map penalty	Transition reward	MHD
50	0	0	3.1183
50	**−0.5**	**−20**	**1.5229**

5.3   Cross Model Comparison

Comparing the best models from above in Table 5, it seems like value iteration does the best because it takes far fewer iterations. However in each iteration of value iteration, every single state is updated once, whereas in the other two methods using batch q-learning, only the states appearing in the randomly samples batch are updated. For both, we used a batch size of 100. This shows that the Q-network was able to learn a sensible policy with much fewer steps of explorations than the tabular Q-Learning method.

Table 5. Method comparison

Method	Iterations	HD
Action value table	1000000	2.1859
Q-Network MLP	50000	1.5229
Value iteration	200	1.6816

5.4   RL Policy Visualizations

Another simple way to validate the policies learned from the three methods is to visually inspect their value functions. The value function is defined as $V(s) = \max_a Q(s, a)$. This is often easier to look at than the action value function, for which there are more plots. What we expect to see is a large value at the end of the demonstrated sequence. In addition, we should also see a policy that follows along the gradient of the value, constantly seeking out adjacent states with higher values. Ideally, we want a policy that respects the properties of the environment, and does not try to move through walls or other obstacles. The second image in Fig. 2 shows the portions of the point cloud that are considered obstacles in the environment. As we can see from Fig. 3, the value iteration policy looks much cleaner, while many of the details are lost in the Q-Network policy.

Fig. 2. Left: Mapping sequence shot of the kitchen. Middle: 3D point cloud reconstruction of the kitchen. Right: Birds-eye view of kitchen point cloud with objects and obstacles highlighted.

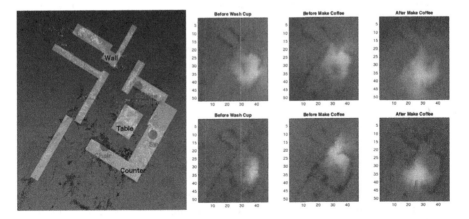

Fig. 3. Left: Location of obstacles and items in the environment. Right: Value function visualizations for the tabular value iteration approach and the deep Q-Network approach. Notice how for each stage of the activity, there is a high value associated with the location where the next action must be performed.

Also notice that the Q-network policy has more noisy artifacts. We also see that in both cases having a very high reward seems to be overpowering the penalty for moving through walls close by the goals.

Figure 4 shows the path generated directly from the policy distribution. Each row represents a point in time. The first column represents the **Has Dirty Cup** stage, the second is the **Has Clean Cup** stage, and the third is the **Has Coffee** stage. As we can see, the policy sensibly moves around the table in the center of the kitchen, first to the sink to wash the cup, then to the coffee machine, and finally finishes at the chair.

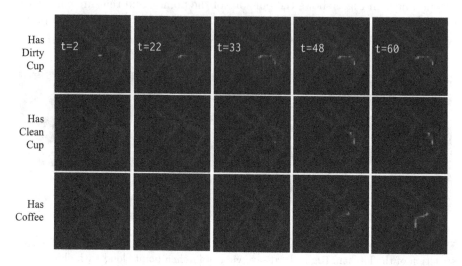

Fig. 4. Path generated by Q-Network. Each column shows the possible states (location and activity stage) at a single timestep, and the light blue represents the probability of being in that state. We can see that at the first timestep, the probability is concentrated at where the agent starts, in the **Has Dirty Cup** stage. The path then makes its way around the table over to the sink, and then we have a probability of being in **Has Clean Cup** stage. (Color figure online)

6 Conclusion

Reinforcement learning is an effective strategy for long term activity forecasting, as it permits us to encode common sense prior knowledge in the reward function definition. By encoding common sense terms into the reward function, we can leverage minimal example sequence data in the training algorithm. We demonstrate the effectiveness of this method on our First-Person Office Behaviour Dataset, and show that using deep Q-Learning to take advantage of low level environmental features gives the forecasting algorithm greater accuracy.

One issue with our approach is our manual search for an optimal reward function. Instead of doing this manually, the next logical step would be to use inverse optimal control (IOC). This would make it possible to learn the reward function constants (R_{end}, R_{wall}, and R_{act}). Currently, there are many parameters that need to be tuned manually, and this would only increase if the reward function were to become more complex. There has also been recent research into deep IOC. This is better suited for our problem setting for the same reasons as deep reinforcement learning. This would allow the IOC algorithm to discover which features in the environment are higher value by leveraging the point cloud structure.

Acknowledgement. This research was funded in part by a grant from the Pennsylvania Department of Healths Commonwealth Universal Research Enhancement Program and CREST, JST.

Appendix

Employing deep Q-Learning introduces several complications, such as divergence and convergence speed. As mentioned in [17], employing delayed updates and memory replay (batch q-learning) helps to control divergence quite significantly.

When using a table to record the action-value function it is possible to update the value for any specific state-action pair without modifying any other values. This is not possible with a neural network, as all the hidden units are connected, and gradient descent will update all parameters in some small way. This raises the issue when updating the q-network, that previous iterations' updates may be modified to incorrect values. One technique we employed to provide more stability to the learning process was to sample terminal states and non-terminal states separately during the batch updates. Since the target value for the terminal states do not depend on a network output, they provide stability to the learning process. However since the terminal states are far fewer than the regular states, they do not get selected as often during the batch updates. Sampling them separately ensures that some terminal states are always used to train the q-network on each iteration, lessening the effect other updates have on the values for the terminal states.

References

1. Huang, D.-A., Kitani, K.M.: Action-reaction: forecasting the dynamics of human interaction. In: Fleet, D., Pajdla, T., Schiele, B., Tuytelaars, T. (eds.) ECCV 2014. LNCS, vol. 8695, pp. 489–504. Springer, Heidelberg (2014). doi:10.1007/978-3-319-10584-0_32
2. Vondrick, C., Pirsiavash, H., Torralba, A.: Anticipating the future by watching unlabeled video. arXiv preprint arXiv:1504.08023 (2015)
3. Pellegrini, S., Ess, A., Schindler, K., Van Gool, L.: You'll never walk alone: modeling social behavior for multi-target tracking. In: 2009 IEEE 12th International Conference on Computer Vision, pp. 261–268. IEEE (2009)
4. Walker, J., Gupta, A., Hebert, M.: Patch to the future: unsupervised visual prediction. In: 2014 IEEE Conference on Computer Vision and Pattern Recognition (CVPR), pp. 3302–3309. IEEE (2014)

5. Kitani, K.M., Ziebart, B.D., Bagnell, J.A., Hebert, M.: Activity forecasting. In: Fitzgibbon, A., Lazebnik, S., Perona, P., Sato, Y., Schmid, C. (eds.) ECCV 2012. LNCS, vol. 7575, pp. 201–214. Springer, Heidelberg (2012). doi:10.1007/978-3-642-33765-9_15

6. Karasev, V., Ayvaci, A., Heisele, B., Soatto, S.: Intent-aware long-term prediction of pedestrian motion. In: Proceedings of the IEEE International Conference on Robotics and Automation (2016)

7. Xie, D., Todorovic, S., Zhu, S.C.: Inferring. In: Proceedings of the IEEE International Conference on Computer Vision, pp. 2224–2231 (2013)

8. Huang, D.A., Farahmand, A.M., Kitani, K.M., Bagnell, J.A.: Approximate Max-Ent inverse optimal control and its application for mental simulation of human interactions (2015)

9. Tesauro, G.: TD-Gammon, a self-teaching backgammon program, achieves master-level play. In: Communications of the ACM (1994)

10. Kakade, S., Langford, J.: Approximately optimal approximate reinforcement learning

11. Farahmand, A.M., Szepesvári, C., Munos, R.: Error propagation for approximate policy and value iteration. In: Advances in Neural Information Processing Systems, pp. 568–576 (2010)

12. Munos, R., Szepesvári, C.: Finite-time bounds for fitted value iteration. J. Mach. Learn. Res. **9**, 815–857 (2008)

13. Riedmiller, M.: Neural fitted Q iteration – first experiences with a data efficient neural reinforcement learning method. In: Gama, J., Camacho, R., Brazdil, P.B., Jorge, A.M., Torgo, L. (eds.) ECML 2005. LNCS (LNAI), vol. 3720, pp. 317–328. Springer, Heidelberg (2005). doi:10.1007/11564096_32

14. Farahmand, A.M., Ghavamzadeh, M., Szepesvári, C., Mannor, S.: Regularized fitted Q-iteration for planning in continuous-space Markovian decision problems. In: American Control Conference, ACC 2009, pp. 725–730. IEEE (2009)

15. Sutton, R.S., Barto, A.G.: Introduction to Reinforcement Learning, 1st edn. MIT Press, Cambridge (1998)

16. Adam, S., Busoniu, L., Babuska, R.: Experience replay for real-time reinforcement learning control. IEEE Trans. Syst. Man Cybern. Part C (Appl. Rev.) **42**, 201–212 (2012)

17. Mnih, V., Kavukcuoglu, K., Silver, D., Rusu, A.A., Veness, J., Bellemare, M.G., Graves, A., Riedmiller, M., Fidjeland, A.K., Ostrovski, G., et al.: Human-level control through deep reinforcement learning. Nature **518**, 529–533 (2015)

18. Wu, C.: VisualSFM: a visual structure from motion system (2011)

19. Wu, C., Agarwal, S., Curless, B., Seitz, S.M.: Multicore bundle adjustment. In: 2011 IEEE Conference on Computer Vision and Pattern Recognition (CVPR), pp. 3057–3064. IEEE (2011)

20. Wu, C.: SiftGPU: a GPU implementation of scale invariant feature transform (SIFT) (2007)

Video Summarization Using Deep Semantic Features

Mayu Otani[1(✉)], Yuta Nakashima[1], Esa Rahtu[2],
Janne Heikkilä[2], and Naokazu Yokoya[1]

[1] Graduate School of Information Science,
Nara Institute of Science and Technology, Ikoma, Japan
{otani.mayu.ob9,n-yuta,yokoya}@is.naist.jp
[2] Center for Machine Vision and Signal Analysis, University of Oulu, Oulu, Finland
{erahtu,jth}@ee.oulu.fi

Abstract. This paper presents a video summarization technique for an Internet video to provide a quick way to overview its content. This is a challenging problem because finding important or informative parts of the original video requires to understand its content. Furthermore the content of Internet videos is very diverse, ranging from home videos to documentaries, which makes video summarization much more tough as prior knowledge is almost not available. To tackle this problem, we propose to use deep video features that can encode various levels of content semantics, including objects, actions, and scenes, improving the efficiency of standard video summarization techniques. For this, we design a deep neural network that maps videos as well as descriptions to a common semantic space and jointly trained it with associated pairs of videos and descriptions. To generate a video summary, we extract the deep features from each segment of the original video and apply a clustering-based summarization technique to them. We evaluate our video summaries using the SumMe dataset as well as baseline approaches. The results demonstrated the advantages of incorporating our deep semantic features in a video summarization technique.

1 Introduction

With the proliferation of devices for capturing and watching videos, video hosting services have gained an enormous number of users. According to [1] for example, almost one third of the people online use YouTube to upload or review videos. This increasing popularity of Internet videos has accelerated the demand for efficient video retrieval. Current video retrieval engines usually rely on various types of metadata, including title, user tags, descriptions, and thumbnails, to find videos, which is usually given by video owners. However, such metadata may not

Electronic supplementary material The online version of this chapter (doi:10. 1007/978-3-319-54193-8_23) contains supplementary material, which is available to authorized users.

© Springer International Publishing AG 2017
S.-H. Lai et al. (Eds.): ACCV 2016, Part V, LNCS 10115, pp. 361–377, 2017.
DOI: 10.1007/978-3-319-54193-8_23

be very descriptive to represent the entire content of a video. Moreover, titles and tags are completely up to video owners and so their semantic granularity can vary video by video, or such metadata can even be irrelevant to the content. Consequently users need to review retrieved videos, at least partially, to get rough ideas on their content.

One potential remedy for this comprehensibility problem in video retrieval results is to adopt video summarization, which generates a compact representation of a given video. By providing such summaries as video retrieval results, the users can easily and quickly find desired videos. Video summarization has been one of the major areas in the computer vision and multimedia fields, and a wide range of techniques have been proposed for various goals. Among them, ideal video summarization tailored for the comprehensibility problem should include video content that is essential to tell the story in the entire video. At the same time, it also needs to avoid inclusion of semantically unimportant or redundant content.

To this end, many existing approaches for video summarization extract short video segments based on a variety of criteria that are designed to find essential parts with small redundancy. Examples of such approaches include sampling some exemplars from a set of video segments based on visual features [2,3] and detecting occurrences of unseen content [4]. These approaches mostly rely on low-level visual features, e.g., color histogram, SIFT [5], and HOG [6], which are usually deemed far from the semantics. Some recent approaches utilize higher-level features including objects and identities of people. Their results are promising, but they cannot handle various concepts except a predefined set of concepts, while an Internet video consists of various levels of semantic concepts, such as objects, actions, and scenes. Enumerating all possible concepts as well as designing concept detectors are almost infeasible, which makes video summarization challenging.

This paper presents a novel approach for video summarization. Our approach enjoys recent advent of deep neural networks (DNNs). Our approach segments the original videos into short video segments, for each of which we calculate deep features in a high-dimensional, continuous semantic space using a DNN. We then sample a subset of video segments such that the sampled segments are semantically representative of the entire video content and are not redundant. For sampling such segments, we define an objective function that evaluates representativeness and redundancy of sampled segments. After sampling video segments, we simply concatenate them in the temporal order to generate a video summary (Fig. 1).

To capture various levels of semantics in the original video, deep features play the most important role. Several types of deep features have been proposed recently using convlutional neural networks (CNNs) [7,8]. These deep features are basically trained for a certain classification task, which predicts class labels of a certain domain, such as objects and actions. Being different from these deep features, our deep features need to encode a diversity of concepts to handle a wide range of Internet video contents. To obtain such deep features, we design a DNN to map videos and descriptions to the semantic space and train it with a dataset

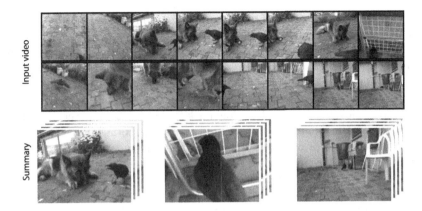

Fig. 1. An example of an input video and a generated video summary. The same content (*i.e.*, the dog) repeatedly appears in the input video in different appearances or background, which may be semantically redundant. Our video summary successfully reduces such redundant video segments, thanks to our deep features encoding higher-level semantics.

consisting of videos and their associated descriptions. Such a dataset contains descriptions like "a man is playing the guitar on stage," which includes various levels of semantic concepts, such as objects ("man", "guitar"), actions ("play"), and a scene ("on stage"). Our DNN is jointly trained using such a dataset so that a pair of a video and its associated sentence gives a smaller Euclidean distance in the semantic space. We use this DNN to obtain our deep features; therefore, our deep features well capture various levels of semantic concepts.

The contribution of this work can be summarized as follows:

– We develop deep features for representing an original input video. In order to obtain features that capture higher level semantics and are well generalized to various concepts, our approach learns video features using their associated descriptions. By jointly training the DNN using videos and descriptions in recently released large-scale video-description dataset [9], we obtain deep features capable of encoding sentence-level semantics.
– We leverage the deep features for generating a video summary. To the best of our knowledge, this is the first attempt to use jointly trained deep features for the video summarization task.
– We represent a video using deep features in a semantic space, which can be a powerful tool for various tasks like video description generation and video retrieval.
– We quantitatively demonstrate that our deep features benefit the video summarization task, comparing ours to deep features extracted using VGG [10].

2   Related Work

Video Summarization. The difficulty in video summarization lies in the definition of "important" video segments to be included in a summary and their extraction. At the early stage of video summarization research, most approaches focus on a certain genre of videos. For example, the importance of a video segment in broadcasting sports program may be easily defined based on the event happening in that segment according to the rules of the sports [11]. Furthermore, a game of some sports (*e.g.*, baseball and American football) has a specific structure that can facilitate important segment extraction. Similarly, characters that appear in movies are also used as domain knowledge [12]. For these domains, various types of metadata (*e.g.*, a textual record of scoring in a game, movie scripts, and closed captions) help to generate video summaries [11–13]. Egocentric videos are another interesting example of video domains, for which a video summarization approach using a certain set of predefined objects as a type of domain knowledge has been proposed [14]. More recent approaches in this direction adopt supervised learning techniques to embody domain knowledge. For example, Potapov *et al.* [15] proposed to summarize a video focusing on a specific event and used an event classifier's confidence score as the importance of a video segment. Such approaches, however, are almost impossible to generalize to other genres because they heavily depend on domain knowledge.

In the last few years, video summarization has been addressed in an unsupervised fashion or without using any domain knowledge. Such approaches introduce the importance of video segment by using various types of criteria and cast video summarization into an optimization problem involving these criteria. Yang *et al.* [16] proposed to utilize an auto-encoder, in which its encoder converts an input video's features into a more compact one, and the decoder then reconstructs the input. The auto-encoder is trained with Internet videos in the same topic. According to the intuition that the decoder can well reconstruct features from videos with frequently appearing content, they assess the segment importance based on the reconstruction errors. Another innovative approach was presented by Zhao *et al.*, which finds a video summary that well reconstructs the rest of the original video. The diversity of segments included in a video summary is an important criterion and many approaches use various definitions of the diversity [3,17,18].

These approaches used various criteria in the objective function, but their contributions have been determined heuristically. Gygli *et al.* added some supervised flavor to these approaches for learning each criterion's weight [19,20]. One major problem of these approaches is that such datasets do not scale because manually creating good video summaries is cumbersome for people.

Canonical views of visual concepts can be an indicator of important video segments, and several existing work uses this intuition for generating a video summary [21–23]. These approaches basically find canonical views in a given video, assuming that results of image or video retrieval using the video's title or keywords as query contain canonical views. Although a group of images or videos retrieved for the given video can effectively predict the importance of video segments, retrieving these images/videos for every input video is expensive and can be difficult because there are only a few relevant images/videos for rare concepts.

For the goal of summarizing Internet videos, we employ a simple algorithm for segment extraction. This is very different from the above approaches that use a sophisticated segment extraction method relying on low-level visual features with manually created video summaries or topic specific data. Due to the dependency of low-level visual features, they do not distinguish semantically identical concepts with different appearances caused by different viewpoints or lighting conditions, and consequently result in semantically redundant video summaries. Instead of designing such a sophisticated algorithm, we focus on designing good features to represent the original video with richer semantics, which can be viewed as the counterpart of sentences' semantics.

Representation Learning. Recent research efforts on CNNs have revealed that the activations of a higher layer of a CNN can be powerful visual features [8,24], and CNN-based image/video representations have been explored for various tasks including classification [8], image/video retrieval [17,25], and video summarization [16,20]. Some approaches learn deep features or metrics between a pair of inputs, possibly in different modalities, using a Siamese network [26,27]. Kiros *et al.* [28] proposed to retrieve image using sentence queries and vice versa by mapping images and sentences into a common semantic space. For doing this, they jointly trained the mappings using video-description pairs and the contrastive loss such that positive pairs (*i.e.*, an image and a relevant sentence) and negative pairs (*i.e.*, an image and a randomly selected irrelevant sentence) give smaller and larger Euclidean distances in the semantic space, respectively.

Inspired by Kiros *et al.*'s work, we develop a common semantic space, which is also jointly trained with pairs of videos and associated sentences (or descriptions). With this joint training, our deep features are expected to encode sentence level semantics, rather than word-or object-level ones. Such deep semantic features can boost the performance of a standard algorithm for important video segment extraction, *i.e.*, clustering-based one, empowering them to cope with higher-level semantics.

3   Approach

Figure 2 shows an overview of our approach for video summarization. We first extract uniform length video segments from the input video in a temporal sliding window manner and compute their deep semantic features using a trained DNN. Inspired by [30], we represent the input video as a sequence of deep features in the semantic space, each of which corresponds to a video segment, as shown in Fig. 3. This representation can encode the semantic transition of the video and thus can be useful for various tasks including video retrieval, video description generation, etc. In Fig. 3, some clusters can be observed, each of which are expected to contain semantically similar video segments. Based on this assumption, our approach picks out a subset of video segments by optimizing an objective function involving the representativeness of the subset.

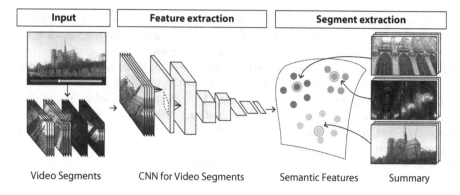

Fig. 2. Our approach for video summarization using deep semantic features. We extract uniform length video segments from an input video. The segments are fed to a CNN for feature extraction and mapped to points in a semantic space. We generate a video summary by sampling video segments that correspond to cluster centers in the semantic space.

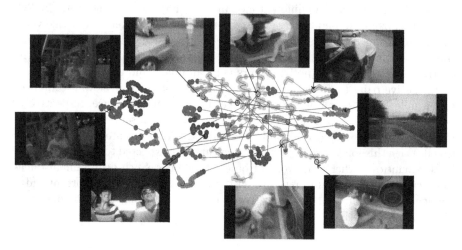

Fig. 3. A two-dimensional plot of our deep features calculated from a video, where we reduce the deep features' dimensionality with t-SNE [29]. Some deep features are represented by the corresponding video segments' keyframes, and the edges connecting deep features represent temporal adjacency of video segments. The colors of deep features indicate clusters obtained by k-means, *i.e.*, points with the same color belong to the same cluster. (Color figure online)

The efficiency of the deep features is crucial in our approach. To obtain good deep features that can capture higher-level semantics, we use the DNN shown in Fig. 4, consisting of two sub-networks to map a video and a description to a common semantic space and jointly train them using a large-scale dataset of videos and their associated descriptions (a sentence). The video sub-network

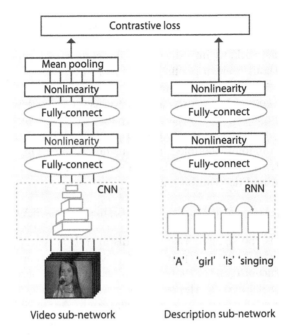

Fig. 4. The network architecture. Video segments and descriptions are encoded into vectors in the same size. Both sub-network for videos and descriptions are trained jointly by minimizing the contrastive loss.

basically is a CNN, and the sentence sub-network is a recurrent neural network (RNN) with some additional layers. We use the contrastive loss function [26] for training, which tries to bring a video and its associated description (a positive pair) closer (*i.e.*, a small Euclidean distance in the semantic space) while a video and a randomly sampled irrelevant description (a negative pair) farther. Being different from other visual features using a CNN trained to predict labels of a certain domain [7,8], our deep features are trained with sentences. Consequently, they are expected to contain sentence-level semantics, including objects, actions, and scenes.

3.1 Learning Deep Features

To cope with higher-level semantics, we jointly train the DNN shown in Fig. 4 with pairs of videos and sentences, and we use its video sub-network for extracting deep features. The video sub-network is a modified version of VGG [10], which is renowned for a good classification performance. In our video sub-network, VGG's classification ("fc8") layer is replaced with two fully-connected layers with hyperbolic tangent (tanh) nonlinearity, which is followed by a mean pooling layer to fuse different frames in a video segment. Let $V = \{v_i | i = 1, \ldots, M\}$ be a video segment, where v_i represents frame i. We feed the frames to the video sub-network and compute a video representation $X \in \mathbb{R}^d$.

For the sentence sub-network, we use skip-thought vector by Kiros *et al.* [28], which encodes a sentence into 4800-dimensional vectors with an RNN. Similarly to the video sub-network, we introduce two fully-connected layers with tanh non-linearity (but without a mean pooling layer) as in Fig. 4 to calculate a sentence representation $Y \in \mathbb{R}^d$ from a sentence S.

For training these sub-networks jointly, we use a video-description dataset (*e.g.*, [9]). We sample positive and negative pairs, where a positive pair consists of a video segment and its associated description, and a negative pair consists of a video and a randomly sampled irrelevant description. Our DNN is trained with the contrastive loss [26], which is defined using extracted features (X_n, Y_n) for the n-th video and description pair as:

$$\text{loss}(X_n, Y_n) = t_n d(X_n, Y_n) + (1 - t_n) \max(0, \alpha - d(X_n, Y_n)), \qquad (1)$$

where $d(X_n, Y_n)$ is the squared Euclidean distance between X_n and Y_n in the semantic space, and $t_n = 1$ if pair (X_n, Y_n) is positive, and $t_n = 0$, otherwise. This loss encourages associated video segment and description to have smaller Euclidean distance in the semantic space, and irrelevant ones to have larger distance. α is a hyperparameter to penalizes irrelevant video segment and description pairs whose Euclidean distance is smaller than α. In our approach, we compute Euclidean distance of positive pairs with initial DNNs before training and employ the largest distance among them as α. This enable most pairs to

Fig. 5. Two-dimensional deep feature embedding with keyframes of corresponding videos, where the feature dimensionality is reduced with t-SNE. The videos located on each colored ellipsis show similar content, *e.g.*, cars and driving people (blue), sports (green), talking people (orange), and cooking (pink). (Color figure online)

be used to update the parameters at the begining of the training. Our DNNs for videos and descriptions can be optimized using the backpropagation technique.

Figure 5 shows a 2D plot of learned deep features, in which the dimensionality of the semantic space is reduced using t-SNE [29] and a keyframe of each video segment is placed at the corresponding position. This plot demonstrates that our deep neural net successfully locates semantically relevant videos at closer points. For example, the group of videos around the upper left area (pink) contains cooking videos, and another group on the lower left (green) shows various sports videos. For video summarization, we use the deep features to represent a video segment.

3.2   Generating Video Summary

Figure 3 shows a two-dimensional plot of deep features from a video, whose dimensionality is reduced again using t-SNE. This example illustrates that a standard method for video summarization, *e.g.*, based on clustering, works well because, thanks to our deep features, video segments with a similar content are concentrated in the semantic space. From this observation, we generate a video summary given an input video by solving the k-medoids problem [20].

In the k-medoids problem, we find a subset $\mathcal{S} = \{S_k | k = 1, \ldots, K\}$ of video segments, which are cluster centers that minimize the sum of the Euclidean distance of all video segments to their nearest cluster centers $S_k \in \mathcal{S}$ and K is a given parameter to determine the length of the video summary. Letting $\mathcal{X} = \{X_j | j = 1, \ldots, L\}$ be a set of deep features extracted from all video segments in the input video, k-medoids finds a subset $\mathcal{S} \subset \mathcal{X}$, that minimizes the objective function defined as:

$$F(\mathcal{S}) = \sum_{X \in \mathcal{X}} \min_{S \in \mathcal{S}} \|X - S\|_2^2. \tag{2}$$

The optimal subset

$$\mathcal{S}^* = \operatorname*{argmin}_{S} F(\mathcal{S}) \tag{3}$$

includes the most representative segments in clusters. As shown in Fig. 5, our video sub-network maps segments with similar semantics to closer points in the semantic space; therefore we can expect that the segments in a cluster have semantically similar content and subset $\mathcal{S}^*$ consequently includes most representative and diverse video segments. The segments in $\mathcal{S}^*$ are concatenated in the temporal order to generate a video summary.

3.3   Implementation Detail

Deep Feature Computation. We uniformly extracted 5-second video segments in a temporal sliding window manner, where the window was shifted by 1 second. Each segment V was re-sampled at 1 frame per second, so V has five frames (*i.e.*, $M = 5$). The activations of VGG's "fc7" layer consists of 4,096 units. We set the unit size of the two fully connected layers to 1,000 and 300 respectively, which means our deep feature is a 300-dimensional vector. For the description sub-network, the fully-connected layers on top of the RNN have the same sizes as the video sub-network's. During the training, we fixed the network parameters of VGG and skip-thought, but those of the top two fully-connected layers for both video and description sub-networks were updated. We sampled 20 negative pairs for each positive pair to compute the contrastive loss. Our DNN was trained over the MSR-VTT dataset [9], which consists of 1 M video clips annotated with 20 descriptions for each. We used Adam [31] to optimize the network parameters with the learning rate of 2^{-4} and trained for 4 epochs.

Video Summarization Generation. Given an input video, we sampled 5-second video segments in the same way as the training of our DNN, and extracted a deep feature from each segment. We then minimize the objective function in Eq. (2) with cost-effective lazy forward selection [19,32]. We set the summary length K to be roughly 15% of the input video's length following [19].

4   Experiment

To demonstrate the advantages of incorporating our deep features in video summarization, we evaluated and compared our approach with some baselines. We used the SumMe dataset [19] consisting of 25 videos for evaluation. As the videos in this dataset are either unedited or slightly edited, unimportant or redundant parts are left in the videos. The dataset includes videos with various contents. It also provides manually created video summaries for each video, with which we compare our summaries. We compute the f-measure that evaluates agreement to reference video summaries using the code provided in [19].

4.1   Baselines

We compared our video summaries with following several baselines as well as recent video summarization approaches: (i) **Manually-created** video summaries are a powerful baseline that may be viewed as the upper bound for automatic approaches. The SumMe dataset provides at least 15 manually-created video summaries whose length is 15% of the original video. We computed the average f-measure of each manually-created video summary with letting each of the rest manually-created video summaries as ground truth (*i.e.*, if there are 20 manually-created video summaries, we compute 19 f-measures for each summary in a pairwise manner and calculate their average). We denote the summary with the highest f-measure among all manually-created video summaries by the best-human video summary. (ii) **Uniform sampling** (Uni.) is widely used baseline for video summarization evaluation. (iii) We also compare to video summaries generated in the same approach as ours except that VGG's "fc7" activations were used instead of our deep features, which is referred to as **VGG**-based video summary. (iv) **Attention-based** video summary (Attn.) is a recently proposed video summarization approach using visual attention [33]. (v) **Interestingness-based** video summary (Intr.) refers to a supervised approach [19], where the weights of multiple objectives are optimized using the SumMe dataset.

4.2 Results

Several examples of video summaries generated with our approach are shown in Fig. 6, along with ratio of annotators who agreed to include each video segments in their manually-created video summary. The peaks of the blue lines indicate that the corresponding video segments were frequently selected to create a video summary. These blue lines demonstrate that human annotators were consistent in some extent. Also we observe that the video segments selected by our approach (green areas) are correlated to the blue lines. This suggests that our approach is consistent with the human annotators.

The results of the quantitative evaluation are shown in the Table 1. In this table, we report the minimum, average, and maximum f-measure scores of manually-created video summaries. Compared to VGG-based summary, ours significantly improved the scores. Our video summaries achieved 58.8% of the

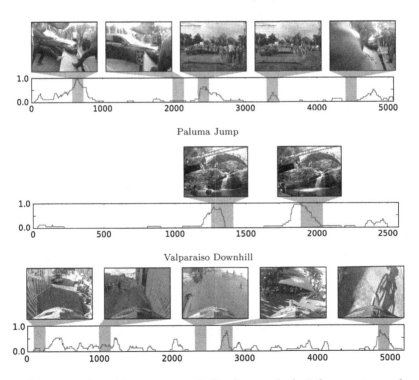

Fig. 6. Segments selected by our approach. Keyframes of selected segments are shown. The green areas in the graphs indicate selected segments. The blue lines represents the ratio of annotators who selected the segment for their manually-created summary. (Color figure online)

Table 1. F-measures of manually-created video summaries and computational approaches (our approach and baselines, higher is better). Since there are multiple manually-created video summaries for each original video and thus multiple f-measures, we show their minimum, mean, and maximum. The best score among the computational approaches are highlighted.

Video	Manually created			Computational approaches				
	Min.	Avg.	Max.	Uni.	VGG	Attn.	Intr.	Ours
Air Force One	0.185	0.332	0.457	0.060	0.239	0.215	**0.318**	0.316
Base Jumping	0.113	0.257	0.396	**0.247**	0.062	0.194	0.121	0.077
Bearpark Climbing	0.129	0.208	0.267	0.225	0.134	**0.227**	0.118	0.178
Bike Polo	0.190	0.322	0.436	0.190	0.069	0.076	**0.356**	0.235
Bus in Rock Tunnel	0.126	0.198	0.270	0.114	0.120	0.112	0.135	**0.151**
Car Railcrossing	0.245	0.357	0.454	0.185	0.139	0.064	**0.362**	0.328
Cockpit Landing	0.110	0.279	0.366	0.103	**0.190**	0.116	0.172	0.165
Cooking	0.273	0.379	0.496	0.076	0.285	0.118	0.321	**0.329**
Eiffel Tower	0.233	0.312	0.426	0.142	0.008	0.136	**0.295**	0.174
Excavators River Crossing	0.108	0.303	0.397	0.107	0.030	0.041	**0.189**	0.134
Fire Domino	0.170	0.394	0.517	0.103	0.124	**0.252**	0.130	0.022
Jumps	0.214	0.483	0.569	0.054	0.000	0.243	**0.427**	0.015
Kids Playing in Leaves	0.141	0.289	0.416	0.051	0.243	0.084	0.089	**0.278**
Notre Dame	0.179	0.231	0.287	0.156	0.136	0.138	**0.235**	0.093
Paintball	0.145	0.399	0.503	0.071	0.270	0.281	**0.320**	0.274
Playing on Water Slide	0.139	0.195	0.284	0.075	0.092	0.124	**0.200**	0.183
Saving Dolphines	0.095	0.188	0.242	0.146	0.103	**0.154**	0.145	0.121
Scuba	0.109	0.217	0.302	0.070	0.160	**0.200**	0.184	0.154
St Maarten Landing	0.365	0.496	0.606	0.152	0.153	**0.419**	0.313	0.015
Statue of Liberty	0.096	0.184	0.280	0.184	0.098	0.083	**0.192**	0.143
Uncut Evening Flight	0.206	0.350	0.421	0.074	0.168	**0.299**	0.271	0.168
Valparaiso Downhill	0.148	0.272	0.400	0.083	0.110	0.231	0.242	**0.258**
Car over Camera	0.214	0.346	0.418	0.245	0.048	0.201	**0.372**	0.132
Paluma Jump	0.346	0.509	0.642	0.058	0.056	0.028	0.181	**0.428**
Playing Ball	0.190	0.271	0.364	0.123	0.127	0.140	0.174	**0.194**
Mean f-measure	0.179	0.311	0.409	0.124	0.127	0.167	**0.234**	0.183
Relative to human avg.	0.576	1.000	1.315	0.398	0.408	0.537	**0.752**	0.588
Relative to human max.	0.438	0.760	1.000	0.303	0.310	0.408	**0.572**	0.447

average score of manually-created video summaries, while VGG-based got 40.8%. This result demonstrates the advantage of our deep features for creating video summaries.

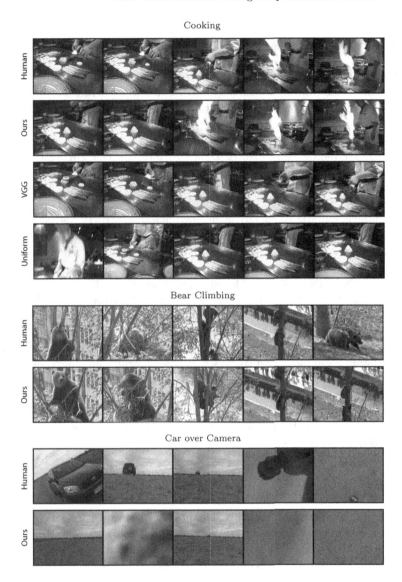

Fig. 7. Uniformly sampled frames of summaries by different approaches. "Human" means the best-human video summary. The full results of "Bear Climbing" and "Car over Camera" are shown in the supplementary material.

One of the recent video summarization approaches, *i.e.*, interestingness-based one [19], got the highest score in this experiment. Note that the interestingness-based approach [19] uses a supervised technique, in which the mixture weights of various criteria in their objective function are optimized over the SumMe dataset. Our video summaries were generated using a relatively simple algorithm to extract a subset of segments; nevertheless, ours outperformed the interestingness-based for some videos, and even got a better mean f-measure score than attention-based.

Our approach got low scores, especially for short videos, such as "Jumps" and "Fire Domino." Since we extract uniform length segments (5 second), in the case of short videos, our approach only extracts a few segments. This may result in a lower f-measure score. This limitation can be solved by extracting shorter video segments or using more sophisticated video segmentation like [12,19].

We also observed that our approach got lower scores than others on the "St Maarten Landing" and "Notre Dame," which are challenging because of long unimportant parts and diversity of content, respectively. For "St Maarten Landing," as our approach is unsupervised, it failed to exclude unimportant segments. For "Notre Dame," generating a summary is difficult because there are too many possible segments to be included in a summary. While our summary shares small parts with manually created summaries, it is a challenging example even for human annotators, which is shown in the low scores of manually-created video summaries.

Figure 7 shows examples of video summaries created with our approach and baselines. The video "Cooking" shows a person cooking some vegetables while doing a performance. Ours and the best-human video summary include the same scene of the performance with fire, while others do not. On the other hand, ours extracts unimportant segments from the video "Car over Camera." The original video is highly redundant with static scenes just showing the ground or the sky, and such scenes make up large clusters in the semantic space even if they are unimportant. As our approach extracts representatives from each cluster, a video with lengthy unimportant parts resulted in a poor video summaries. We believe that this problem can be avoided by using visual cues such as interestingness [34] and objectiveness [35].

5   Conclusion

In this work, we proposed to learn semantic deep features for video summarization and a video summarization approach that extracts a video summary based on the representativeness in the semantic feature space. For deep feature learning, we designed a DNN with two sub-networks for videos and descriptions, which are jointly trained using the contrastive loss. We observed that learned features extracted from videos with similar content make clusters in the semantic space. In our approach, the input video is represented by deep features in the semantic space, and segments corresponding to cluster centers are extracted to generate a video summary. By comparing our summaries to manually created summaries, we shown that the advantage of incorporating our deep features in a video summarization technique. Furthermore, our results even outperformed the worst human created summaries. We expect that the quality of video summaries will be improved by incorporating video segmentation methods. Moreover, our objective function can be extended by considering other criteria used in the area of video summarization, such as interestingness and temporal uniformity.

Acknowledgement. This work is partly supported by JSPS KAKENHI No. 16K16086.

References

1. YouTube.com: Statistics-YouTube (2016). https://www.youtube.com/yt/press/en-GB/statistics.html
2. Gong, Y., Liu, X.: Video summarization using singular value decomposition. In: Proceedings of IEEE Computer Society Conference Computer Vision and Pattern Recognition (CVPR), pp. 174–180 (2000)
3. Gong, B., Chao, W.L., Grauman, K., Sha, F.: Diverse sequential subset selection for supervised video summarization. In: Proceedings of Advances in Neural Information Processing Systems (NIPS), pp. 2069–2077 (2014)
4. Zhao, B., Xing, E.P.: Quasi real-time summarization for consumer videos. In: Proceedings of IEEE Computer Society Conference Computer Vision and Pattern Recognition (CVPR), pp. 2513–2520 (2014)
5. Lowe, D.G.: Distinctive image features from scale invariant keypoints. Int. J. Comput. Vis. **60**, 91–11020042 (2004)
6. Dalal, N., Triggs, B.: Histograms of oriented gradients for human detection. In: Proceedings of IEEE Computer Society Conference Computer Vision and Pattern Recognition (CVPR), pp. 886–893 (2005)
7. Yao, L., Ballas, N., Larochelle, H., Courville, A.: Describing videos by exploiting temporal structure. In: Proceedings of IEEE International Conference Computer Vision (ICCV), pp. 4507–4515 (2015)
8. Donahue, J., Jia, Y., Vinyals, O., Hoffman, J., Zhang, N., Tzeng, E., Darrell, T.: DeCAF: a deep convolutional activation feature for generic visual recognition. In: Proceedings of International Conference Machine Learning (ICML), vol. 32, pp. 647–655 (2014)
9. Xu, J., Mei, T., Yao, T., Rui, Y.: MSR-VTT: a large video description dataset for bridging video and language. In: Proceedings of IEEE Computer Society Conference Computer Vision and Pattern Recognition (CVPR), pp. 5288–5296 (2016)
10. Simonyan, K., Zisserman, A.: Very deep convolutional networks for large-scale image recoginition. In: Proceedings International Conference Learning Representations (ICLR), pp. 14 (2015)
11. Babaguchi, N., Kawai, Y., Ogura, T., Kitahashi, T.: Personalized abstraction of broadcasted American football video by highlight selection. IEEE Trans. Multimed. **6**, 575–586 (2004)
12. Sang, J., Xu, C.: Character-based movie summarization. In: Proceedings of ACM International Conference Multimedia (MM), pp. 855–858 (2010)
13. Evangelopoulos, G., Zlatintsi, A., Potamianos, A., Maragos, P., Rapantzikos, K., Skoumas, G., Avrithis, Y.: Multimodal saliency and fusion for movie summarization based on aural, visual, and textual attention. IEEE Trans. Multimed. **15**, 1553–1568 (2013)
14. Lu, Z., Grauman, K.: Story-driven summarization for egocentric video. In: Proceedings of IEEE Computer Society Conference Computer Vision and Pattern Recognition (CVPR), pp. 2714–2721 (2013)
15. Potapov, D., Douze, M., Harchaoui, Z., Schmid, C.: Category-specific video summarization. In: Fleet, D., Pajdla, T., Schiele, B., Tuytelaars, T. (eds.) ECCV 2014. LNCS, vol. 8694, pp. 540–555. Springer, Heidelberg (2014). doi:10.1007/978-3-319-10599-4_35

16. Yang, H., Wang, B., Lin, S., Wipf, D., Guo, M., Guo, B.: Unsupervised extraction of video highlights via robust recurrent auto-encoders. In: Proceedings of IEEE International Conference Computer Vision (ICCV), pp. 4633–4641 (2015)

17. Xu, J., Mukherjee, L., Li, Y., Warner, J., Rehg, J.M., Singh, V.: Gaze-enabled egocentric video summarization via constrained submodular maximization. In: Proceedings of IEEE Computer Society Conference Computer Vision and Pattern Recognition (CVPR), pp. 2235–2244 (2015)

18. Tschiatschek, S., Iyer, R.K., Wei, H., Bilmes, J.A.: Learning mixtures of submodular functions for image collection summarization. In: Proceedings of Advances in Neural Information Processing Systems (NIPS), pp. 1413–1421 (2014)

19. Gygli, M., Grabner, H., Riemenschneider, H., Gool, L.: Creating summaries from user videos. In: Fleet, D., Pajdla, T., Schiele, B., Tuytelaars, T. (eds.) ECCV 2014. LNCS, vol. 8695, pp. 505–520. Springer, Heidelberg (2014). doi:10.1007/978-3-319-10584-0_33

20. Gygli, M., Grabner, H., van Gool, L.: Video summarization by learning submodular mixtures of objectives. In: Proceedings of IEEE Computer Society Conference Computer Vision and Pattern Recognition (CVPR), pp. 3090–3098 (2015)

21. Song, Y., Vallmitjana, J., Stent, A., Jaimes, A.: TVSum: summarizing web videos using titles. In: Proceedings of IEEE Computer Society Conference Computer Vision and Pattern Recognition (CVPR), pp. 5179–5187 (2015)

22. Khosla, A., Hamid, R., Lin, C.j., Sundaresan, N.: Large-scale video summarization using web-image priors. In: Proceedings of IEEE Computer Society Conference Computer Vision and Pattern Recognition (CVPR), pp. 2698–2705 (2013)

23. Chu, W.S., Jaimes, A.: Video co-summarization: video summarization by visual co-occurrence. In: Proceedings of IEEE Computer Society Conference Computer Vision and Pattern Recognition (CVPR), pp. 3584–3592 (2015)

24. Wang, X., Gupta, A.: Unsupervised learning of visual representations using videos. In: Proceedings of IEEE International Conference Computer Vision (ICCV), pp. 2794–2802 (2015)

25. Frome, A., Corrado, G., Shlens, J.: DeViSE: a deep visual-semantic embedding model. In: Proceedings of Advances in Neural Information Processing Systems (NIPS), pp. 2121–2129 (2013)

26. Chopra, S., Hadsell, R., LeCun, Y.: Learning a similarity metric discriminatively, with application to face verification. In: Proceedings of IEEE Computer Society Conference Computer Vision and Pattern Recognition (CVPR), pp. 539–546 (2005)

27. Lin, T.Y., Belongie, S., Hays, J.: Learning deep representations for ground-to-aerial geolocalization. In: Proceedings of IEEE Computer Society Conference Computer Vision and Pattern Recognition (CVPR), pp. 5007–5015 (2015)

28. Kiros, R., Zhu, Y., Salakhutdinov, R.R., Zemel, R., Urtasun, R., Torralba, A., Fidler, S.: Skip-thought vectors. In: Proceedings of Advances in Neural Information Processing Systems (NIPS), pp. 3276–3284 (2015)

29. Maaten, L.V.D., Hinton, G.E.: Visualizing high-dimensional data using t-SNE. J. Mach. Learn. Res. 9, 2579–2605 (2008)

30. DeMenthon, D., Kobla, V., Doermann, D.: Video summarization by curve simplification. In: Proceedings of ACM International Conference Multimedia (MM), pp. 211–218 (1998)

31. Kingma, D., Ba, J.: Adam: A method for stochastic optimization. In: Proceedings of Internatonal Conference Learning Representations (ICLR), pp. 11 (2015)

32. Leskovec, J., Krause, A., Guestrin, C., Faloutsos, C., VanBriesen, J., Glance, N.: Cost-effective outbreak detection in networks. In: Proceedings of ACM SIGKDD International Conference Knowledge Discovery and Data Mining (KDD), pp. 420–429 (2007)

33. Ejaz, N., Mehmood, I., Wook Baik, S.: Efficient visual attention based framework for extracting key frames from videos. Sig. Process.: Image Commun. **28**, 34–44 (2013)

34. Gygli, M., Grabner, H., Riemenschneider, H., Nater, F., Gool, L.V.: The interestingness of images. In: IEEE International Conference Computer Vision (ICCV), pp. 1633–164 (2013)

35. Alexe, B., Deselaers, T., Ferrari, V.: What is an object? In: Proceedings of IEEE Computer Society Conference Computer Vision and Pattern Recognition (CVPR), pp. 73–80 (2010)

Towards Segmenting Consumer Stereo Videos: Benchmark, Baselines and Ensembles

Wei-Chen Chiu[1(✉)], Fabio Galasso[2], and Mario Fritz[1]

[1] Max Planck Institute for Informatics, Saarland Informatics Campus,
Saarbrücken, Germany
walon@mpi-inf.mpg.de
[2] OSRAM Corporate Technology, Munich, Germany

Abstract. Are we ready to segment *consumer stereo videos*? The amount of this data type is rapidly increasing and encompasses rich information of appearance, motion and depth cues. However, the segmentation of such data is still largely unexplored. First, we propose therefore a new benchmark: videos, annotations and metrics to measure progress on this emerging challenge. Second, we evaluate several state of the art segmentation methods and propose a novel ensemble method based on recent spectral theory. This combines existing image and video segmentation techniques in an *efficient* scheme. Finally, we propose and integrate into this model a novel regressor, learnt to optimize the stereo segmentation performance directly via a differentiable proxy. The regressor makes our segmentation ensemble *adaptive* to each stereo video and outperforms the segmentations of the ensemble as well as a most recent RGB-D segmentation technique.

1 Introduction

We witness a fast growing number of stereo streams on the web, due to the advent of consumer stereo video cameras. Are we ready to expoit the rich cues which stereo videos deliver? Our work focuses on segmentation of such data sources, as it is a common pre-processing step for further analysis such as action [1–3] or scene classification [4].

We propose a new consumer stereo video challenge, to understand the opportunities and foster the research in this new area. The new type of data combines the availability of appearance and motion with the possibility of extracting depth information. Considering consumer videos means addressing a most abundant web data, which is however also very heterogeneous, due to a variety of consumer cameras.

The new consumer stereo video challenge explicitly concerns the semantics of the video. A number of existing benchmarks have offered ground truth depth

Electronic supplementary material The online version of this chapter (doi:10. 1007/978-3-319-54193-8_24) contains supplementary material, which is available to authorized users.

© Springer International Publishing AG 2017
S.-H. Lai et al. (Eds.): ACCV 2016, Part V, LNCS 10115, pp. 378–395, 2017.
DOI: 10.1007/978-3-319-54193-8_24

and motion, recurring to controlled recordings [5] or computer graphics simulations [6]. By contrast, here we address stereo videos *in the wild* and specifically consider the semantics of the data. While this might partly harm analysis (no true depth available), it addresses directly what we are most interested in, the actors and objects in the videos.

We warm-start the challenge with a number of baselines, extending best available image and video segmentations to the consumer stereo videos and their available features, e.g. color, motion and depth. Most baselines perform well on some videos, however none performs well on all. As an example, motion segmentation techniques [7] perform well while the object moves, but encounter difficulty with static video shots. On the other hand, camouflaged (but moving) objects impinge appearance-based image [8] and video [9] segmentation techniques.

Thus motivated, we introduce in Sect. 5 a new efficient segmentation ensemble model, which leverages existing results where they perform best. Furthermore, we introduce in Sect. 6 the framework to learn a regressor which adapts the ensemble model to each particular stereo video. The proposed technique is overviewed in Sect. 4 and demonstrated in Sect. 7. Although *only* combining optimally existing results, our new algorithm outperforms a most recent RGB-D segmentation technique [10].

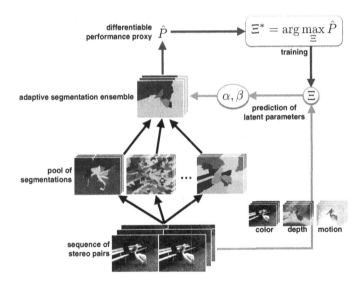

Fig. 1. Overview of the proposed efficient adaptive stereo segmentation technique. Our proposed segmentation ensemble model leverages the best available image and video segmentation results efficiently. A regressor makes the ensemble model adaptive to each stereo video, based on color, depth and motion features. In our novel learning framework, the segmentation performance is optimized via a differentiable proxy. (Color figure online)

2   Related Work

Video Segmentation. Video segmentation has recently received much attention. It strikes how diversely the segmentation problem is defined. [11] looks at
the problem of motion segmentation by using optical-flow long term trajectories.
[12] also uses trajectories (defined densely with superpixel-regions) and looks at
motion but focuses on people motion. [13] considers appearance to generate several image proposals and tracks the most temporally consistent ones with motion.
[9] addresses general unsupervised video segmentation based on appearance and
motion. [14] introduces an unsupervised, geodesic distance based, salient video
object segmentation method. [15] proposes a non-local consensus voting scheme
defined over a graph of similar regions in the video sequence. We note that:
1. the strong diversity of existing algorithms hardly allows combining their
results; 2. none of those techniques may seamlessly generalize to stereo videos.
This work proposes a solution to both aspects.

Depth/Stereo Segmentation. There is a long tradition of work on 3D reconstruction which estimates 3D coordinates, thus depth, from pair or multiple
views [16,17]. These efforts have been recently combined with reasoning on the
object appearance and the physical constraints of the 3D scene in the work of
[18], whereby segmentation proposals are produced for semantic objects. The
underlying assumption of a static scene for these methods does not allow their
extension to stereo video sequences which we consider here. Additionally, the
use of consumer cameras contrasts the high quality images which they generally
require.

Scene Flow and Stereo Videos. Recent work addresses scene flow, the joint
estimation of optical flow and depth, assuming calibrated [19,20] or uncalibrated
cameras [21]. Those do not address segmentation. Elsewhere, video segmentation is addressed by considering RGB-D information [10,22–24], Kinect colour
images with depth. The stereo videos which we consider are not assumed calibrated nor from the same camera. Since we consider consumer cameras, the
videos are further unconstrained in terms of spatio-temporal resolution, zooming, sensitivity and dynamical range, and present challenging motion blur effects
and image latency.

Image/Video Co-Segmentation. Recent researches on image and video co-
segmentation [25–28] tasks provide a way to jointly extract common objects
across multiple images or videos. The general assumption of co-segmentation
problem on commonality of objects in a video set makes it a plausible fit to
stereo video segmentation task when we consider left and right videos of stereo
pairs as two separate sequences in the same set. However, from this perspective
the depth cue in the stereo videos is not explicitly explored which can provide
rich information to outline objects while other features are with ambiguities.

Algorithm Selection and Combination. There has been previous work
which attempted to select the best algorithm from a candidate pool depending on the specific task e.g. for recognition on a budget [29,30] or active learning
[31]. For optical flow, [32] presented a supervised learning approach to predict

the most suitable algorithm, based on the confidence measures of the optical flow estimates. By contrast, Bai et al. [33] shows an object cutout system where the segmentation is done by a set of local classifiers, each adaptively *combining* multiple local features. Our proposed technique not only *combines* different feature cues but also the available segmentation algorithms, weighting their contributions rather than selecting one. We show experimentally that combination provides better results. Our proposed approach relates closely to [8] which efficiently combines image segmentation algorithms within the spectral clustering framework outperforming the original results in the pool. Our extension of [8] is twofold: 1. we generalize from image to stereo video segmentation; 2. we design and learn discriminatively an *adaptive scheme* which allows to combine optimally the pool of video and segmentation algorithms *at each superpixel*, based on the appearance and motion features of the local pixels. Experimentally, the adaptive scheme further improves on the static combination.

3   Consumer Stereo Video Segmentation Challenge (CSVSC)

We launch a Consumer Stereo Video Segmentation Challenge (CSVSC). The new dataset consists of 30 video sequences which we have selected from Youtube based on their heterogeneity. In fact, the footage differs in the number of objects (2–15), the kind of portrayed actors (animals or people), the type of motion (a few challenging stop-and-move scenes and objects entering or exiting the scene), the appearance visual complexity (also in relation to the background, a few objects may be harder to discern) and the distance of the objects from the camera (varying disparity and thus depth). Not less importantly, we have selected videos acquired by different consumer stereo cameras, which implies diverse camera intrinsic parameters, zooms and (as a further challenge) noise degradations such as motion blurs and camera shake. We illustrate a few sample sequences in Fig. 2.

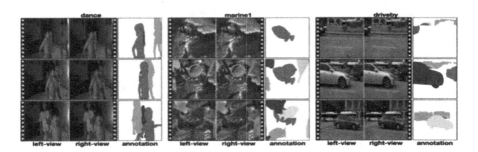

Fig. 2. Sample frames from the Consumer Stereo Video Segmentation Challenge (CSVSC) dataset (left-right views) with the corresponding annotations. The stereo videos differ in content (appearance, motion, number and type of subjects) and in camera characteristics (intrinsic parameters, zoom, noise).

Benchmark Annotation and Metrics. We have gathered human annotations and defined metrics to quantify progress on the new benchmark. In particular, we equidistantly sample 5 frames from the left view of all 30 videos to be labelled (150 frames are labelled in whole benchmark, while stereo videos altogether total to 1738 left-right-pair frames).

As for the metrics, we have considered state-of-the-art image [34] and video segmentation [35] metrics:

Boundary Precision-Recall (BPR). This reflects the per-frame boundary alignment between a video segmentation solution and the human annotations. In particular, BPR indicates the F-measure between recall and precision [34].

Volume Precision-Recall (VPR). This measures the video segmentation property of temporal consistency. As for BPR, VPR also indicates the F-measure between recall and precision [35].

It is of research interest to determine which of the metrics is best for learning. We answer this question in Sect. 7, where we consider BPR and VPR alone or combined by their: **arithmetic mean (AM-BVPR)** or **harmonic mean (HM-BVPR)**, formulated respectively as $\frac{\text{BPR}+\text{VPR}}{2}$ and $\frac{2 \cdot \text{BPR} \cdot \text{VPR}}{\text{BPR}+\text{VPR}}$.

Preparation of Stereo Videos. Not having a ground truth depth may impinge comparison among techniques applied to the dataset. We define therefore an initial set of comparisons among depth-estimation algorithms and make the results available.

We have considered the per-frame rectification of [38] and the stereo matching algorithm of [36], filling-in the missing correspondences with [39]. Furthermore, we have estimated depth by the optical flow algorithm of [37] between the right and left views. We illustrate samples in Fig. 3. Our initial findings are that estimating depth by optical flow leads to best downstream stereo segmentation outputs, which we use therefore in the rest of the paper.

Fig. 3. Sample disparity estimation. The first two columns are the original stereo pair and their rectified images. The top-right picture is the disparity map computed by [36], the bottom-right is the depth map obtained by optical flow [37] between the left and right view.

4   Efficient Adaptive Segmentation of Stereo Videos

We warm start the CSVSC challenge with a basic segmentation ensemble model. To this purpose we first pre-process the stereo videos with a pool of state-of-the-art image and video segmentation algorithms. Then we combine the segmentation outputs with a new efficient segmentation ensemble model (cf. Sect. 5). Finally, we propose the learning framework to adapt the combination parameters of each stereo video (cf. Sect. 6).

Figure 1 gives an overview of our ensemble model:

Pool of Image and Video Segmentations. We select most recent algorithms which are available online. These are used to segment the single frames (image segmentations) and the left views of the stereo videos (video segmentations). This results in a pool of segments which are respectively superpixels and supervoxels.

Efficient Segmentation Ensemble Model. We bring together the pool of segments and connect them to the stereo video voxels. The segmentation ensemble model is represented by a graph and parameterized by α and β's, which weight the contribution of each segmentation method. The model is accurate (voxel-based) but costly. We propose therefore an efficient graph reduction which is exact, i.e. it provides the same solutions as the voxel-based at a lower computational complexity.

Performance-Driven Adaptive Combination. We compute stereo video features from the stereo videos based on color, flow and depth. From these features, we regress the combination parameters α and β's, i.e. we combine optimally the pooled segmentation outputs. To this purpose, we propose a novel regressor Ξ and an inference procedure, to learn the optimal regression parameters ξ from data. For the first time in literature, the regressor parameters ξ are directly optimized according to the final performance measure P (resulting from the graph partitioning and the metric evaluation, cf. Sect. 6.2). We achieve this with a novel differentiable performance proxy $\hat{P}$.

None of the state-of-the-art segmentation algorithms performs well with all of the challenging consumer stereo videos (cf. experiments in Sect. 7). Both the contributions on the ensemble model (Sect. 5) and the performance-driven adaptive combination (Sect. 6) turn out important for better results.

5   Efficient Segmentation Ensemble Model

We propose a graph for bringing together the available video segmentation outputs. Additionally, we propose the use of recent spectral techniques to reduce the voxel graph to one based on tailored superpixels/supervoxels, to improve efficiency without any (proven) compromise on performance. The graph partitioning with spectral clustering provides the segmentation output.

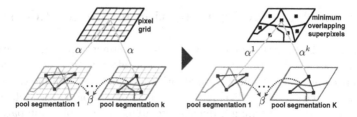

Fig. 4. Proposed video segmentation model. A number of K pooled image and video segmentation outputs are brought together as hypotheses of grouping for the considered video sequence (cf. Sect. 5 for details). We propose to replace the model of [8] (*left*) with a new one (*right*) based on minimally overlapping superpixels, which is provably equivalent but yields better efficiency (cf. Sect. 5.1)

5.1   Unifying Graph

Given a number of video segmentation outputs, we propose to bring all of them together by defining a unifying graph.

Let us consider Fig. 4 *left*. Each video segmentation algorithm provides groupings of the video sequence voxels. In the unifying graph, each pixel is therefore linked to the groupings to which it belongs. For example, one algorithm may compute spatio-temporal tubes (supervoxels) [9], another one may compute image-based superpixels [34]. The video sequence voxels would then be linked to the tube to which they belong (temporally) and to their superpixels (spatially). Altogether, the outputs from the pool of video segmentations provide hypotheses of grouping for the video voxels.

More formally, we define a graph $\mathcal{G} = (\mathcal{V}, \mathcal{E})$ to jointly represent the video and the segmentation outputs. Nodes from the vertex set $\mathcal{V}$ are of two kinds:

Voxels are the video sequence elements which we aim to segment;
Pooled segmentation outputs are the computed spatial- and/or temporal-groupings, providing voxel grouping hypotheses.

Further to being connected to the voxels, the pooled groupings from the same output are also connected to their neighbors, which defines the video volume structure. Edges are therefore of two types:

β-**edges** are between the groupings of each pooled segmentation k; we assume C features (appearance, motion, etc. cf. Sect. 4) and distances based on β^c-weighted features: $w_{I,J}^k = e^{-(\beta^1 d_{I,J}^{k1} + \cdots + \beta^C d_{I,J}^{kC})}$, where $d_{I,J}^{k_c}$ is the distance between superpixels I and J from the k-th pooled output based on c-th feature.

α-**edges** are between the voxels and the grouping that it belongs to. The α^k's encode the trust towards the respective K segmentation algorithm, ideally proportional to its accuracy. In other words, the weights of *alpha* edges correspond to the importance of individual segmentations, thus a higher α^k can be interpreted as a larger contribution to the overall performance, and shows the importance of individual segmentations.

Partitioning graph $\mathcal{G}$ with spectral clustering is computationally demanding as the number of nodes (and edges) depends linearly on the video voxels. The theory of [8] reduces the complexity of a first stage of spectral clustering (the eigendecomposition) but not of the second one (k-means), still of linear complexity in the number of voxels (and thus bottleneck of [8]). We address both with graph reduction in the following Section.

5.2 Improved Efficiency with Graph Reduction

Let us consider again Fig. 4. A huge number of voxels are similar both in appearance and in motion and are therefore grouped in all segmentation outputs. When partitioning the original graph $\mathcal{G}$, these voxels are always segmented together. (The trivial proof leverages their equal edges and therefore eigenvectors).

Rather than considering all voxels, we propose to reduce the original graph $\mathcal{G}$ to one of smaller size $\mathcal{G}^Q$ which is equivalent (provides exactly the same clustering solutions). In $\mathcal{G}^Q$, we basically group all those voxels with equal connections into super-nodes (reweighting their edges equivalently). This reduces the algorithmic complexity, as the spectral clustering (both the eigendecomposition and the k-means) now depends only on the number of super-nodes (which is determined for most pooled segmentation algorithms by the number of objects, rather than voxels).

We identify the voxels with equal connections by intersecting the available segmentation outputs. The result of the intersection is an oversegmentation into superpixels, which can generate all pooled segmentation outputs by merging. We name these *minimally overlapping superpixels*.

More formally, the reduced graph $\mathcal{G}^Q = (\mathcal{V}^Q, \mathcal{E}^Q)$ takes the minimally overlapping superpixels as nodes $\mathcal{V}^Q$ and the following edge weights

$$
w_{IJ}^Q = \begin{cases} \displaystyle\sum_{i \in I} \sum_{j \in J} w_{ij} & \text{if } I \neq J \\ \displaystyle\frac{1}{|I|} \sum_{i \in I} \sum_{j \in J} w_{ij} - \frac{(|I|-1)}{|I|} \sum_{i \in I} \sum_{j \in \mathcal{V} \setminus I} w_{ij} & \text{if } I = J \end{cases} \tag{1}
$$

where $|\cdot|$ indicates the number of pixels within the superpixel, I, J are two minimally overlapping superpixels, and w_{ij} stands for a generic edge of the original graph. According to (1), two pixels i and j are reduced if belonging to the same superpixel, i.e. if $I = J$. These self-edges are of great importance, because spectral clustering normalizes clusters by their accumulated volumes of merged pixels, i.e. summations of merged α's. (Since the superpixel connections are equal for the pixels within the same superpixel by construction, the reduction is exact, cf. [40].)

5.3 Implementation Details

The output segmentation is obtained by graph partitioning $\mathcal{G}^Q$ with spectral clustering [41–43]. In particular, the labels of the minimally overlapping superpixels provide the voxel labels and thus the video segmentation solution.

In this work, we use $K = 6$ image and video segmentation algorithms: (1) The hierarchical image segmentation of [34]. We choose one layer from hierarchy based on best performance on a validation set. We take three segmentation outputs by applying the Simple Linear Iterative Clustering (SLIC) [44] respectively on (2) depth, (3) optical-flow [37] and the (4) LAB-color coded cues, bilaterally filtered for noise removal and edge preservation [45]. (5) Hierarchical graph-based video segmentation (GBH) [9]. We choose one layer from the hierarchy on the validation set. (6) The motion segmentation technique (moseg) of [7].

While the features are computed on the stereo video. The graph is constructed on one of the two views (the left one) of the stereo videos, which is then evaluated for the segmentation quality. The contribution of segmentation outputs is weighted by α. β defines the affinities between superpixels/supervoxels from the same pooled segmentation output, weighting $C = 3$ feature cues based on mean Lab-color, depth and motion.

Note the importance of α's and β's in the graph $\mathcal{G}$ and therefore $\mathcal{G}^Q$. These parameters define how much each pooled segmentation output is trusted and how to compute the similarity among superpixels/supervoxels in these outputs. Such parameters can be defined statically (cf. [8]) or adjusted dynamically in a data-dependent fashion, as we propose in the next Section.

6   Performance-Driven Adaptive Combination

We propose a regressor Ξ to estimate the optimal segmentation ensemble parameters α and β from the appearance-, motion- and depth-based features of the stereo videos. (Cf. Fig. 1 where the regressor is given by the red arrows). Furthermore, we propose a novel inference framework to learn the regressor parameters ξ from the training stereo videos. (Cf. Fig. 1 where the training is represented with blue arrows). A new differentiable performance proxy $\hat{P}$ enables optimization driven by the stereo video segmentation performance measure P.

6.1   Adaptive Combination by Regression

Let us define a regressor Ξ, with parameters ξ. Ξ takes as input a set of features $\mathcal{F}$ computed from the stereo video and outputs the parameters α and β for the ensemble segmentation model (i.e. the coefficients to optimally combine K segmentation outputs from the pool based on C features, cf. Sect. 5.1). Intuitively, the regressor should select the best segmentation outputs from the pool, based on the stereo video content. This would imply, for example, a larger trust towards image- rather than motion-segmentation outputs, for those stereo videos where no motion is present.

While any type of regressor could be adopted, here in our approach we employ a second order regressor Ξ which we parameterize by a matrix B. Overall, α and β are computed as:

$$(\alpha^1, \ldots, \alpha^K, \beta^1, \ldots, \beta^C) = \Xi(\mathcal{F}; \xi) = \mathcal{F}^T B \mathcal{F} \qquad (2)$$

where B is learnt with least squares and the input-output pairs of $\mathcal{F}$ and (α, β), and ξ are the regression coefficients contained in B. We consider in $\mathcal{F}$ features based on appearance, motion and depth. A large feature set is important to allow the regressor to understand the type of stereo video (dynamic, static, textured etc.) For each feature, we compute therefore histograms, means, medians, variances and entropies. We would leave the learning framework to choose from the right feature, i.e. training the best regressor Ξ. This should ideally consider the system performance P for optimization or the tractable differentiable proxy which we discuss next.

6.2 Performance-Driven Regressor Learning by Differentiable Proxies

Let us consider Fig. 1. The α and β, regressed by Ξ according to features $\mathcal{F}$, correspond to a stereo video segmentation performance P. During *training*, we seek to optimize Ξ for the maximum segmentation performance P:

$$\hat{\Xi} = \max_{\Xi} P(\Xi(\mathcal{F})) \tag{3}$$

There are two main obstacles to our goal. First, typical video segmentation performance metrics are not differentiable and therefore do not lend themselves to directly optimizing an overall performance. To address this, we propose a differentiable performance proxy $\hat{P}$ in Sect. 6.2.

Second, α and β are not part of the objective (3) and have to be considered *latent*. In Sect. 6.2, we define therefore an EM-based strategy to jointly learn the regressor Ξ, α and β. An overview of our training procedure is given in Algorithm 1.

Algorithm 1. Joint learning of the regressor Ξ and latent combination weights α, β

Require: $\forall$ training videos with initial set of parameters (α, β) and stereo video features $\mathcal{F}$
1: **repeat**
2:     Given the current estimates of (α, β), train the Ξ which regresses them from $\mathcal{F}$
3:     **for all** training video **do**
4:         predict $(\alpha'', \beta'') = \Xi(\mathcal{F})$;
5:         use (α'', β'') as initialization for $(\hat{\alpha}, \hat{\beta}) = \arg\max_{\alpha, \beta} \hat{P}(\alpha, \beta)$;
6:         update (α, β) for the training video by $(\hat{\alpha}, \hat{\beta})$;
7:     **end for**
8: **until** Convergence or max. iterations exceeded

Metric Specific Performance Proxy. In image segmentation, performance is generally measured by boundary precision recall (BPR) and its associated best F-measure [34]. In video segmentation, benchmarks additionally include volume precision recall (VPR) metrics [35]. Both these performance measures are plausible P, but neither of them is differentiable, which complicates optimization. (We experiment on various performance measures in Sect. 7.)

We propose to estimate a differentiable performance proxy $\hat{P}$ which approximates the true performance P. We do so by a second order approximation parameterized by the matrix Y. Taking χ a vector of features which are sufficient to represent the stereo video (at least as far as the estimation of (α, β) is concerned, will be described in the next paragraph) we have:

$$(\hat{\alpha}, \hat{\beta}) = \arg\max_{\alpha, \beta} \hat{P}(\alpha, \beta) = \arg\max_{\alpha, \beta} \chi^\top Y \chi \tag{4}$$

We perform training by sampling α and β, computing then input-output pairs of vector χ and the real performance values P, and finally fitting the parameter matrix Y.

Stereo Video Representation by Spectral Properties. We are motivated by prior work on supervised learning in spectral clustering [46–48] to represent the stereo videos by their spectral properties. In particular, we draw on [46] and consider the normalized-cut cost NCut (of the similarity graph, based on the training set groundtruth labelling) and its lower bound Trace_R. Our representation vector is therefore $\chi = [\alpha, \beta, \text{NCut}, \text{Trace}_R]^\top$.

In more details, given the indicator matrix $E = \{e_r\}_{r=1\cdots R}$ where $e_r \in \mathbb{R}^{N^m}$, $e_r(i) = 1$ if superpixel i belongs to r-th cluster otherwise $= 0$, and N^m is the number of superpixels, we have:

$$\text{NCut}(\alpha, \beta, E) = \sum_{r=1}^{R} \frac{e_r^\top (D - W) e_r}{e_r^\top D e_r}$$

$$\text{Trace}_R(\alpha, \beta) = R - \sum_{r=1}^{R} \lambda_r(L) \tag{5}$$

where $D = \mathbf{diag}(W\mathbf{1})$ is the degree matrix of W and $\lambda_r(L)$ is the r-th eigenvalues of the generalized Laplacian matrix $L = D^{-1} \cdot W$ of the similarity matrix W.

Derivatives of Performance Proxy. Our performance proxy $\hat{P}$ is now differentiable. For gradient descent optimization, we use its derivatives w.r.t. parameters $\theta \in \{\alpha^k, \beta^c\}$:

$$\frac{\partial \chi^\top Y \chi}{\partial \theta} = \frac{\partial \chi^\top}{\partial \theta} (Y + Y^\top) \chi \quad \forall \theta \in \{\alpha^k, \beta^c\} \tag{6}$$

The derivatives of NCut and $\text{Trace}_R \in \chi$ are:

$$\frac{\partial (\text{NCut})}{\partial \theta} = \sum_{r=1}^{R} \frac{-e_r^\top \frac{\partial W}{\partial \theta} e_r e_r^\top D e_r + e_r^\top W e_r e_r^\top \frac{\partial D}{\partial \theta} e_r}{(e_r^\top D e_r)^2}$$

$$\frac{\partial (\text{Trace}_R)}{\partial \theta} = \text{trace}(V^\top \frac{\partial L(\theta)}{\partial \theta} V) \tag{7}$$

where V denotes the subspace spanned by the first R eigenvectors of L. Note that W, D and L are all parameterized by α and β. By plain chain rule differentiation, we compute the gradients of the differentiable proxy $\hat{P}$ in closed form. (Cf. All gradients are presented in the supplementary material.)

Joint Learning of Regressor and Latent Parameter Combinations. As stated in Eq. 3, we are interested in optimizing the performance P w.r.t. the regressor Ξ and therefore the ensemble combination parameters α and β have to be treated as latent variables. As described in Algorithm 1, we solve this by an EM-type optimization scheme in which we iterate finding optimal parameters α and β and predicting new α and β parameters based on the re-fitted regressor Ξ.

Intuitively, this scheme strikes a balance between the generalization capabilities of the regressor and optimal parameters α and β. We found this to be particular important, as in many cases a wide range of parameters leads to good results. Fixing the best parameters as a learning target, leads to a more difficult regression and overall worse performance. The metric specific performance proxy is continuously updated by using the samples in a small neighborhood in order to improve the local approximation of the desired metric P.

6.3 Implementation Details

As already noted, the computation of NCut at training involves the ground truth annotations. In particular, the NCut for the entire video requires all frames labeled, while ours and most segmentation datasets [35,49] only offer sparse labeling. Aggregating dense optical flow over time allows to connect the sparsely annotated frames. The spatial and temporal connections of these labeled frames are then used for the NCut computation.

Our representation vector χ in (4) consists of $[\alpha, \beta, \mathrm{NCut}, \mathrm{Trace}_R]$. We have empirically found that this combination improves of the individual parts and subsets by 5% and therefore we use the full vector in the following experiments.

In order to increase the number of examples for our training procedure, we divide each video into subsequences so that each of them contains two frames with groundtruth.

7 Experimental Results

We evaluate our proposed **efficient and adaptive stereo video segmentation** algorithm (**EASVS**) on the CSVSC benchmark. In particular, first we test the pooled segmentation outputs, then we compare EASVS against relevant state-of-the-art on stereo video sequences, finally we present an in-depth analysis of EASVS.

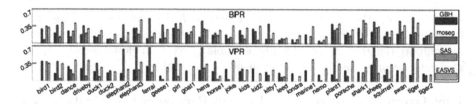

Fig. 5. Results of the considered video segmentation algorithms (GBH [9], moseg [7], SAS [8]) and our proposed EASVS on the CSVSC stereo video sequences using BPR and VPR. No considered method performs consistently well on all videos. moseg may achieve high performance of stereo videos with large and distinctive motion such as "elephants3" and "hens" but underperforms when motion is not strong, e.g. "marine1". Complementary features are given by GBH. SAS combines statically (cf. segmentation ensemble model of Sect. 5.1) the two video segmentation techniques as well as the pooled image segments but also underperforms, because a static combination cannot address the variety of the stereo videos.

7.1   Video Segmentations and Their (Static) Ensemble

Among the pooled segmentations (details in Sect. 5.3), we have included two state-of-the-art video segmentation techniques: the motion segmentation algorithm of [7] (moseg) and the graph-based hierarchical video segmentation method of [9] (GBH).

In Fig. 5, we illustrate performance of each of moseg and GBH on all stereo video sequences. (Cf. detailed comments in the figure caption.) As expected, none of the two performs satisfactorily on all sequences. Rather, they have in most cases complementary performance, e.g. moseg aiming for motion segmentation takes the lead on sequences with evident motion and good optical flow estimates; while GBH overtakes when spatio-temporal appearance cues are more peculiar in the visual objects.

A third technique illustrated in Fig. 5 is the segmentation by aggregating superpixel method of [8] (SAS). This is an interesting baseline for our proposed algorithm. SAS is based on a static combination of pooled segmentation outputs. We extend its original image-based formulation to stereo videos by including into its pool the GBH and moseg video segmentation methods, as we illustrate in Sect. 5.1.

Figure 5 clearly states that a static combination does not suffice to address the segmentation of stereo videos. By contrast, quite surprisingly, trying to always pool *all* video and image segmentation output *with the same contributing weights* turns out to harm performance.

7.2   EASVS and the State-of-the-art

Our adaptive combination of pooled segmentation outputs poses the question of which measure to use for learning. As mentioned in Sect. 3, the BPR and VPR measures may push for adaptive algorithms with better boundaries or temporally-consistent volumes. Averaging BPR and VPR may balance the two

Table 1. Results on the CSVSC benchmark. The proposed EASVS outperforms the baselines from video segmentation, static ensemble, RGB-D video segmentation, and video co-segmentation for all metrics. See detailed discussion in Sect. 7.2.

Stereo video segmentation	BPR	VPR	AM-BVPR	HM-BVPR
GBH [9]	0.187	0.208	0.198	0.198
moseg [7]	0.247	0.285	0.266	0.264
SAS [8]	0.184	0.087	0.135	0.118
4D-seg [10]	0.128	0.146	0.137	0.120
VideoCoSeg [28]	0.238	0.140	0.189	0.169
Proposed EASVS	**0.301**	**0.296**	**0.295**	**0.288**
Oracle	0.371	0.505	0.423	0.428

aspects, which we may achieve by arithmetic (AM-BVPR) or harmonic mean (HM-BVPR).

In Table 1, we illustrate performance of EASVS against moseg, GBH and SAS, measured according to the four available metrics (BPR, VPR, AM-BVPR, HM-BVPR). For EASVS, the measured performance statistic has also been respectively used for learning the adaptive ensemble segmentation model. (Since our approach involves learning, we address train/test splits with three-fold cross validation and the results are averaged on three folds). The results in the table match the intuition that only an adaptive combination can successfully address all videos. Furthermore, our proposed EASVS outperforms a recent depth video segmentation method [10] (4D-seg) by more than 50% on all measures, as well as a recent video co-segmentation algorithm [28] that we run on each video stereo pair by 65%. This is confirmed by the qualitative examples shown in Table 3.

We delve further into the understanding of the potential result improvements within the EASVS framework with an oracle. In more details, we allow our algorithm to estimate the optimal segmentation-pool combination-parameters (α and β) by accessing the ground truth performance measure P for each stereo video sequence. The higher oracle performance by up to 70% (with the current representation and quadratic regressors) anticipate future improvements with richer models and more data.

7.3   Deeper Analysis of EASVS

In Table 2, we provide additional insights into EASVS. First, we experiment with (1) no depth and (2) fixed depth contribution. The performance drops by 11.5%

Table 2. Analysis of the proposed EASVS, which shows the importance of the depth cue as well as the proposed adaptive strategy. See Sect. 7.3 for detailed discussion.

	No depth	Fixed depth	Fixed α	Fixed β	Fixed both α, β	Proposed EASVS
HM-BVPR	0.254	0.276	0.270	0.276	0.272	**0.288**

Table 3. Examples of the proposed EASVS optimized for different metrics compared to the state-of-the-art algorithms. Note how GBH outlines the object boundaries but tends to over-segment, while moseg produces under-segmentations and fails to extract objects without significant motion. The static combination scheme SAS cannot strike good compromised parameters across all videos, which results in degraded results. 4D-seg [10] is a clear leap forward but suffers from some of the drawbacks of GBH. Our EASVS benefits the learning framework and the adaptive nature for a better output.

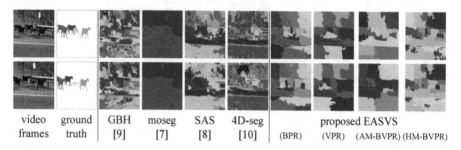

video frames	ground truth	GBH [9]	moseg [7]	SAS [8]	4D-seg [10]	proposed EASVS (BPR)	(VPR)	(AM-BVPR)	(HM-BVPR)

for (1) and 4.2% for (2) in HM-BVPR. This shows the importance of the depth cue within the full system, which benefits for videos with motion or appearance ambiguities. Additionally, this speaks in favor a the adaptive strategy. (Cf. 4D-seg [10] also leverages depth but cannot reach the same performance as the adaptive depth combination.)

Second, we fix the combination parameters (3) α, (4) β, and (5) both α, β to the single best values determined on the training set, therefore limiting the system adaptivity. The performance drops by 6.3%, 4.2%, and 5.6% respectively. (Please note that although conceptually fixing both α and β is the same as the SAS baseline, but for SAS we use default parameters in the code of [8].) Once again, we find that adaptivity is therefore crucial for the performance of our system and that both adaptive aspects are strictly needed: weighting the pooled segmentation (α) and measuring similarity of the resulting superpixels (β).

8   Conclusions

We have considered the emerging topic of consumer stereo cameras and proposed a benchmark to evaluate progress for the task of segmentation with this interesting type of data. The dataset is challenging and it includes diverse visual cues and camera setups. None of the existing segmentation algorithms can perform well in all conditions.

Furthermore, we have introduced a novel efficient and adaptive stereo video segmentation algorithm. Our method is capable of combining optimally a pool of segmentation outputs from a number of "expert" algorithms. The quality of results highlights that combining single algorithms is promising and that research on such a framework is perfectly orthogonal to pushing performance in the single niches, e.g. motion segmentation, image segmentation, supervoxelization etc.

References

1. Le, Q.V., Zou, W.Y., Yeung, S.Y., Ng, A.Y.: Learning hierarchical invariant spatio-temporal features for action recognition with independent subspace analysis. In: CVPR (2011)
2. Oneata, D., Revaud, J., Verbeek, J., Schmid, C.: Spatio-temporal object detection proposals. In: Fleet, D., Pajdla, T., Schiele, B., Tuytelaars, T. (eds.) ECCV 2014. LNCS, vol. 8691, pp. 737–752. Springer, Heidelberg (2014). doi:10.1007/978-3-319-10578-9_48
3. Taralova, E.H., Torre, F., Hebert, M.: Motion words for videos. In: Fleet, D., Pajdla, T., Schiele, B., Tuytelaars, T. (eds.) ECCV 2014. LNCS, vol. 8689, pp. 725–740. Springer, Heidelberg (2014). doi:10.1007/978-3-319-10590-1_47
4. Raza, S.H., Grundmann, M., Essa, I.: Geometric context from video. In: CVPR (2013)
5. Scharstein, D., Hirschmüller, H., Kitajima, Y., Krathwohl, G., Nešić, N., Wang, X., Westling, P.: High-resolution stereo datasets with subpixel-accurate ground truth. In: Jiang, X., Hornegger, J., Koch, R. (eds.) GCPR 2014. LNCS, vol. 8753, pp. 31–42. Springer, Heidelberg (2014). doi:10.1007/978-3-319-11752-2_3
6. Butler, D.J., Wulff, J., Stanley, G.B., Black, M.J.: A naturalistic open source movie for optical flow evaluation. In: Fitzgibbon, A., Lazebnik, S., Perona, P., Sato, Y., Schmid, C. (eds.) ECCV 2012. LNCS, vol. 7577, pp. 611–625. Springer, Heidelberg (2012). doi:10.1007/978-3-642-33783-3_44
7. Ochs, P., Brox, T.: Object segmentation in video: a hierarchical variational approach for turning point trajectories into dense regions. In: ICCV (2011)
8. Li, Z., Wu, X.M., Chang, S.F.: Segmentation using superpixels: a bipartite graph partitioning approach. In: CVPR (2012)
9. Grundmann, M., Kwatra, V., Han, M., Essa, I.: Efficient hierarchical graph-based video segmentation. In: CVPR (2010)
10. Hickson, S., Birchfield, S., Essa, I., Christensen, H.: Efficient hierarchical graph-based segmentation of RGBD videos. In: CVPR (2014)
11. Brox, T., Malik, J.: Object segmentation by long term analysis of point trajectories. In: Daniilidis, K., Maragos, P., Paragios, N. (eds.) ECCV 2010. LNCS, vol. 6315, pp. 282–295. Springer, Heidelberg (2010). doi:10.1007/978-3-642-15555-0_21
12. Galasso, F., Iwasaki, M., Nobori, K., Cipolla, R.: Spatio-temporal clustering of probabilistic region trajectories. In: ICCV (2011)
13. Li, F., Kim, T., Humayun, A., Tsai, D., Rehg, J.M.: Video segmentation by tracking many figure-ground segments. In: ICCV (2013)
14. Wang, W., Shen, J., Porikli, F.: Saliency-aware geodesic video object segmentation. In: CVPR (2015)
15. Faktor, A., Irani, M.: Video segmentation by non-local consensus voting. In: BMVC (2014)
16. Kanade, T., Okutomi, M.: A stereo matching algorithm with an adaptive window: theory and experiment. TPAMI 16, 920–932 (1994)
17. Scharstein, D., Szeliski, R.: A taxonomy and evaluation of dense two-frame stereo correspondence algorithms. IJCV 47, 7–42 (2002)
18. Bleyer, M., Rhemann, C., Rother, C.: Extracting 3D scene-consistent object proposals and depth from stereo images. In: Fitzgibbon, A., Lazebnik, S., Perona, P., Sato, Y., Schmid, C. (eds.) ECCV 2012. LNCS, vol. 7576, pp. 467–481. Springer, Heidelberg (2012). doi:10.1007/978-3-642-33715-4_34

19. Huguet, F., Devernay, F.: A variational method for scene flow estimation from stereo sequences. In: ICCV (2007)
20. Basha, T., Moses, Y., Kiryati, N.: Multi-view scene flow estimation: a view centered variational approach. IJCV **101**, 6–21 (2013)
21. Vogel, C., Schindler, K., Roth, S.: 3D scene flow estimation with a rigid motion prior. In: ICCV (2011)
22. den Bergh, M.V., Gool, L.J.V.: Real-time stereo and flow-based video segmentation with superpixels. In: WACV (2012)
23. Weikersdorfer, D., Schick, A., Cremers, D.: Depth-adaptive superpixels for RGB-D video segmentation. In: ICIP (2013)
24. Held, D., Guillory, D., Rebsamen, B., Thrun, S., Savarese, S.: A probabilistic framework for real-time 3D segmentation using spatial, temporal, and semantic cues. In: RSS (2016)
25. Kim, G., Xing, E.P.: On multiple foreground cosegmentation. In: CVPR(2012)
26. Joulin, A., Bach, F., Ponce, J.: Multi-class cosegmentation. In: CVPR (2012)
27. Fu, H., Xu, D., Zhang, B., Lin, S.: Object-based multiple foreground video co-segmentation. In: CVPR (2014)
28. Chiu, W.C., Fritz, M.: Multi-class video co-segmentation with a generative multi-video model. In: CVPR (2013)
29. Karayev, S., Baumgartner, T., Fritz, M., Darrell, T.: Timely object recognition. In: Advances in Neural Information Processing Systems (NIPS) (2012)
30. Karayev, S., Fritz, M., Darrell, T.: Anytime recognition of objects and scenes. In: CVPR (2014)
31. Ebert, S., Fritz, M., Schiele, B.: Ralf: A reinforced active learning formulation for object class recognition. In: CVPR (2012)
32. Mac Aodha, O., Brostow, G.J., Pollefeys, M.: Segmenting video into classes of algorithm-suitability. In: CVPR (2010)
33. Bai, X., Wang, J., Simons, D., Sapiro, G.: Video SnapCut: robust video object cutout using localized classifiers. In: SIGGRAPH (2009)
34. Arbelaez, P., Maire, M., Fowlkes, C., Malik, J.: Contour detection and hierarchical image segmentation. TPAMI **33**, 898–916 (2011)
35. Galasso, F., Nagaraja, N.S., Cardenas, T.J., Brox, T., Schiele, B.: A unified video segmentation benchmark: annotation, metrics and analysis. In: ICCV (2013)
36. Geiger, A., Roser, M., Urtasun, R.: Efficient large-scale stereo matching. In: Kimmel, R., Klette, R., Sugimoto, A. (eds.) ACCV 2010. LNCS, vol. 6492, pp. 25–38. Springer, Heidelberg (2011). doi:10.1007/978-3-642-19315-6_3
37. Zach, C., Pock, T., Bischof, H.: A duality based approach for realtime TV-L^1 optical flow. In: Hamprecht, Fred, A., Schnörr, Christoph, Jähne, Bernd (eds.) DAGM 2007. LNCS, vol. 4713, pp. 214–223. Springer, Heidelberg (2007). doi:10.1007/978-3-540-74936-3_22
38. Fusiello, A., Irsara, L.: Quasi-Euclidean uncalibrated epipolar rectification. In: ICPR (2008)
39. Janoch, A., Karayev, S., Jia, Y., Barron, J.T., Fritz, M., Saenko, K., Darrell, T.: A category-level 3-D object dataset: putting the kinect to work. In: IEEE Workshop on Consumer Depth Cameras for Computer Vision (2011)
40. Galasso, F., Keuper, M., Brox, T., Schiele, B.: Spectral graph reduction for efficient image and streaming video segmentation. In: CVPR (2014)
41. Ng, A.Y., Jordan, M.I., Weiss, Y., et al.: On spectral clustering: analysis and an algorithm. In: NIPS (2002)
42. Shi, J., Malik, J.: Normalized cuts and image segmentation. TPAMI **22**, 888–905 (2000)

43. von Luxburg, U.: A tutorial on spectral clustering. Stat. Comput. **17**, 395–416 (2007)

44. Achanta, R., Shaji, A., Smith, K., Lucchi, A., Fua, P., Süsstrunk, S.: SLIC superpixels compared to state-of-the-art superpixel methods. TPAMI **34**, 2274–2282 (2012)

45. Zhang, Q., Shen, X., Xu, L., Jia, J.: Rolling guidance filter. In: Fleet, D., Pajdla, T., Schiele, B., Tuytelaars, T. (eds.) ECCV 2014. LNCS, vol. 8691, pp. 815–830. Springer, Heidelberg (2014). doi:10.1007/978-3-319-10578-9_53

46. Meilă, M., Shortreed, S., Xu, L.: Regularized spectral learning. In: AISTATS (2005)

47. Jordan, F., Bach, F.: Learning spectral clustering. In: NIPS (2004)

48. Ionescu, C., Vantzosy, O., Sminchisescu, C.: Matrix backpropagation for deep networks with structured layers. In: ICCV (2015)

49. Ochs, P., Malik, J., Brox, T.: Segmentation of moving objects by long term video analysis. TPAMI **36**, 1187–1200 (2014)

No-Reference Video Shakiness Quality Assessment

Zhaoxiong Cui and Tingting Jiang[✉]

National Engineering Laboratory for Video Technology,
Cooperative Medianet Innovation Center, School of EECS,
Peking University, Beijing 100871, China
ttjiang@pku.edu.cn

Abstract. Video shakiness is a common problem for videos captured by hand-hold devices. How to evaluate the influence of video shakiness on human perception and design an objective quality assessment model is a challenging problem. In this work, we first conduct subjective experiments and construct a data-set with human scores. Then we extract a set of motion features related to video shakiness based on frequency analysis. Feature selection is applied on the extracted features and an objective model is learned based on the data-set. The experimental results show that the proposed model predicts video shakiness consistently with human perception and it can be applied to evaluating the existing video stabilization methods.

1 Introduction

With the development of digital video capture devices, such as smart phones or wearable devices, more and more people are able to take videos in daily life and upload these videos to the social media. Compared to traditional broadcast videos, these handy videos usually are not perfect because most of them are taken by amateurs. For example, due to the lack of tripods, many videos encounter the problem of shakiness. If the shakiness is severe, it will influence the video quality perceived by people. Therefore, understanding the subjective perception of human to video shakiness is important for many video applications, *e.g.*, video editing, bootleg detection. Furthermore, how to design an objective assessment model for video shakiness which is consistent with subjective perception is a challenging problem. That is, given an input video, it is expected to output a shakiness score which is consistent with human perception.

Video shakiness has been extensively studied by many researchers from different perspectives. Some works [1–4] take the amount of camera motion into account. The underlying assumption is that the larger the camera motion is, the more shaky the video is. However, this assumption is not always true. For example, if the camera moves constantly, even if the motion is large, it would not affect the video quality that much. On the other hand, if the camera moves up and down frequently, even if the motion is small, it will be annoying for the audience. Therefore, there are several methods proposed based on the frequency

© Springer International Publishing AG 2017
S.-H. Lai et al. (Eds.): ACCV 2016, Part V, LNCS 10115, pp. 396–411, 2017.
DOI: 10.1007/978-3-319-54193-8_25

analysis [5–8]. They apply different filters on the motion signals and design frequency-based models.

In this paper, we first conduct subjective experiments and construct a data-set which can provide ground truth for the design of object assessment models. Second, based on this subjective data-set, we propose a frequency-based model in order to objectively evaluate the video quality with respect to shakiness. Specifically, we extract motion signals (including translation, rotation and scaling) from videos and then apply frequency band decomposition on each signal. Later these frequency-related features from videos are selected by a genetic algorithm and an objective video shakiness assessment model is learned by support vector regression method (SVR). The experimental results show that our objective assessment model can predict the video shakiness score more consistently with subjective scores than previous work.

Besides the above subjective experiments and objective model design, another contribution of our work is that we apply the proposed video shakiness assessment model on evaluating video stabilization methods. By comparing the shakiness scores given by the proposed model before and after the video stabilization, we can objectively compare the improvements of different stabilization methods, while this comparison was usually performed by human eyes subjectively before. This demonstrates one application of our method.

The rest of the paper is organized as follows. Section 2 introduces the related work. Section 3 explains the subjective experiments and Sect. 4 shows the feature extraction for video shakiness. The objective model learning and experimental results are shown in Sect. 5. Section 6 demonstrates its application of evaluating performance of video stabilization algorithms. Finally conclusion is given in Sect. 7.

2   Related Work

2.1   Video Quality Assessment

According to the availability of reference videos, video quality assessment (VQA) can be classified as three kinds: full-reference (FR) VQA, reduced-reference(RR) VQA and no-reference(NR) VQA. Among these works, NR-VQA is most challenging because no reference video information can be used. To solve this problem, many methods have been proposed. For example, Bovik et al. [9,10] extract video features and apply machine learning methods in order to design a general-purpose VQA model. Our work also belongs to NR-VQA, but we are specifically interested in video quality regarding shakiness.

2.2   Video Shakiness Analysis

Most previous work on video shakiness analysis can be classified as two categories in general: one is based on camera motion without filtering and the other is based on frequency analysis.

As for the former category, the underlying assumption is that the degree of video shakiness depends on the amount of camera motion only. For home video editing, Girgensohn et al. [1] compute a numerical "unsuitability score" based on a weighted average of horizontal and vertical pan. According to the unsuitability score, videos can be classified as four categories. Besides pan, Mei et al.[2] represent the camera motion as three independent components(pan, tilt and zoom) and proposes a "jerkiness factor" for each frame as follows:

$$S_i = \max\{(\omega_p P + \omega_T T)/(\omega_p + \omega_T), Z\} \tag{1}$$

where S_i denotes the jerkiness factor for i-th frame, P is pan, T is tilt, Z is zoom. P, T, Z are all normalized to $[0, 1]$, ω_p and ω_q are weighting factors. A video's jerkiness is defined as the average of frame-level factors. These two works are cited by Xia et al. [3] and used as a component as a general video quality assessment system for web videos with weighting parameters as $\omega_p = 1, \omega_q = 0.75$. Similarly, Hoshen et al. [4] defines the shakiness of t-th frame $Q_{stab}(t)$, as the average square displacement of all feature points between adjacent frames, i.e.,

$$Q_{stab}(t) = \sqrt{(dx(t))^2 + (dy(t))^2} \tag{2}$$

where $dx(t)$ and $dy(t)$ denote the horizontal and vertical movement of this frame. The above methods all take the amount of camera motion between frames as the indicator of video shakiness, but ignore that different frequency components contained by the camera motion have different influences on human perception.

To address this issue, the latter category of previous work applies frequency analysis on motion signals from videos. For example, Shrestha et al. [5] and Campanella et al. [6] apply a FIR filter on the translation of video frames and then take the difference between the filtered signal and original signal, which corresponds to the high-frequency component, as the amount of shakiness. Alam et al. [7] and Saini et al. [8] take similar approaches but median filter is used. Although these works realize the importance of frequency decomposition, they only exploit the high-frequency component and discard other frequency components. In addition, the influence of frame rates on the filtering is ignored.

Besides these two categories, there are some previous work using other methods. For example, Yan et al. [11] compare the movement vectors between adjacent frames. If the angle between the two vectors is larger than $\pi/2$, they think this frame contains shakiness. In order to detect bootleg automatically, Visentini-Scarzanella et al. [12] retrieve the inter-frame motion trajectories with feature tracking techniques and then compute a normalized cross-correlation matrix based on the similarities between the high-frequency components of the tracked features' trajectories. Bootleg classification is based on the comparison between the correlation distribution and the trained models. However, these two works do not give quantitative metrics for video shakiness evaluation.

It is worth noting that all the above works do not consider video watching conditions, such as the screen size and watching distance. And these models are not verified by subjective experiments devoted to video shakiness.

3 Subjective VQA Experiment

3.1 Test Sequences

We selected 4 queries, "scenery", "animal", "vehicle" and "sport", designed to retrieve the top ranked, high-definition real-world videos in four respective categories from youku.com. In November 2015, we issued the four queries to video search engine, soku.com, and collected all retrieved videos. All original videos we collected are encoded by H264/AVC codec, with target bit-rate 1600 kbps, all in .flv format. For the sake of compatibility with our test platform, we converted the videos into .webm format encoded by VP8 codec. We used FFmpeg libvpx library for trans-coding, and set the quality parameters of output videos good enough (crf = 4, targetbitrate = 2 Mbps) to guarantee the fidelity. Then we cropped the videos into 512 sequences as our data-set. Each sequence lasts 10 s, and most (>99%) of the sequences are cropped within one shot to avoid the influence of scene switching between shots. Sequences with other severe distortions, like blurring and color distortion, were eliminated to avoid the masking effects. Numbers of the sequences in each categories are listed below (Table 1):

Table 1. Size of our data-set

Category	Number of sequences
Scenery	35
Animal	134
Vehicle	297
Sport	46
Total	512

As recommended in ITU-T Recommendation P.910 [13], we calculated the Spatial Information (SI) and Temporal Information (TI) of video sequences. SI and TI metrics quantify spatial and temporal perceptual information content of a given sequence. As shown in Fig. 1, the sequences span a large portion of spatial-temporal information plane, which implied a good variety of our data-set.

Among 512 video sequences in the data-set, 2 sequences falling at the extremes of the shakiness quality scale (one for the best quality, the other for the worst) were chosen for anchoring. Anchoring sequences were displayed with shakiness quality labeled to indicate the range boundaries of shakiness intensity. For the purpose of training, another 10 sequences were randomly selected as dummy (or stabilizing) presentations. Dummy presentations were adopted to familiarize the participants with the experiment process and to stabilize their opinion. The remaining 500 sequences were used as real presentations.

For the session division, the real presentations were divided into 4 parts (125 for each part). The session division, display orders of the dummy presentations, and display orders of the real presentations, were randomized for each observer to avoid the influence by the order of presentations.

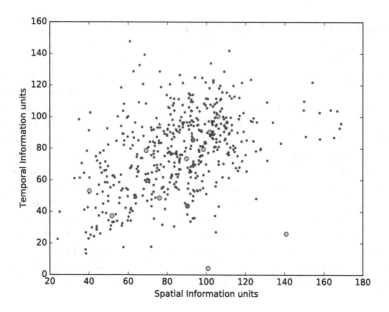

Fig. 1. Spatial-temporal plot for our test data-set: red circles for 2 anchoring sequences; green circles for 10 dummy presentations; blue dots for 500 real presentations (Color figure online).

3.2   Test Protocol

Test Environment. The test sequences were displayed on a Dell UltraSharp U2414H 23.8-inch light-emitting diode liquid crystal display (LED-LCD) monitor (1920 × 1080 at 60 Hz). At default factory settings, the U2414H was set to 75% brightness, which we measured at $254\,cd/m^2$. The contrast ratio was 853:1 and the viewing angle reached 178-degree. Other room illumination was low. A mini-DisplayPort video signal output from a HP folio 9470 m laptop computer was adopted as signal source. The distance between the observer and the monitor was held at about 85 cm which is about three times the height of the monitor.

Observers. Twenty adults, including 9 female and 11 male, aged between 19 and 22, took part in the experiment. All of them were undergraduate college students, 13 majored in Computer Science, 5 in Electronics Engineering, 1 in Physics and 1 in Maths. 4 observers were practitioners in related fields (Computer Vision, Computer Graphics, or Image Processing), and the remaining 16 observers had no related expertise. No observers had experience with video quality assessment study, and no observers were, or had been, directly involved in this study. All observers reported normal visual acuity and normal color vision.

Voting Method. The Single Stimulus (SS) non-categorical judgement method, with numerical scaling, was adopted for this experiment. In our SS method, an

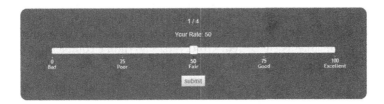

Fig. 2. The voting panel

observer is presented with a video sequence, and then asked to evaluate the shakiness of the sequence by drawing a slider on a numerical scale from 0 to 100. 0 means "bad" in quality, or shaking violently, and 100 represents "excellent" in quality, or shaking unnoticeably. We labeled 5 ITU-R semantics of quality [14] with respective scores at two ends of scale and three intermediate points ("Bad" at 0, "Poor" at 25, "Fair" at 50, "Good" at 75 and "Excellent" at 100) for reference of more specific quality levels (see Fig. 2).

3.3   Experiment Procedure

An experiment contains 4 sessions in total. At the beginning of each session, anchoring sequences were presented first, followed by dummy presentations, then real presentations. Breaks were allowed between three phases. Although assessment trials in real and dummy presentations are just the same, subjective assessment data (voting scores) issued from real presentations were saved and collected after experiment, but results for dummy presentations were not processed (Figs. 3 and 4).

In an assessment trial, a 10-second sequence faded in, presented and faded out. After that, voting panel faded in, the observer was asked to evaluate the video. Then voting panel faded out when evaluation submitted. The rating time was given at least 5 s, assuring that the observer voted carefully and adjacent stimuli were well isolated. The duration of a fade-in or a fade-out was set to 500 ms, which provided comfortable transitions between tasks.

Observers were carefully introduced to the voting method, the grading scale, the sequence and timing at the beginning of experiment. A session lasted about

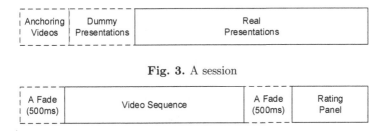

Fig. 3. A session

Fig. 4. A trial

half an hour, which meets the requirement prescribed by [14]. Observers were allowed to rest for a while between sessions. Usually an observer completed all 4 sessions at one time as suggested. Observers who didn't complete at one time were introduced again before their next session.

3.4   Results

A total number of 10,000 (500 sequences by 20 observers) voting scores were processed. As recommended in [14], the outlier detection for observers were imposed, but no outlier was detected.

The Mean Opinion Score (MOS) value of i-th sequence is defined as

$$MOS_i = \frac{1}{K} \sum_k S_{ki}, \tag{3}$$

where S_{ki} is score of sequence i voted by observer k, K is the number of observers. A higher MOS indicates better subjective shakiness quality of a sequence.

4   Feature Extraction

In this section, we design a no-reference video quality metric to predict perceived shakiness quality of web videos. Firstly, the global motion, namely the motion between adjacent frames, is extracted from the video sequence. Next, we transform the translation into the *deflection angle*, directly relating to the signal perceived by human visual system (HVS). Thereafter, the motion signals are decomposed into sub-bands, which contain frequency components of different levels. In the end, the statistics of each sub-band of the motion signals are calculated, as the features we designed for the video shakiness quality.

4.1   Global Motion Estimation

Global motion is defined as the geometrical transformation between adjacent video frames. It also indicates the motion of camera. Here we describe the global motion with a similarity transformation model, with four parameters $[d_x, d_y, \theta, \rho]$, corresponding to pan, tilt, rotation and isotropic scaling. Assuming (x_1, y_1) is the coordinates (with respect to the center of the frame) of a point in current frame F_t, and (x_2, y_2) is the coordinates of the corresponding point in next frame F_{t+1}, global motion can be illustrated by

$$\begin{bmatrix} x_2 \\ y_2 \end{bmatrix} = \rho \begin{bmatrix} \cos\theta & -\sin\theta \\ \sin\theta & \cos\theta \end{bmatrix} \begin{bmatrix} x_1 \\ y_1 \end{bmatrix} + \begin{bmatrix} d_x \\ d_y \end{bmatrix}. \tag{4}$$

Generally, there are two types of global motion estimation (GME) approaches: feature-based methods and featureless methods. Feature-based methods (e.g. [15–18]) utilize geometric features extracted in frames, such as Harris corners (see [19]), and then estimate the motion by matching the corresponding features between adjacent frames. Featureless approaches directly estimate the global motion from all pixels on each frame. Usually, feature-based approaches are fast and accurate, however fragile. On the contrary, though time-consuming, featureless approaches are usually robust. In our task, web videos contained complex and intensive motion. So robustness is necessary. We adopted an FFT-based featureless approach in [20–22], measuring translation (d_x, d_y), rotation θ and scaling ρ directly from the spectrum correlation between frames. During our test on web videos, this FFT-based approach reaches a satisfying robustness and accuracy, with an acceptable time cost.

4.2   Perceptual Modeling

To properly measure the influence of global motion perceived by the viewer, we need to model the global motion signal in a physical meaning, and consider its impact on human visual system (HVS).

In previous section, the translation signals (d_x, d_y) between adjacent frames are estimated, in pixel unit. However, we need physics quantities directly related to the stimulus received by HVS. Considering the viewing condition, including viewing distance and display size, translation (d_x, d_y) shall be transformed into *deflection angle* (α_x, α_y), *i.e.*,

$$\alpha_{x,y}(t) = \arctan\left(\frac{L_d d_{x,y}(t)}{Zs}\right) \approx \frac{L_d d_{x,y}(t)}{Zs} \tag{5}$$

where $d_{x,y}(t)$ is translation at frame t in pixel unit, L_d is the diagonal length of the display monitor, Z is the viewing distance, and $s = \sqrt{h^2 + w^2}$, where h, w is the height and the width of the video frame in pixel unit, respectively. Deflection angle indicates the *shift* of viewing angle caused by the translation between the two frames (see Fig. 5).

It is noticed in [23] that the subjective sensation of motion is proportional to the logarithm of the stimulus intensity, *i.e.*, velocity (Weber-Fechner law [24]). So we take the logarithm of $\alpha_{x,y}(t)$, called *logarithm of deflection angle*, as

$$l_{x,y}(t) = \log \alpha_{x,y}(t). \tag{6}$$

Rotation signal $\theta(t)$ and scaling signal $\rho(t)$ are used directly. This is because HVS perception of the rotation and scaling signals are not directly influenced by viewing condition.

4.3   Sub-band Decomposition

There is evidence that different frequency compositions have different impact on HVS perception [25], more specifically on shakiness perception. So we decompose the signals into three different frequency sub-bands: low band for $(0, 3\,\text{Hz})$, mid band for $(3\,\text{Hz}, 6\,\text{Hz})$ and high band for $(6\,\text{Hz}, 9\,\text{Hz})$. Decomposition is done by filtering the original signal by three respective filters, *i.e.*,

$$S_{\text{l,m,h}}(t) = S(t) * h_{\text{l,m,h}}(t) \tag{7}$$

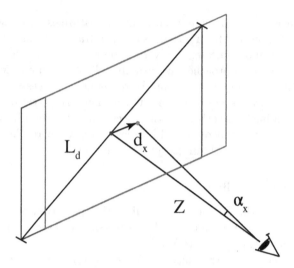

Fig. 5. Illustration of the deflection angle α_x.

where $S(t)$ is the original signal, $S_{l,m,h}(t)$ are filtered low-, mid- and high-band signals, and $h_{l,m,h}(t)$ represent the corresponding impulse response functions of the three filters, and $*$ denotes the convolution operation. An illustration of the band decomposition is shown in Fig. 6.

Ideal filters are adopted in this decomposition work. The ideal filters keep frequency components only in an interval of frequency. The frequency response $H(f)$ of ideal filters are

$$H(f) = \begin{cases} 1 & f_t < f \le f_h \\ 0 & \text{else} \end{cases} \tag{8}$$

where $H(f)$ is the Fourier transform of impulse response $h(t)$.

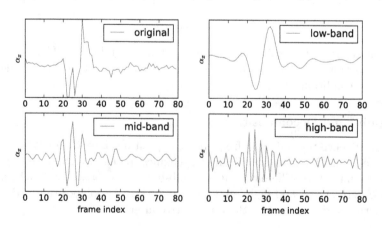

Fig. 6. Band decomposition result of $\alpha_x(t)$ signal of a video sequence.

Table 2. Sub-bands of motion signals

Original	$\alpha_x(t)$	$\alpha_y(t)$	$l_x(t)$	$l_y(t)$	$\theta(t)$	$\rho(t)$
Low band	$\alpha_{x,l}(t)$	$\alpha_{y,l}(t)$	$l_{x,l}(t)$	$l_{y,l}(t)$	$\theta_l(t)$	$\rho_l(t)$
Mid band	$\alpha_{x,m}(t)$	$\alpha_{y,m}(t)$	$l_{x,m}(t)$	$l_{y,m}(t)$	$\theta_m(t)$	$\rho_m(t)$
High band	$\alpha_{x,h}(t)$	$\alpha_{y,h}(t)$	$l_{x,h}(t)$	$l_{y,h}(t)$	$\theta_h(t)$	$\rho_h(t)$

The six motion signals $\alpha_x(t), \alpha_y(t), l_x(t), l_y(t), \theta(t), \rho(t)$ are extracted from every video sequence. Decompositions are done for each of them. As a result, the six signals are decomposed into 18 sub-bands (see Table 2).

4.4 Statistics

Finally, statistical features that capture the impact of motion signals on HVS are extracted from each sub-band. Suppose that $s_i(t)$ is one of the sub-band signals, and T is the total number of frames of the video sequence, we estimate the first to the fourth central or standardized moments of $s_i(t)$, $i.e.$,

$$\text{Mean } s_i^1 = \sum_t s(t)/T$$
$$\text{Variance } s_i^2 = \sum_t [s(t) - s_i^1]^2/T$$
$$\text{Skewness } s_i^3 = \sum_t [s(t) - s_i^1]^3/[T(s_i^2)^{3/2}]$$
$$\text{Flatness } s_i^4 = \sum_t [s(t) - s_i^1]^4/[T(s_i^2)^2]$$

In summary, for each video sequence, six motion signals are extracted, and decomposed into 18 sub-bands. In the next step, four statistics are estimated from each sub-band. In total, 72 ($6 \times 3 \times 4$) feature values are calculated from each video sequence.

5 Objective Experiment

In this section, we validate the performance of the extracted features, and obtain an objective no-reference video shakiness metric. We run the cross-validation test on the features, and validate the performance of the features by SROCC [26]. This cross-validation process is used for feature selection, and an optimal subset of features is obtained. We also compare our approach with other related works.

5.1 Cross Validation

We use a hold-out cross validation to evaluate the performance of the features. In each iteration, the data-set is randomly split into two parts, training set (90%) and validation set (10%). On the training set, a SVR model is trained, and then tested by the validation set. Then the performance of features is validated by calculating SROCC between the subjective MOS and the output of the SVR model on the validation set.

LIBSVM [27] is used for SVR training. We adopt a ν-SVR with RBF kernel to get the optimal result of training. Given a set of features, this train-test process is repeated on the data-set 1000 times randomly. The median SROCC is calculated as the final performance of the set of features.

5.2   Feature Selection

In the previous section, 72 feature values are calculated from one video sequence. To get the best performance, an optimal subset of features must be chosen where SVR performs the best.

We resort to a wrapper model feature selection using a genetic algorithm (GA) [28]. Each subset of features is regarded as a genome, represented by a 72-bit number x. $x(i) = 1$ denotes the feature i is chosen and $x(i) = 0$ denotes the feature is not chosen. The fitness of genome x is determined by the median-SROCC of a 1000-times cross validation with the corresponding feature subset:

$$\text{fitness}(x) = \begin{cases} \text{SROCC}_{1000} - \text{P} & \text{SROCC}_{1000} \geq \text{P} \\ 0 & \text{SROCC}_{1000} < \text{P} \end{cases} \tag{9}$$

where P denotes the pressure of the evolution. During each generation, genomes with larger fitness are more likely to be selected to breed a new generation. More specifically, genomes with 0 fitness would never be chosen. So P determines the minimum fitness allowed in the evolution. The population of the next generation is generated by both crossover and mutation of the selected genomes (see [28]).

We adopt a two-step solution to find the optimal feature subset. In the first step, initialize the genomes by randomly choosing $x(i)$ for each i, setting $P = 0.8$, and run the genetic algorithm for 100 generations. In the second step, initialize the genomes by the genomes of the last generation in the first step, setting $P = 0.85$, and run the genetic algorithm again. The first run picks out a group of genomes with high fitness (SROCC). The second run imposes a more strict restriction, and purifies the genomes to be optimal. Finally the genome with the highest fitness in the last 5 generations during the second GA run is selected to be the optimal subset of the features. From the feature selection result, we find the translation is more important than rotation and zooming, and the low-rank moments of middle- and high-bands are more significant.

5.3   Results

The GA finally chooses an optimal set of 32 features from the 72 features. This optimal feature set performs a good result in cross-validation, with median-SROCC reaching 0.8767 (90% data for training, 10% for testing). Fig. 7 shows the scatter plot of MOS versus objective scores.

By adjusting the portion of training data, the relationship between the algorithm's performance and the amount of training data can be investigated. Starting with 1%, we gradually increased the portion of training data, and got a curve as shown in Fig. 8. With training portion exceeding 20% (train with only 100

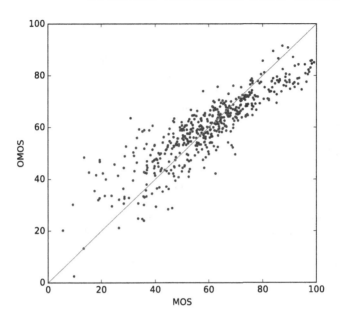

Fig. 7. MOS and objective scores (OMOS)

video sequences), median-SROCC reaches 0.8. When training portion exceeds 40%, the SROCC will become stable. It shows that our approach can reach a good performance with small amount of training data. The generalization ability of our algorithm is excellent.

We compare our approach with related works [3,4,6], as well as only SI and TI features (trained with SVR). See Table 3. Note that the authors of the related works have not yet shared the source code, so we implement their works and test on our data-set by ourselves. The result shows that, our approach outperformed all the state-of-the-art methods.

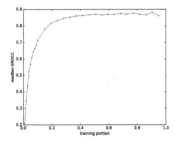

Fig. 8. The relationship between the performance and the training portion.

Table 3. Results

Method	SROCC
SI + TI (SVR)	0.4945
[4]	0.6506
[3]	0.6534
[6]	0.7669
Our approach	**0.8767**

6   Benchmark for Stabilization Algorithms

Stabilization algorithms are designed to eliminate shakiness artifacts in videos. However, objective benchmark to test the performance of stabilization algorithms do not exist. As an application of our shakiness VQA approach, we propose a method to evaluate stabilization algorithms, by means of the NR-VQA model we learned.

Suppose V_i to be i-th video with shakiness artifact, $O(V_i)$ to be the original score given by our shakiness NR-VQA model, and suppose V_i^k to be the video stabilized by k-th stabilization algorithm, $O(V_i^k)$ to be the shakiness score of the stabilized video. Then $O(V_i^k) - O(V_i)$ is called the *enhancement* E_i^k of the stabilization algorithm k on the video V_i.

It is supposed that, the shakiness score will increase after stabilization, *i.e.*, $E_i^k > 0$. Unfortunately, E_i^k may also decrease after stabilization. For instance, if a video without shakiness artifact is stabilized, it is possible that stabilization algorithm unwillingly introduces a motion artifact to the video. In such cases, E_i^k will be less than zero, and we call the video quality is *degenerated*.

To evaluate the performance of a certain stabilization algorithm k, we define the following two indexes:

1. Average Enhancement E^k: the enhancement of stabilization algorithm k on the given data-set.

$$E^k = \frac{1}{N} \sum_i \left(O(V_i^k) - O(V_i) \right). \tag{10}$$

2. Degeneration Frequency P_d^k: the frequency of degeneration in videos stabilized by algorithm k on the given data-set.

$$P_d^k = \frac{1}{N} \sum_i I(O(V_i^k) < O(V_i)). \tag{11}$$

N is the amount of videos in the data-set. I is the indicator function: $I(A) = 1$ when A is true, otherwise $I(A) = 0$.

We stabilize all videos in our data-set, by three popular stabilization tools: Microsoft Project Oxford Video API [29], proDAD Mercalli 2.0 [30] and Adobe After Effect CC 2015 (VX deformation stabilizer) [31]. Then, we score original videos and stabilization videos by our NR-VQA model. We plot the scores of stabilization videos of three algorithms, in reference of the original scores, see Fig. 9. As shown in the figure, after stabilization the scores of videos increase generally. This shows the effect of stabilization algorithms. It is also observed that the enhancement of low-quality videos is more significant than that of high-quality videos. Moreover, indeed some videos degenerate after stabilization, and high-quality videos degenerate more frequently, exactly as expected.

Therefore, we calculate E^k and P_d^k for each algorithm separately in videos of three different quality levels: high-quality level (videos with the highest

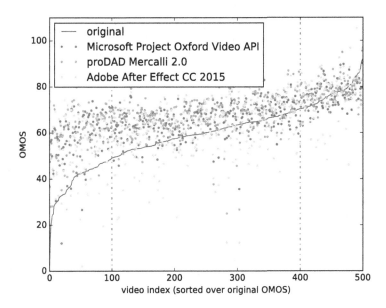

Fig. 9. Objective scores of stabilized and original videos. Dash lines indicate the boundary of video quality levels.

Table 4. P_d and E indexes of three stabilization algorithms

Stabilization algorithm	Microsoft project oxford video API		proDAD Mercalli		Adobe after effects CC 2015	
Quality level	P_d	E	P_d	E	P_d	E
Low	0.060	18.997	0.010	19.607	0.140	14.075
Mid	0.160	6.151	0.170	6.857	0.367	2.246
High	0.530	−0.137	0.440	0.860	0.450	−0.166
Overall	0.214	7.459	0.192	8.208	0.338	4.129

$100\ O(V_i)$), low-quality level (videos with the lowest $100\ O(V_i)$), and mid-quality level (the other 300 videos). From the following table, it can be seen that proDAD Mercalli 2.0 performs best (Table 4).

7   Conclusion

We propose a new method for video shakiness quality assessment. First, we construct a data-set based on subjective experiments. Second, based on this data-set we extract video features and learn an objective model to predict video quality in terms of shakiness. The proposed model has been validated on the constructed data-set and used to evaluate the performance of existing video stabilization methods.

Acknowledgment. This work was partially supported by National Basic Research Program of China (973 Program) under contract 2015CB351803 and NSFC under contracts 61572042, 61390514, 61421062, 61210005, 61527084, as well as the grant from Microsoft Research-Asia.

References

1. Girgensohn, A., Boreczky, J., Chiu, P., Doherty, J., Foote, J., Golovchinsky, G., Uchihashi, S., Wilcox, L.: A semi-automatic approach to home video editing. In: ACM Symposium on User Interface Software and Technology, pp. 81–89 (2000)
2. Mei, T., Hua, X.S., Zhu, C.Z., Zhou, H.Q., Li, S.: Home video visual quality assessment with spatiotemporal factors. IEEE Trans. Circuits Syst. Video Technol. **17**, 699–706 (2007)
3. Xia, T., Mei, T., Hua, G., Zhang, Y.D., Hua, X.S.: Visual quality assessment for web videos. J. Vis. Commun. Image Represent. **21**, 826–837 (2010)
4. Hoshen, Y., Ben-Artzi, G., Peleg, S.: Wisdom of the crowd in egocentric video curation. In: Computer Vision and Pattern Recognition Workshops, pp. 587–593 (2014)
5. Shrestha, P., Weda, H., Barbieri, M., With, P.H.N.D.: Video Quality Analysis for Concert Video Mashup Generation. Springer, Heidelberg (2010)
6. Campanella, M., Barbieri, M.: Edit while watching: home video editing made easy. In: Electronic Imaging 2007, vol. 6506, pp. 65060L-1–65060L-10 (2007)
7. Alam, K.M., Saini, M., Ahmed, D.T., Saddik, A.E.: VeDi: a vehicular crowd-sourced video social network for VANETs. In: IEEE Conference on Local Computer Networks Workshops (LCN Workshops), pp. 738–745 (2014)
8. Saini, M.K., Gadde, R., Yan, S., Wei, T.O.: MoViMash: online mobile video mashup. In: ACM International Conference on Multimedia, pp. 139–148 (2012)
9. Mittal, A., Saad, M.A., Bovik, A.C.: A completely blind video integrity oracle. IEEE Trans. Image Process. **25**, 289–300 (2016)
10. Saad, M.A., Bovik, A.C., Charrier, C.: Blind prediction of natural video quality. IEEE Trans. Image Process. **23**, 1352–1365 (2014)
11. Yan, W.Q., Kankanhalli, M.S.: Detection and removal of lighting and shaking artifacts in home videos. In: Proceeding of ACM Multimedia, pp. 107–116 (2002)
12. Visentini-Scarzanella, M., Dragotti, P.L.: Video jitter analysis for automatic bootleg detection. In: IEEE International Workshop on Multimedia Signal Processing, pp. 101–106 (2012)
13. ITU-T P.910: Subjective video quality assessment methods for multimedia applications (1999)
14. ITU-R, BT.500-13: Methodology for the subjective assessment of the quality of television pictures. International Telecommunications Union, Technical report (2012)
15. Perez, P., Garcia, N.: Robust and accurate registration of images with unknown relative orientation and exposure. In: International Conference on Image Processing, vol. 3, pp. 1104–1107. IEEE (2005)
16. Torr, P.H.S., Zisserman, A.: Feature based methods for structure and motion estimation. In: Triggs, B., Zisserman, A., Szeliski, R. (eds.) IWVA 1999. LNCS, vol. 1883, pp. 278–294. Springer, Heidelberg (2000). doi:10.1007/3-540-44480-7_19
17. Huang, J.C., Hsieh, W.S.: Automatic feature-based global motion estimation in video sequences. IEEE Trans. Consum. Electron. **50**, 911–915 (2004)

18. Ryu, Y.G., Chung, M.J.: Robust online digital image stabilization based on point-feature trajectory without accumulative global motion estimation. IEEE Signal Process. Lett. **19**, 223–226 (2012)
19. Harris, C., Stephens, M.: A combined corner and edge detector. In: Alvey Vision Conference, vol. 15, p. 50 (1988)
20. Kuglin, C.D.: The phase correlation image alignment method. In: Proceeding of International Conference Cybernetics and Society, pp. 163–165 (1975)
21. Reddy, B.S., Chatterji, B.N.: An FFT-based technique for translation, rotation, and scale-invariant image registration. IEEE Trans. Image Process. **5**, 1266–1271 (1996)
22. Wolberg, G., Zokai, S.: Robust image registration using log-polar transform. In: International Conference on Image Processing, vol. 1, pp. 493–496. IEEE (2000)
23. Wang, Z., Li, Q.: Video quality assessment using a statistical model of human visual speed perception. J. Opt. Soc. Am. A **24**, B61–B69 (2007)
24. Hecht, S.: The visual discrimination of intensity and the Weber-Fechner law. J. Gen. Physiol. **7**, 235–267 (1924)
25. Winkler, S.: Issues in vision modeling for perceptual video quality assessment. Sig. Process. **78**, 231–252 (1999)
26. Wang, Z., Sheikh, H.R., Bovik, A.C.: Objective video quality assessment. In: The Handbook of Video Databases: Design and Applications, pp. 1041–1078 (2003)
27. Chang, C.C., Lin, C.J.: Libsvm: a library for support vector machines. ACM Trans. Intell. Syst. Technol. **2**, 389–396 (2011)
28. Yang, J., Honavar, V.: Feature subset selection using a genetic algorithm. IEEE Intell. Syst. Appl. **13**, 44–49 (1998)
29. Microsoft: cognitive services - video API. https://www.microsoft.com/cognitive-services/en-us/video-api
30. proDAD: mercalli v2. www.prodad.com/Home-29756,l-us.html
31. Adobe: after effects CC. http://www.adobe.com/products/aftereffects.html

Learning to Extract Motion from Videos in Convolutional Neural Networks

Damien Teney[1][✉] and Martial Hebert[2]

[1] The University of Adelaide, Adelaide, Australia
damien.teney@adelaide.edu.au
[2] Carnegie Mellon University, Pittsburgh, USA

Abstract. This paper shows how to extract dense optical flow from videos with a convolutional neural network (CNN). The proposed model constitutes a potential building block for deeper architectures to allow using motion without resorting to an external algorithm, *e.g.* for recognition in videos. We derive our network architecture from signal processing principles to provide desired invariances to image contrast, phase and texture. We constrain weights within the network to enforce strict rotation invariance and substantially reduce the number of parameters to learn. We demonstrate end-to-end training on only 8 sequences of the Middlebury dataset, orders of magnitude less than competing CNN-based motion estimation methods, and obtain comparable performance to classical methods on the Middlebury benchmark. Importantly, our method outputs a distributed representation of motion that allows representing multiple, transparent motions, and dynamic textures. Our contributions on network design and rotation invariance offer insights non-specific to motion estimation.

1 Introduction

The success of convolutional neural networks (CNNs) on image-based tasks, from object recognition to semantic segmentation or geometry prediction, has inspired similar developments with videos. Example applications include activity recognition [1–3], scene classification [4], or semantic segmentation of scenes with dynamic textures [5]. The appeal of CNNs is to be trainable end-to-end, *i.e.* taking raw pixel values as input, and learning their mapping to the output of choice, identifying appropriate intermediate representations in the layers of the network. The natural application of this paradigm to videos involves a 3D volume of pixels as input, made of stacked consecutive frames. The direct application of existing architectures on such inputs has shown mixed results. An alternative is to first extract optical flow or dense trajectories with an external algorithm [1–3] and feed the CNN with this information in addition to the pixel values. The success of this approach can be explained by the intrinsically different nature of spatial and temporal components, now separated during a preprocessing. The extraction of motion regardless of image contents is not trivial, and has been addressed by the long-standing line of successful optical flow

© Springer International Publishing AG 2017
S.-H. Lai et al. (Eds.): ACCV 2016, Part V, LNCS 10115, pp. 412–428, 2017.
DOI: 10.1007/978-3-319-54193-8_26

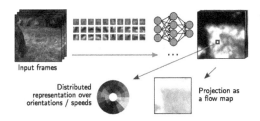

Input frames

Distributed
representation over
orientations / speeds

Projection as
a flow map

Fig. 1. The proposed CNN takes raw pixels as input and produces features representing evidence for motion at various speeds and orientations. These can be projected as a traditional optical flow map. First-layer kernels (pictured) typically identify translating patterns in the image.

algorithms. Conversely, the end-to-end training of CNNs on videos for high-level tasks has shown limited capability for identifying intermediate representations of motion. In this paper, we show that specifically training a CNN to extract optical flow can be achieved once some key principles are taken into account (Fig. 1).

We leverage signal processing principles and how motion manifests itself in the frequency domain of a spatiotemporal signal (Sect. 3) to derive convolutions, pooling and non-linear operations able to map input pixels to a representation of motion (Sect. 4). The resulting network is designed as a building block for deeper architectures addressing higher-level tasks. The current practice of extracting optical flow as an independent preprocessing might be suboptimal, as the assumptions of an optical flow algorithm may not hold for a particular end application. The proposed approach would potentially allow to fine-tune the motion representation with the whole model in a deep learning setting.

The proposed network outputs a distributed representation of motion. The penultimate layer comprises, for each pixel, a population of neurons selective to various orientations and speeds. These can represent a multimodal distribution of activity at a single spatial location, and represent non-rigid, overlapping, and transparent motions that traditional optical flow usually cannot. The distributed representation can be used as a high-dimensional feature by subsequent applications, or decoded into a traditional map of the dominant flow. The latter allows training and evaluation with existing optical flow datasets.

The motivation for the proposed approach is not direct competition with existing optical flow algorithms. The aim is to enable using CNNs with videos in a more principled way, using pixel input for both spatial and temporal components. Instead of considering the flow of a complete scene as the end-goal, we rather wish to identify local motion cues without committing to an early scene interpretation. This contrasts with modern optical flow approaches, which often perform implicit or explicit tracking and/or segmentation. In particular, we avoid motion smoothness and rigidness priors, and spatial scene-level reasoning. This makes the features produced by the network suitable to characterize situations that break such assumptions, *e.g.* with transparent phenomena and dynamic textures [5,6]. As a downside, our evaluation on the Sintel benchmark shows inferior performance to state-of-the-art techniques. This confirms that a *scene-level* interpretation of motion requires such priors and higher-level reasoning. In particular, we do not perform explicit feature tracking and long-range matching, which are the highlights of the best performers on this dataset (*e.g.* [7,8]).

The contributions of this paper are fourfold. (1) We derive, from signal processing principles, a CNN able to learn mapping pixels to optical flow. (2) We train this CNN end-to-end on videos with ground truth flow, then demonstrate performance on the Middlebury benchmarks comparable to classical methods. (3) We show how to enforce strict rotation invariance within a CNN through weight-sharing, and demonstrate significant benefit for training on a small dataset with no need for data augmentation. (4) We show that the distributed representation of motion produced within our network is more versatile than traditional flow maps, able to represent phenomena such as dynamic textures and multiple, overlapping motions. Our pretrained network is available as a building block for deeper architectures addressing higher-level tasks. Its training is significantly less complex than competing methods [9] while providing similar or superior accuracy on the Middlebury benchmark.

2   Related Work

Learning Spatiotemporal Features. Several recent works have used CNNs for classification and recognition in videos. Karpathy *et al.* [10] consider the large-scale classification of videos, but obtain only a modest improvement over single-frame input. Simonyan and Zisserman [1] propose a CNN for activity recognition in which appearance and motion are processed in two separate streams. The temporal stream is fed with optical flow computed by a separate algorithm. The advantage of using separate processing of spatial and temporal information was further examined in [2,3]. We propose to integrate the identification of motion into such networks, eliminating the need for a separate algorithm, and potentially allowing to fine-tune the representation of motion. Tran *et al.* [4] proposed an architecture to extract general-purpose features from videos, which can be used for classification and recognition. Their deep network captures high-level concepts that integrate both motion and appearance. In comparison, our work focuses on the extraction of motion alone, *i.e.* independently of appearance, as motivated by the two-stream approach [1]. Learning spatiotemporal features outside of CNN architectures was considered earlier. Le *et al.* [11] used independent subspace analysis to identify filter-based features. Konda and Memisevic [12] learned motion filters together with depth from videos. Their model is based on the classical energy-based motion model, similarly to ours. Taylor *et al.* [13] used restricted Boltzmann machines to learn unsupervised motion features. Their representation can be used to derive the latent optical flow, but was shown to capture richer information, useful *e.g.* for activity recognition. The high-dimensional features produced by our network bear similar benefits. Earlier work by Olshausen [14] learned sparse decompositions of video signals, identifying features resembling the filters learned in our approach. In comparison to all works mentioned above, we focus on the extraction of motion independently of spatial appearance, whereas decompositions such as in [14] result in representations that confound appearance and temporal information.

Extraction of Optical Flow. The estimation of optical flow has been studied for several decades. The basis for many of today's methods dates back to the seminal work of Horn and Schunk [15]. The flow is computed as the minimizer of data and smoothness terms. The former relies on the conservation of a measurable image feature (typically corresponding to the assumption of brightness constancy) and the latter models priors such as motion smoothness. Many works proposed improvements to these two terms (see [16] for a recent survey). Heeger [17] proposed a completely different approach, applying spatiotemporal filters to the input frames to sample their frequency contents. This method naturally applies to sequences of more than two frames, and relies on a bank of hand-designed filters (typically, Gaussian derivatives or spatiotemporal Gabor filters [18]). Subsequent improvements [19–21] focused on the design of those filters. They must balance the sampling of narrow regions of the frequency spectrum, i.e., to accurately estimate motion speed and orientation, while retaining the ability to precisely locate the stimuli in image. A practical consequence of this tradeoff is the typically blurry flow maps produced by the method. This historically played in favor of the more popular approach of Horn and Schunk. Another downside of the filter-based approach was the computational expense of convolutions. Our work revisits Heeger's approach, motivated by two key points. On the one hand, applying spatiotemporal filters naturally falls in the paradigm of convolutional neural networks, which are currently of particular interest for analyzing videos. On the other hand, modern advances can overcome the two initial burdens of the filter-based approach by (1) learning the filters using backpropagation, and (2) leveraging GPU implementations of convolutions.

Recently, Fisher *et al.* proposed another CNN-based method named Flownet [9]. They obtain very good results on optical flow benchmarks. In comparison to our work, they train a much deeper network that requires tens of thousands of training images. Our architecture is derived from signal processing principles, which contains fewer weights by several orders of magnitude. This allows training on much smaller datasets. The final results in [9] also include a variational refinement, essentially using the CNN to initialize a traditional flow estimation. Our procedure is entirely formulated as a CNN. This potentially allows fine-tuning when integrated into a deeper architecture.

A number of recent works [5,6,22,23] studied the use of spatiotemporal filters to characterize motion in *e.g.* transparent and semi-transparent phenomena, and dynamic textures such as a swirling cloud of smoke, reflections on water, or swaying vegetation. These works highlighted the potential of filter-based features, and the need for motion representations – such as those produced by our approach – that go beyond displacement (flow) maps.

Invariances in CNNs. One of our technical contributions is to enforce rotation invariance by tying groups of weights together. In contrast to weight *sharing* which involves weights at different layers, this applies to weights of a same layer.

Encouraging or enforcing invariances in neural networks has been approached in several ways. The convolutional paradigm ensure translation invariance by reusing weights between spatial locations. Other schemes of weight sharing were proposed in a simple model in [24], and later in [25] with a method to learn which weights to share. No published work discussed the implementation of strict rotation invariance in CNN to our knowledge. In [26,27], schemes akin to ensemble methods were proposed for invariance to geometric transformations. In [28] and more recently in [29], a network first predicts a parametric transformation, used to rotate or warp the image to a canonical state before further processing. Our approach, in addition to the actual invariance, has the benefit of reducing the number of weights to learn and facilitates the training. Soft invariances, *e.g.* to contrast can be encouraged by specific operations, such as local response normalization (LRN, *e.g.* in [30]). We also use a number of such operations. Note that this paper abuses of the term "invariance" for cases more accurately involving *equi*variance or *co*variance [31].

3 Filter-Based Motion Estimation

Our rationale for estimating motion with a convolutional architecture is based on the motion energy model [18]. Classical implementations [17,21,32] are based on convolutions with hardcoded spatiotemporal filters (*e.g.* 3D Gabors or Gaussian derivatives). The convolution of a signal with a kernel in the spatiotemporal domain corresponds to a multiplication of their spectra in the frequency domain. Convolutions with a bank of bandpass filters produce measurements of energy in these bands, which are then suitable for frequency analysis of the signal. A pattern moving in a video with a constant speed and orientation manifests itself as a plane in the frequency domain [18,33], and the signal energy entirely lies within this plane. It passes through the origin, and its tilt corresponds to the motion orientation and speed in the image domain. Classical implementations have used various schemes to identify the best-fitting plane to the energy measurements. In our model, we learn the spatiotemporal filters together with additional layers to decode their responses into the optical flow. Importantly, transparent patterns in an image moving with different directions or speeds correspond to distinct planes in the frequency domain. The same principle can thus identify multiple, overlapping motions.

4 Proposed Network Architecture

Our network is fully convolutional, i.e., it does not contain any fully connected layers. Each location in the output flow map is linked to a spatially-limited receptive field in the input, and each pixel of a training sequence can thus be seen as a unique training point. We describe below each layers of our network in their feedforward order. We use x_{ijk}^ℓ to refer to the scalar value at coordinates (i, j, k) in the 3D tensor obtained by evaluating the ℓth layer. Indices i and j refer to spatial dimensions, k to feature channels. We use a colon (:) to refer to all elements along a dimension. We denote with $*_{2D}$ and $*$ convolutions in two and three dimensions, respectively. In contrast to CNNs used for image recognition, the desired output here is dense, *i.e.* a 2D flow vector for every pixel. To achieve this, all convolutions use a $1\,px$ stride and the pooling (Eq. 5) a $2\,px$ stride, all with appropriate padding. The output is thus at half the resolution of the input. Our experiments use bilinear upsampling (except otherwise noted) to obtain flow fields at the original resolution.

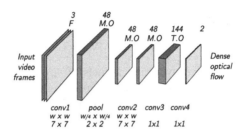

Fig. 2. The proposed neural network takes raw pixels of F video frames as input, and outputs a dense optical flow at half its resolution. The network comprises 2 convolutional layers, 1 pooling layer, and 2 pixelwise weights (1×1 convolutions). Dimensions of receptive fields and feature maps are shown, with numbers chosen in our implementation in gray. Normalizations and non-linearities are not shown.

4.1   Network Input

The input to the network is the $H \times W \times F$ volume of pixels made by stacking F successive grayscale frames of a video (typically, $F=3$). The desired network output is the flow between frames $\lceil F/2 \rceil$ and $\lceil F/2 \rceil+1$.

$$x^1_{::k} \;=\; k\text{th frame of the video} \quad \forall\, k=1\ldots F. \tag{1}$$

4.2   Invariance to Brightness and Contrast

The estimated motion should be insensitive to additive changes of brightness of the input. Since the subsequent processing will be local, instead of subtracting the average brightness over the whole image, we subtract the local low-frequency component:

$$x^{\ell+1}_{::k} \;=\; x^{\ell}_{::k} \;-\; \left(x^{\ell}_{::k} \;*_{2D}\; \mathcal{H}^0\right) \quad \forall\, k, \tag{2}$$

where $\mathcal{H}^0$ denotes a fixed, 2D Gaussian kernel of standard deviation $w/3$. Note that this operation could also be written as a convolution with a center-surround filter. We then ensure invariance to local contrast changes with a normalization using the standard deviation in local neighbourhoods:

$$x^{\ell+1}_{ijk} \;=\; x^{\ell}_{ijk} \Big/ \operatorname*{std}_{i',j'\in\Omega(i,j)} \left(x^{\ell}_{i'j'k}\right) \quad \forall\, i,j,k, \tag{3}$$

where $\Omega(i,j)$ refers to the square region of length w centered on (i,j).

The two above operations proved essential to learn subsequent filters with little training data. They can be seen as a local equivalent to a typical image-wide whitening [34]. This formulation is better suited to the subsequent local processing of a fully convolutional network.

4.3   Motion Detection

This key operation convolves the volume of pixels with learned 3D kernels. They can be interpreted as spatiotemporal filters that respond to various patterns moving at different speeds.

$$x_{::k}^{\ell+1} = x_{:::}^{\ell} * \mathcal{H}_k^1 + b_k^1 \quad \forall\, k=1\ldots MO, \tag{4}$$

where $\mathcal{H}_k^1$ are MO learned convolution kernels of size $w \times w \times F$, and b_k^1 the associated biases. The constants M and O respectively fix the number of independent kernels and the number of orientations explicitly represented within the network (Sect. 4.7).

4.4   Invariance to Local Image Phase

The learned kernels used as motion detectors above typically respond to lines and edges in the image. The estimated motion should however be independent of such image structure. Classical models [17,20] account for this using pairs of quadrature filters, though this is not trivial to enforce with our learned filters. Instead, we approximate a phase-invariant response as follows. (1) The response of the convolution is rectified by pointwise squaring. Responses out of phase by 180° (*e.g.* dark-bright and bright-dark transitions) then give a same response. (2) We apply a spatial max-pooling. The wavenumber of the pattern captured by our kernels of size w is at least $2/w$ cycles/px. The worst-case phase shift of 90° then corresponds to a spatial shift of $w/4$ px. We maxpool responses over windows of size $w/4$ with a fixed stride of 2, and thus approximate a phase-invariant response at the price of a lowered resolution.

$$x_{ijk}^{\ell+1} = \max_{i',j'\in\Omega'(2i,2j)} \left(x_{i'j'k}^{\ell}{}^2\right) \quad \forall\, i,j,k, \tag{5}$$

where $\Omega'(2i,2j)$ refers to the square region of length $\lceil w/4 \rceil$ centered on $(2i,2j)$.

4.5   Invariance to Local Image Structure

The estimated motion should be independent from the amount and type of texture in the image. To account for intensity differences of patterns at different orientations at any particular location (*e.g.* a grid pattern of horizontal lines crossing fainter vertical ones), we normalize the responses by their sum over all orientations [17]:

$$x_{ijk}^{\ell+1} = x_{ijk}^{\ell} \Big/ \Big(\sum_{k'} x_{ijk'}^{\ell} + \epsilon \Big) \quad \forall\, i,j,k, \tag{6}$$

where ϵ is a small constant to avoid divisions by a small value in low-texture areas of the image. The sum is performed over feature channels k' that correspond to the O variants of k at all orientations (see Sect. 4.7).

To account for the aperture problem, we allow local interaction by introducing an additional convolutional layer with MO learned kernels $\mathcal{H}^2$ of size $w \times w \times MO$.

$$x^{\ell+1}_{::k} = x^{\ell}_{:::} * \mathcal{H}^2_k + b^2_k \quad \forall\, k = 1 \ldots MO, \tag{7}$$

$$x^{\ell+1}_{ijk} = \max(x^{\ell}_{ijk}, 0). \tag{8}$$

The classical hardcoded models typically use here 2D convolutions with Gaussian kernels. Our experiments showed that supervised training lead to similar kernels, although slightly non-isotropic, and modeling some cross-channel and center-surround interactions.

4.6 Decoding into Flow Vectors

The features maps at this point represent evidence for different types of motion at every pixel. This evidence is now decoded with a hidden layer, a softmax nonlinearity and a linear output layer:

$$x^{\ell+1}_{::k} = x^{\ell}_{:::} * \mathcal{H}^3_k + b^3_k \quad \forall\, k = 1 \ldots TO \tag{9}$$

$$x^{\ell+1}_{ijk} = e^{x^{\ell}_{ijk}} / \sum_{k'} e^{x^{\ell}_{ijk'}} \quad \forall\, i, j, k = 1 \ldots TO \tag{10}$$

$$x^{\ell+1}_{::k} = x^{\ell}_{:::} * \mathcal{H}^4_k + b^4_k \quad \forall\, k = \{1, 2\}, \tag{11}$$

where T is a constant that fixes the number of hidden units. The decoding is performed pixelwise, $i.e.$ $\mathcal{H}^3_k$ and $\mathcal{H}^3_k$ are 1×1. Intuitively, the activations of the hidden layer (Eq. 9) represent scores for motions at S and O discrete speeds and orientations, of which the softmax picks out the highest. Assuming a unimodal distribution of scores ($i.e.$ a single motion at any pixel), the output layer interpolates these scores and maps them to a 2D flow vector for every pixel (see Sect. 6).

4.7 Invariance to In-Plane Rotations

In our context, rotation invariance implies that a rotated input must produce a correspondingly rotated output. Note the contrast with image recognition where rotated inputs should give a *same* output. All of our learned weights (Eqs. 4–9) are split into groups corresponding to discrete orientations. The key is to enforce these groups of weights to be equivalent, $i.e.$ so that they make the same use of features from the preceding layer at the same *relative* orientations. In addition, convolutional kernels need to be 2D rotations of each other. These strict requirements allow us to maintain only a single version of the weights at a canonical orientation, and generate the others when evaluating the network (see Fig. 4). During training with backpropagation, the gradients are aligned with this canonical orientation and averaged to update the single version of the weights.

Formally, let us consider a convolutional layer[1] $\ell+1$. The feature maps x^ℓ (respectively $x^{\ell+1}$) are split into O (P) groups of M (N) channels. For example, in Eq. 7, $O = P$ and $M = N$. The groups of channels correspond to regular orientations θ_i^ℓ $(\theta_j^{\ell+1})$ in $[0, 2\pi[$. Considering the convolution weights $\mathcal{H}$ and their slice h_{imjn} the 2D kernel acting on the input (respectively output) channel of orientation θ_i^ℓ $(\theta_j^{\ell+1})$, we constrain the weights as follows:

$$h_{imjn} = \underset{\theta_j^{\ell+1} - \theta_{j'}^{\ell+1}}{\text{rotate2D}} \left(h_{i'mj'n} \right) \qquad (12)$$

$$\forall \, i, i', j, j', m, n \quad \text{s.t.} \quad cos(\theta_{j'}^{\ell+1} - \theta_{i'}^\ell) = cos(\theta_j^{\ell+1} - \theta_i^\ell) \qquad (13)$$

Equation 12 ensures that convolution kernels are rotated versions of each other (implemented with bilinear interpolation) and Eq. 13 ensures that the same weights are applied to input channels representing a same *relative* orientation with respect to a given output channel of the layer. In other words, the weights are shifted between each group so as to act similarly on channels representing the same relative orientations (see Fig. 4). In Eqs. 7 and 9, $N = N'$. In Eq. 4, $N=1$. In Eq. 11, $N=2$ with $\theta^{\ell+1} = \{0, \pi/2\}$.

It follows that the number of convolution kernels to explicitly maintain is reduced from $OPMN$ to $\lceil O/2 \rceil MN$. In Eq. 7 for example, in our implementation with $O = P = 12$ orientations, this amounts to a decrease by a factor 24. It allows training on small amounts of data with lower risk of overfitting. This also negates the need to artificially augment the dataset with rotations and flips.

5   Multiscale Processing

Equations 1–11 form a complete network that maps pixels to a dense flow field. However, the detection of large motion speeds is limited by the small effective receptive field of the output units, due to the limited number of convolutional layers. We remedy this in two ways (Fig. 3) without increasing the number of weights to train. First, we apply the network (Eqs. 1–7) on multiple downsized

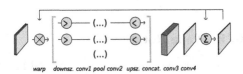

warp   downsz. conv1 pool conv2 upsz. concat. conv3 conv4

Fig. 3. We bring two modifications to the basic network (Fig. 2) for multiscale processing. First, we apply the network on several downsampled versions of the input, concatenating the feature maps before the decoding stage. Second, we add a recurrent connection to warp the input frames according to the estimated flow, iterating its evaluation for a fixed number of steps. This is inspired by the classical "coarse-to-fine" strategy for optical flow, and designed to estimate motions larger than the receptive field of the network output units.

---

[1] The general formulation applies to convolutional layers as well as to our pixelwise weights (Eqs. 9, 11), in which case the 2D rotation of the kernel has no effect.

versions of the input frames. The feature maps are brought back to a common resolution by bilinear upsampling and concatenated before the decoding stage (Eqs. 9–11).

Second, we add a recurrent connection to the network that warps the input frames according to the current estimate of the flow. The evaluation runs through the recurrent connection for a fixed number of steps. This is inspired by the classical coarse-to-fine warping strategy [35]. It allows the model to approximate the flow iteratively. Note that the recurrent connection to the warping layer is not a strictly linear operation contrary to typical recurrent neural networks. Training is performed by backpropagation through the unfolded recurrent iterations, but not through the recurrent connection itself. Since the result of each iteration is summed to give the final output, we can append and sum the same loss (Sect. 6) at the end of each unfolded iteration.

6   Implementation and Training

We train the weights $(\mathcal{H}^1, \ldots, \mathcal{H}^4)$ and biases $(b^1, \ldots, b^4)$ end-to-end with back-propagation. Even though the network ultimately performs a regression to flow vectors, we found more effective to train it first for *classification*. First, we pick a number TO of flow vectors uniformly in the distribution of training flows. We train the network for nearest-neighbour classification over these possible outputs, with a logarithmic loss over Eq. 10. Second, we add the linear output layer (Eq. 11) and initialize the rows of $\mathcal{H}^4_k$ with the vectors used for classification. Second, the network is fine-tuned with a Euclidean ("end-point error") loss over the decoded vectors. That fine-tuning has a marginal effect in practice. The softmax values are practically unimodal and sum to one by construction, hence they can directly interpolate the vectors used for classification with a linear operation. We believe the 2-step training is helpful because the Euclidean loss alone does not reflect well the quality (*e.g.* smoothness) of the flow. The classification loss guides the optimization towards a better optimum.

Considering the above, the softmax values (Eq. 10) constitute a distributed representation of motion, where each dimension corresponds to a different orientation and speed. Feature maps at this layer can encode multimodal distributions over this representation and represent patterns that optical flow cannot.

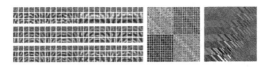

Fig. 4. Convolutional kernels $\mathcal{H}^1$, $\mathcal{H}^2$, and pixelwise weights $\mathcal{H}^3$ learned on the Middlebury dataset (selection). The obvious structure is enforced by constraints that ensure rotation invariance of the overall network.

7   Experimental Evaluation

We present three sets of experiments. (1) We evaluate the non-standard design choices of the proposed architecture through ablative analysis. (2) We compare our performance versus existing methods on the Middlebury and Sintel benchmarks. (3) We demonstrate applicability to dynamic textures and multiple transparent motions that goes beyond traditional optical flow. Code and trained models are available on the author's website [36].

7.1   Ablative Analysis

Our ablative analysis uses the public Middlebury dataset, with the sequences split into halves for training (Grove2, RubberWhale, Urban3) and testing (Grove3, Dimetrodon, Hydrangea). For each run, we modify the network in one particular way, retrain it from scratch, and report its performance on the test set (Table 1). We observe that all preprocessing and normalization steps (2–3) have a positive impact, and some are even necessary for the training to converge at all (at least with the small dataset used in these experiments). Operations inspired by the classical implementations of the motion energy model, i.e., (8) rectification by squaring of filter responses and (6) normalization across orientations, proved beneficial as well. This shows the benefit for our principled approach to network design. We perform a comparison to the classical, hardcoded filter-approach approach [17] by setting the first filters to Gaussian derivatives (4). Although learned filters are visually somewhat similar (Fig. 4), learned kernels clearly perform better. This confirms the general benefit of a data-driven approach to motion estimation.

Table 1. We evaluate every non-standard design choice by retraining a modified network. N.C. denotes networks for which the training did not converge within a reasonable number of iterations. See discussion in Sect. 7.1.

	EPE (px)	AAE $(°)$
Full model, $F=3$, $O=12$, 10 scales, 1 rec.iter.	0.66	6.9
(1) Number of input frames $F=2/5$	0.67 0.90	6.8 8.7
(2) No center-surround filter	N.C.	N.C.
(3) No local normalization	0.67	6.9
(4) Hard-coded $\mathcal{H}_k^1$: Gauss. deriv.	0.78	8.4
(5) No L1-normalization over orientations	N.C.	N.C.
(6) No pooling for phase invariance	0.93	11.2
(7) ReLU after conv1 (default: squaring)	N.C.	N.C.
(8) No constraints for rotation invariance	N.C.	N.C.
(9) Number of orientations $O=6/8/16$	1.65 0.72 0.65	26.6 8.1 6.6
(10) Loss: classification/regression	0.66/0.83	7.0 8.7
(default: 2-step, classification (logarithmic) then regression (Euclidean))		
(11) Number of scales: 4/8/16	0.69 0.66 0.67	6.9 6.8 6.9
(12) Recurrent iterations: 2/3/4/5	0.57 0.55 0.56 0.57	6.5 6.4 6.7 6.8

7.2 Performance on Optical Flow Benchmarks

We use the Middlebury and Sintel benchmarks for evaluation of networks trained from scratch on their respective training sets. A network trained on the smaller Middlebury dataset performs decently on Sintel sequences with small motions. However, most include much faster motions that had to be retrained for.

Middlebury. Flow maps estimated by our method on the Middlebury dataset [37] are generally smooth and accurate (see Fig. 5). Most errors occur near boundaries of objects that become, or cause occlusions. Although our flow maps remain generally more blurry than those of state-of-the-art methods, some fine details are remarkably well preserved (*e.g.* Fig. 5, second row). This blurriness, or imprecision in the spatial localization of motions, is a well-known drawback of filter-based motion estimation. Convolutional kernels of smaller extent would be desirable to provide better localization, but the extent of its response in the frequency domain (Sect. 3) would correspondingly increase, which would imply a coarser sampling of the signal spectrum and lesser accuracy in motion direction and speed. Comparisons with existing methods show performance on the level of classical methods. We obtain much better performance than the recent implementation of Solari *et al.* [20] of a filter-based method with no learning.

Sintel. The MPI Sintel dataset [38] contains computer-generated scenes of a movie provided in "clean" and "final" versions, the latter including atmospheric effects, reflections, and defocus/motion blur. Flow maps estimated by our method (Fig. 5) are often smooth. Flows in scenes with small motions are usually accurate, but they lack details at the objects' borders and near small image details. Although this is partly alleviated by our recurrent iterative processing within the network, large errors remain in scenes with fast motions. This is reflected by a poor quantitative performance (Table 2). Additional insights can

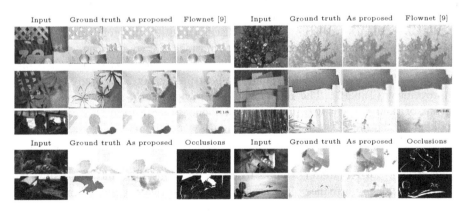

Fig. 5. Estimated flow on sequences from the Middlebury (top two rows) and Sintel (others) datasets. Most failure cases (*e.g.* bottom row) occur near (dis)occlusions, which are common on the Sintel dataset due to large motions of small objects and the small relative camera field-of-view. (Color figure online)

Table 2. Comparison with existing algorithms on the Middlebury and Sintel benchmarks. We report average end-point errors (EPE, in pixels) average angular errors (AAE, in gray, in degrees), and execution times per frame on Sintel. Numbers in parentheses correspond to the sets used for training the model.

Method	Middlebury		Sintel private	
	Public	Private	Clean	Final
FlowNetS [9]	1.09 13.28	–	4.44	7.76
FlowNetS + refinement [9]	0.33 3.87	0.47 4.58	4.07	7.22
DeepFlow [7]	0.21 3.24	0.42 4.22	4.56	7.21
LDOF [8]	0.45 4.97	0.56 4.55	6.42	9.12
Classic++ [39]	0.28 –	0.41 3.92	8.72	9.96
FFV1MT [20]	0.95 9.96	1.24 11.66	–	–
Proposed	(0.45) (5.47)	0.70 6.41	9.36	10.04
Proposed + refinement as [9]	(0.35) (4.10)	0.58 5.22	9.47	10.14

be gained by examining the flow maps (Fig. 5). Errors arise not on the estimates of large motions, but on their localization, in particular near zones of (dis)occlusions. This is obvious *e.g.* in Fig. 5, last row, with a thin wing flapping over a blank sky. Although the actual motion (in yellow) is detected, it spills on both sides of the thin wing structure. Since such occlusons are caused by large motions, they result in a large penalty in EPE. As argued before [38], good overall performance in such situations clearly require reasoning over larger spatial and/or temporal extent than the local motions cues that our method was designed around.

Interesting comparisons can be made with the competing approach Flownet [9]. It uses a more standard deep architecture with numerous convolutional and pooling layers. It also includes a variational refinement to improve the precision of motion estimates from the CNN. As discussed in [9], this refinement cannot correct large errors of correspondingly large motions. For comparison, we applied this same post-processing to our own results. Our results right off the CNN on Middlebury are already accurate, and the post-processing brings only marginal improvement (Table 2). The refinement on Flownet has a stronger effect: the output of their CNN is less precise, and it benefits more from this refinement. Looking at the Sintel dataset, the situation is very different. The main metric (the average EPE) is dominated by large motions, which are the weak point of the filter-based principles (Sect. 3) that we rely on. Flownet is particularly good at long-range matching thanks to its deep architecture, and this results in vastly superior performance. As stated above, the refinement is of little use with large motions, and brings minimal improvement to either method on Sintel. In conclusion, the different design choices in Flownet and our approach seem complimentary in different regimes. It would be interesting to investigate how to combine their strengths.

7.3   Applicability to Transparent Motions and Dynamic Textures

We tested the applicability of our method on scenes that are challenging (dynamic textures) or impossible (transparencies) to handle with traditional optical flow methods. There are no established benchmarks related to motion estimation and dynamic textures. Recent works [5,6,22,23] that highlighted the potential of filter-based motion features in such situation focused on applications such as segmentation [5,23] or scene recognition [6,22]. In Fig. 7, we show scenes containing dynamic textures (water, steam) from which we identified the dominant motion. The flow estimated by a typical method [40] is typically noisy and/or inaccurate, as the usual assumption of brightness constancy does not hold (*e.g.* flickering effect on the water surface, changes of brightness/transparency of the steam, etc.). Although no ground truth is available for these scenes, the flow estimated by our methods is more reliable in comparison. We then demonstrate

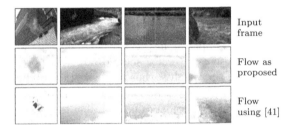

Fig. 6. The extraction of optical flow on dynamic textures is challenging for traditional methods, as transparencies (*e.g.* with steam, left) or flicker (*e.g.* on water ripples, middle right) violate the typical assumption of brightness constancy. The core of our approach relies on the analysis of the frequency contents of the video, and produces more stable and reliable motion estimates.

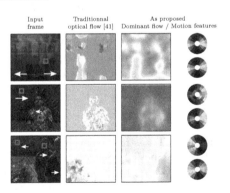

Fig. 7. In scenes with multiple, transparent motions, traditional optical flow methods fail and typically produce incoherent results. Our method identifies a more stable dominant motion. More importantly, our higher-dimensional motion descriptor can capture multiple motions at single locations (red squares; white arrows indicate approximate ground truth direction of motion; see text for details). (Color figure online)

the ability of our distributed representation to capture multiple motions at a single location (transparencies and semi-transparencies), thus going beyond the optical flow representation of pixelwise displacements. We show features in Fig. 7 from three sequences. The first depicts two alpha-blended (in equal proportions) textures moving in opposite directions, thus simulating transparency. The other two depict persons moving behind a fence in directions different than the fence itself [22]. Feature vectors from different locations in the image are visualized as radial bins (orientations) of concentric rings (speeds). Larger values (brighter bins) indicate motion evidence. As expected, areas with simple translations produce one major peak, whereas areas with transparencies produce correspondingly more complex, multimodal distributions. These experiments used a model trained on the Middlebury dataset (Fig. 6).

8   Conclusions

We showed how to identify optical flow entirely within a convolutional neural network. By reasoning about required invariances and by using signal processing principles, we designed a simple architecture that can be trained end-to-end, from pixels to dense flow fields. We also showed how to enforce strict rotation invariance by constraining the weights, thus reducing the number of parameters and enabling training on small datasets. The resulting network performs on the Middlebury benchmark with performance comparable to classical methods, but inferior to the best engineered methods.

We believe the approach presented here bears two major advantages over existing optical flow algorithms. First, building upon the classical motion energy model, our approach is able to produce high-dimensional features that can capture non-rigid, transparent, or superimposed motions, which traditional optical flow cannot represent. Second, it constitutes a method for motion estimation formulated entirely as a shallow, easily-trainable CNN, without requiring any post-processing. Its potential is to be used as a building block in deeper architectures (*e.g.* for activity or object recognition in videos) offering the possibility for fine-tuning the representation of motion. The potential of these two aspects will deserve further exploration and should be addressed in future work.

References

1. Simonyan, K., Zisserman, A.: Two-stream convolutional networks for action recognition in videos. CoRR NIPS Spotlight Session abs/1406.2199 (2014)
2. Wu, Z., Wang, X., Jiang, Y.G., Ye, H., Xue, X.: Modeling spatial-temporal clues in a hybrid deep learning framework for video classification. In: ACM Multimedia Conference (2015)
3. Ye, H., Wu, Z., Zhao, R.W., Wang, X., Jiang, Y.G., Xue, X.: Evaluating two-stream CNN for video classification. In: ACM on International Conference on Multimedia Retrieval (ICMR) (2015)
4. Tran, D., Bourdev, L.D., Fergus, R., Torresani, L., Paluri, M.: C3D: generic features for video analysis. CoRR abs/1412.0767 (2014)

5. Teney, D., Brown, M.: Segmentation of dynamic scenes with distributions of spatiotemporally oriented energies. In: British Machine Vision Conference (BMVC) (2014)

6. Derpanis, K.G., Wildes, R.P.: Spacetime texture representation and recognition based on a spatiotemporal orientation analysis. IEEE Trans. Pattern Anal. Mach. Intell. (PAMI) **34**, 1193–1205 (2012)

7. Weinzaepfel, P., Revaud, J., Harchaoui, Z., Schmid, C.: DeepFlow: large displacement optical flow with deep matching. In: International Conference on Computer Vision (ICCV) (2013)

8. Brox, T., Malik, J.: Large displacement optical flow: descriptor matching in variational motion estimation. IEEE Trans. Pattern Anal. Mach. Intell. (PAMI) **33**, 500–513 (2011)

9. Fischer, P., Dosovitskiy, A., Ilg, E., Häusser, P., Hazirbas, C., Golkov, V., van der Smagt, P., Cremers, D., Brox, T.: Flownet: learning optical flow with convolutional networks. CoRR abs/1504.06852 (2015)

10. Karpathy, A., Toderici, G., Shetty, S., Leung, T., Sukthankar, R., Fei-Fei, L.: Large-scale video classification with convolutional neural networks. In: IEEE Conference on Computer Vision and Pattern Recognition (CVPR) (2014)

11. Le, Q.V., Zou, W.Y., Yeung, S.Y., Ng, A.Y.: Learning hierarchical invariant spatiotemporal features for action recognition with independent subspace analysis. In: IEEE Conference on Computer Vision and Pattern Recognition (CVPR) (2011)

12. Konda, K.R., Memisevic, R.: Unsupervised learning of depth and motion. CoRR abs/1312.3429 (2013)

13. Taylor, G.W., Fergus, R., LeCun, Y., Bregler, C.: Convolutional learning of spatiotemporal features. In: Daniilidis, K., Maragos, P., Paragios, N. (eds.) ECCV 2010. LNCS, vol. 6316, pp. 140–153. Springer, Heidelberg (2010). doi:10.1007/978-3-642-15567-3_11

14. Olshausen, B.: Learning sparse, overcomplete representations of time-varying natural images. In: ICIP, vol. 1, pp. I-41 (2003)

15. Horn, B.K.P., Schunck, B.G.: Determining optical flow. Artif. Intell. **11**, 185–203 (1981)

16. Fortun, D., Bouthemy, P., Kervrann, C.: Optical flow modeling and computation: a survey. Comput. Vis. Image Underst. (CVIU) **393**, 1–21 (2015)

17. Heeger, D.J.: Model for the extraction of image flow. J. Opt. Soc. Am. A **4**, 1455–1471 (1987)

18. Adelson, E.H., Bergen, J.: Spatiotemporal energy models for the perception of motion. J. Opt. Soc. Am. A **2**, 284–299 (1985)

19. Rust, N.C., Mante, V., Simoncelli, E.P., Movshon, J.A.: How MT cells analyze the motion of visual patterns. Nature Neurosci. **9**, 1421–1431 (2006)

20. Solari, F., Chessa, M., Medathati, N., Kornprobst, P.: What can we expect from a V1-MT feedforward architecture for optical flow estimation? Signal Process.: Image Commun. **39**, 342–354 (2015)

21. Ulman, V.: Improving accuracy of optical flow of heeger's original method on biomedical images. In: Campilho, A., Kamel, M. (eds.) ICIAR 2010. LNCS, vol. 6111, pp. 263–273. Springer, Heidelberg (2010). doi:10.1007/978-3-642-13772-3_27

22. Derpanis, K.G., Wildes, R.P.: The structure of multiplicative motions in natural imagery. IEEE Trans. Pattern Anal. Mach. Intell. (PAMI) **32**, 1310–1316 (2010)

23. Teney, D., Brown, M., Kit, D., Hall, P.: Learning similarity metrics for dynamic scene segmentation. In: IEEE Conference on Computer Vision and Pattern Recognition (CVPR) (2015)

24. Fasel, B., Gatica-Perez, D.: Rotation-invariant neoperceptron. In: International Conference on Pattern Recognition (ICPR) (2006)
25. Le, Q.V., Ngiam, J., Chen, Z., Chia, D., Koh, P.W., Ng, A.Y.: Tiled convolutional neural networks. In: Advances in Neural Information Processing Systems (NIPS) (2010)
26. Dieleman, S., Willett, K.W., Dambre, J.: Rotation-invariant convolutional neural networks for galaxy morphology prediction. CoRR abs/1503.07077 (2015)
27. Laptev, D., Buhmann, J.M.: Transformation-invariant convolutional jungles. In: IEEE Conference on Computer Vision and Pattern Recognition (CVPR) (2015)
28. Rowley, H., Baluja, S., Kanade, T.: Rotation invariant neural network-based face detection. In: IEEE Conference on Computer Vision and Pattern Recognition (CVPR) (1998)
29. Jaderberg, M., Simonyan, K., Zisserman, A., Kavukcuoglu, K.: Spatial transformer networks. In: Advances in Neural Information Processing Systems (NIPS) (2015)
30. Krizhevsky, A., Sutskever, I., Hinton, G.E.: Imagenet classification with deep convolutional neural networks. In: Advances in Neural Information Processing Systems (NIPS) (2012)
31. Jayaraman, D., Grauman, K.: Learning image representations equivariant to egomotion. CoRR abs/1505.02206 (2015)
32. Niyogi, S.A.: Fitting models to distributed representations of vision. In: International Joint Conference on Artificial Intelligence, San Francisco, CA, USA, pp. 3–9. Morgan Kaufmann Publishers Inc. (1995)
33. Fleet, D., Jepson, A.: Computation of component image velocity from local phase information. Int. J. Comput. Vis. (IJCV) 5, 77–104 (1990)
34. LeCun, Y.A., Bottou, L., Orr, G.B., Müller, K.-R.: Efficient BackProp. In: Montavon, G., Orr, G.B., Müller, K.-R. (eds.) Neural Networks: Tricks of the Trade. LNCS, vol. 7700, pp. 9–48. Springer, Heidelberg (2012). doi:10.1007/978-3-642-35289-8_3
35. Memin, E., Perez, P.: A multigrid approach for hierarchical motion estimation. In: IEEE Intenational Conference on Computer Vision (ICCV) (1998)
36. Anonymous: Website to be provided upon acceptance of the paper. http://damienteney.info/cnnFlow.htm
37. Baker, S., Scharstein, D., Lewis, J., Roth, S., Black, M.J., Szeliski, R.: A database and evaluation methodology for optical flow. In: International Conference on Computer Vision (ICCV) (2007)
38. Butler, D.J., Wulff, J., Stanley, G.B., Black, M.J.: A naturalistic open source movie for optical flow evaluation. In: Fitzgibbon, A., Lazebnik, S., Perona, P., Sato, Y., Schmid, C. (eds.) ECCV 2012. LNCS, vol. 7577, pp. 611–625. Springer, Heidelberg (2012). doi:10.1007/978-3-642-33783-3_44
39. Sun, D., Roth, S., Black, M.J.: Secrets of optical flow estimation and their principles. In: IEEE Conference on Computer Vision and Pattern Recognition (CVPR), pp. 2432–2439. IEEE (2010)
40. Brox, T., Bruhn, A., Papenberg, N., Weickert, J.: High accuracy optical flow estimation based on a theory for warping. In: Pajdla, T., Matas, J. (eds.) ECCV 2004. LNCS, vol. 3024, pp. 25–36. Springer, Heidelberg (2004). doi:10.1007/978-3-540-24673-2_3

Author Index